从零开始学 Hadoop大数据分析

（视频教学版）

温春水 毕洁馨 ◎ 编著

机械工业出版社
China Machine Press

图书在版编目（CIP）数据

从零开始学Hadoop大数据分析：视频教学版/温春水，毕洁馨编著.—北京：机械工业出版社，2019.2（2021.1重印）

ISBN 978-7-111-61931-4

Ⅰ. 从… Ⅱ. ①温… ②毕… Ⅲ. 数据处理软件 Ⅳ. TP274

中国版本图书馆CIP数据核字（2019）第025726号

本书全面介绍了Hadoop大数据分析的基础知识、14个核心组件模块及4个项目实战案例。为了帮助读者高效、直观地学习，作者特意为本书录制了20小时同步配套教学视频。

本书共19章，分为3篇。第1篇Hadoop基础知识，涵盖大数据概述、Hadoop的安装与配置、Hadoop分布式文件系统及基于Hadoop 3的HDFS高可用等相关内容；第2篇Hadoop核心技术，涵盖的内容有Hadoop的分布式协调服务——ZooKeeper；分布式离线计算框架——MapReduce；Hadoop的集群资源管理系统——YARN；Hadoop的数据仓库框架——Hive；大数据快速读写——HBase；海量日志采集工具——Flume；Hadoop和关系型数据库间的数据传输工具——Sqoop；分布式消息队列——Kafka；开源内存数据库——Redis；Ambari和CDH；快速且通用的集群计算系统——Spark。第3篇Hadoop项目案例实战，主要介绍了基于电商产品的大数据业务分析系统、用户画像分析、基于个性化的视频推荐系统及电信离网用户挽留4个项目实战案例，以提高读者的大数据项目开发水平。

本书内容全面，实用性强，适合作为Hadoop大数据分析与挖掘的入门读物，也可作为Java程序员的进阶读物。另外，本书还特别适合想要提高大数据项目开发水平的人员阅读。对于专业的培训机构和相关院校而言，本书也是一本不可多得的教学用书。

从零开始学Hadoop大数据分析（视频教学版）

出版发行：机械工业出版社（北京市西城区百万庄大街22号　邮政编码：100037）	
责任编辑：欧振旭　李华君	责任校对：姚志娟
印　　刷：北京捷迅佳彩印刷有限公司	版　　次：2021年1月第1版第4次印刷
开　　本：186mm×240mm　1/16	印　　张：23
书　　号：ISBN 978-7-111-61931-4	定　　价：89.00元
客服电话：（010）88361066　88379833　68326294	投稿热线：（010）88379604
华章网站：http://www.hzbook.com	读者信箱：hzit@hzbook.com

版权所有·侵权必究
封底无防伪标均为盗版
本书法律顾问：北京大成律师事务所　韩光/邹晓东

前言

随着互联网的发展，人们日常工作和生活中产生的数据越来越多，伴随着信息的爆炸，大数据应运而生。分布式集群对大量数据的存储和分析处理有极大优势，因此 Hadoop 的各种技术得到了广泛应用和普及。大数据项目的开发除了需要扎实的理论基础外，还需要掌握 Hadoop 的搭建环境和运行部署方法，这样才能在大数据技术领域有更强的竞争力和职业发展前景。

目前市场上关于 Hadoop 的原理介绍和环境搭建的图书不少，但是真正从实战出发，通过"理论讲解→环境搭建→项目案例实战"这种符合初学者学习规律的科学编排体系的图书却不多。本书便是基于这一编排体系而写，以实战为主旨，通过 Hadoop 的 14 个基础组件的相关模块和 4 个完整的项目实战案例，让读者在理解大数据原理的同时，完成 Hadoop 的环境搭建，并亲自动手实现书中的实战案例，提高开发水平和项目实战能力。

本书可以帮助大数据开发人员充分了解当下流行的大数据技术和应用方法，从而在大数据项目中能更加自信、高效地完成项目开发。书中为有意涉猎大数据领域的人提供了详尽的指导，让他们能够更快、更好地掌握大数据的核心技术，并应用于项目实践，从而脱颖而出，顺利进军大数据行业。另外，本书也为大数据项目开发小组提供了可参考和借鉴的选拔大数据人才的技术标准。

本书特色

1．提供了20小时同步配套教学视频，高效、直观

为了便于读者高效、直观地学习，笔者专门为本书重点内容录制了 20 小时同步配套教学视频。读者可以一边看书，一边结合教学视频进行学习，取得更好的学习效果。

2．对Hadoop开发做了基础上的准备

本书从一开始就对大数据的应用、特点和 Hadoop 的起源与发展做了基本介绍，并简

要介绍了大数据的技术框架及 Hadoop 的核心构件，然后详细介绍了 Hadoop 的安装和配置步骤，便于读者理解后续章节中介绍的各种组件和案例。

3．全面涵盖Hadoop的各种核心技术

本书介绍了 Hadoop 的核心构件 HDFS 和 MapReduce，并详细介绍了基于存储和计算的 YARN、Hive、HBase、Flume、Sqoop、Kafka 和 Redis 等大数据技术的原理、环境搭建步骤和整合应用示例。

4．模块驱动，实用性强

本书介绍了 Hadoop 开发的 14 个典型模块，有很强的实用性。这些模块都是 Hadoop 开发经常要用到的模块，开发人员可以随时查阅和参考。

5．详解4个高价值项目实战案例

本书介绍了 4 个项目实战案例，这些案例来源于大数据实际项目，有较高的参考价值和实际应用价值。这些案例用不同的大数据整合技术实现，读者稍加修改即可用于自己的实际项目中。通过这些实战案例，可以让读者对书中介绍的相关理论知识和技术细节有更加透彻的理解。

6．提供完善的售后服务

本书提供了专门的售后服务邮箱：hzbook2017@163.com。读者在阅读本书的过程中有任何疑问都可以通过该邮箱获得帮助。

7．提供教学PPT，方便老师教学和学生学习

笔者专门为本书制作了专业的教学 PPT，以方便相关院校的教学人员讲课时使用；读者也可以通过教学 PPT，来提纲挈领地掌握书中的内容脉络。

本书内容

第1篇　Hadoop基础知识（第1~4章）

第 1 章初识 Hadoop，介绍了大数据的特点和在各行业的应用；阐述了大数据和云计算、物联网之间的关系；讲述了 Hadoop 的起源、发展和意义。

第 2 章 Hadoop 的安装与配置，介绍了 Hadoop 安装与配置的相关知识，主要包括虚拟机的创建、克隆服务器、SSH 免密码登录、JDK 安装、Hadoop 环境变量配置及 Hadoop 分布式安装等。

第3章 Hadoop 分布式文件系统，主要介绍了 Hadoop 的分布式文件系统，包括 HDFS 的核心概念、读写文件的流程，以及 HDFS 基于 Shell 和 Java API 的操作。

第4章基于 Hadoop 3 的 HDFS 高可用，主要介绍了 Hadoop 3.x 的发展和 HDFS 的高可用实现原理，以及如何基于 Hadoop 3 搭建完全分布式和 NameNode 的高可用。

第2篇　Hadoop核心技术（第5~15章）

第5章 Hadoop 的分布式协调服务——ZooKeeper，介绍了 ZooKeeper 的核心概念，包括 Session、数据节点（Znode）、版本、Watcher 和 ACL 等；还介绍了 ZooKeeper 的安装步骤、服务器端和客户端的相关命令，以及 Java API 访问 ZooKeeper 的多种操作。

第6章分布式离线计算框架——MapReduce，主要介绍了 MapReduce 的原理和应用知识，包括 MapReduce 的特点、应用场景、执行原理和测试实例。

第7章 Hadoop 的集群资源管理系统——YARN，比较了 YARN 和 MapReduce 的异同，并介绍了 YARN 集群资源管理系统的基本架构、工作流程和环境搭建步骤等。

第8章 Hadoop 的数据仓库框架——Hive，介绍了 Hive 的理论基础，以及 Hive 和数据库的异同、Hive 设计目的与应用、Hive 运行框架及执行原理；完成了 Hive 的环境搭建、内部表的创建、外部表的创建及数据操作；另外，还介绍了如何通过 Java 访问 Hive 及 Hive 的优化等相关内容。

第9章大数据快速读写——HBase，介绍了 HBase 列式数据库的体系架构、执行原理及安装步骤，还介绍了通过 Shell 操作 HBase，以及基于 Java API 访问 HBase 实现数据增加和查询的相关内容。

第10章海量日志采集工具——Flume，主要介绍了 Flume 的概念、特点、架构，以及其主要组件 Event、Client、Agent、Source、Channel 和 Sink 的作用，并详细介绍了本地读取和配置设置的分日期储存和自动读取实例。

第11章 Hadoop 和关系型数据库间的数据传输工具——Sqoop，主要介绍了数据采集工具 Sqoop 的运行机制、安装和配置，以及 Sqoop 的导入和导出实例。

第12章分布式消息队列——Kafka，介绍了在大数据背景下的分布式消息队列 Kafka 的相关知识，主要包括 Kafka 的基本概念、核心组件、Kafka 集群安装及应用案例等。

第13章开源的内存数据库——Redis，介绍了 Redis 的核心概念、特点、安装和配置步骤及基于客户端登录 Redis 实例；还介绍了 Redis 的数据类型，包括 String、List、Hash 和 Set 等。

第14章 Ambari 和 CDH，主要介绍了 Ambari 和 CDH 的基本概念及其特点，并详细介绍了 Ambari 和 CDH 的安装步骤。

第15章快速且通用的集群计算系统——Spark，主要介绍了 Spark 的核心概念和运行机制，涉及 Spark 分布式集群的安装、平台搭建和应用案例等。

第3篇　Hadoop项目案例实战（第16~19章）

第 16 章基于电商产品的大数据业务分析系统实战，通过一个项目实战案例，详细介绍了数据采集、数据存储、数据清洗、数据转化、数据分析及最终数据的展现过程。

第 17 章用户画像分析实战，通过一个项目实战案例，详细介绍了项目背景、数据采集、数据预处理、模型构建、数据分析等项目开发的过程，并对项目核心代码做了详细解读和部署运行。

第 18 章基于个性化的视频推荐系统实战，通过一个项目实战案例，详细介绍了推荐系统的基本概念、协同过滤推荐算法、项目架构、模型构建的详细过程，并对相关核心代码做了详细解读。

第 19 章电信离网用户挽留实战，通过一个项目实战案例，详细介绍了数据挖掘标准流程中的商业理解、数据理解、数据准备、建模、评估和部署这 6 个步骤，并利用代码实现了数据建模、评估和部署，最终得到用户离网预警清单，有效防止用户流失。

本书配套资源及获取方式

本书提供了以下配套资源：
- 20 小时配套教学视频；
- 实例源代码文件；
- 教学 PPT。

这些资源需要读者自行下载。请在华章公司的网站（www.hzbook.com）上搜索到本书，然后单击"资料下载"按钮，即可在本书页面上找到"配套资源"下载链接。

读者也可以在微信上搜索并关注"程序员的足迹"公众帐号，然后发送"大数据"，即可得到本书配套资源的下载链接。

适合阅读本书的读者

- 需要全面学习 Hadoop 大数据技术的人员；
- Java 程序员；
- 大数据开发工程师；
- 需要提高大数据项目开发水平的人员；

- 大数据开发项目经理;
- 专业培训机构的学员;
- 对大数据技术感兴趣的学生;
- 需要一本案头必备查询手册的人员。

阅读建议

- 没有 Hadoop 技术基础的读者,建议从第 1 章开始顺次阅读并搭建环境,演练每一个实例。
- 有一定 Hadoop 框架基础的读者,可以根据实际情况有重点地选择阅读相关章节和项目案例。
- 对于每一个实例和项目案例,读者可以先自己思考一下实现的思路,然后再详细阅读,这样学习效果会更好。
- 对于重点内容,建议读者先看一遍教学视频,对相关内容有个基本了解,然后再详细阅读书中的内容,会更加事半功倍。

本书作者

本书由温春水和毕洁馨编写。感谢在本书编写过程中提供过帮助的各位编辑!

由于作者的水平所限,加之成书时间较为仓促,书中可能还存在一些疏漏和不当之处,敬请各位读者斧正。联系我们请发电子邮件到 hzbook2017@163.com。

编著者

目录

前言

第1篇　Hadoop 基础知识

第1章　初识 Hadoop ... 2
1.1　大数据初探 ... 2
1.1.1　大数据技术 ... 2
1.1.2　大数据技术框架 ... 3
1.1.3　大数据的特点 ... 3
1.1.4　大数据在各个行业中的应用 ... 4
1.1.5　大数据计算模式 ... 4
1.1.6　大数据与云计算、物联网的关系 ... 4
1.2　Hadoop 简介 ... 5
1.2.1　Hadoop 应用现状 ... 6
1.2.2　Hadoop 简介与意义 ... 6
1.3　小结 ... 6

第2章　Hadoop 的安装与配置 ... 7
2.1　虚拟机的创建 ... 7
2.2　安装 Linux 系统 ... 10
2.3　配置网络信息 ... 11
2.4　克隆服务器 ... 12
2.5　SSH 免密码登录 ... 13
2.6　安装和配置 JDK ... 15
2.6.1　上传安装包 ... 15
2.6.2　安装 JDK ... 16
2.6.3　配置环境变量 ... 16
2.7　Hadoop 环境变量配置 ... 16
2.7.1　解压缩 Hadoop 压缩包 ... 17
2.7.2　配置 Hadoop 的 bin 和 sbin 文件夹到环境变量中 ... 17

		2.7.3 修改/etc/hadoop/hadoop-env.sh	17
2.8	Hadoop 分布式安装		17
	2.8.1	伪分布式安装	17
	2.8.2	完全分布式安装	19
2.9	小结		21

第 3 章 Hadoop 分布式文件系统 ... 22

- 3.1 DFS 介绍 ... 22
  - 3.1.1 什么是 DFS ... 22
  - 3.1.2 DFS 的结构 ... 22
- 3.2 HDFS 介绍 ... 23
  - 3.2.1 HDFS 的概念及体系结构 ... 23
  - 3.2.2 HDFS 的设计 ... 23
  - 3.2.3 HDFS 的优点和缺点 ... 24
  - 3.2.4 HDFS 的执行原理 ... 24
  - 3.2.5 HDFS 的核心概念 ... 25
  - 3.2.6 HDFS 读文件流程 ... 27
  - 3.2.7 HDFS 写文件流程 ... 28
  - 3.2.8 Block 的副本放置策略 ... 29
- 3.3 Hadoop 中 HDFS 的常用命令 ... 30
  - 3.3.1 对文件的操作 ... 30
  - 3.3.2 管理与更新 ... 31
- 3.4 HDFS 的应用 ... 31
  - 3.4.1 基于 Shell 的操作 ... 31
  - 3.4.2 基于 Java API 的操作 ... 33
  - 3.4.3 创建文件夹 ... 34
  - 3.4.4 递归显示文件 ... 34
  - 3.4.5 文件上传 ... 35
  - 3.4.6 文件下载 ... 35
- 3.5 小结 ... 36

第 4 章 基于 Hadoop 3 的 HDFS 高可用 ... 37

- 4.1 Hadoop 3.x 的发展 ... 37
  - 4.1.1 Hadoop 3 新特性 ... 37
  - 4.1.2 Hadoop 3 HDFS 集群架构 ... 38
- 4.2 Hadoop 3 HDFS 完全分布式搭建 ... 39
  - 4.2.1 安装 JDK ... 40
  - 4.2.2 配置 JDK 环境变量 ... 40
  - 4.2.3 配置免密码登录 ... 40

		4.2.4 配置 IP 和主机名字映射关系	41
		4.2.5 SSH 免密码登录设置	41
		4.2.6 配置 Hadoop 3.1.0	42
	4.3	什么是 HDFS 高可用	47
		4.3.1 HDFS 高可用实现原理	47
		4.3.2 HDFS 高可用实现	48
	4.4	搭建 HDFS 高可用	50
		4.4.1 配置 ZooKeeper	50
		4.4.2 配置 Hadoop 配置文件	52
		4.4.3 将配置文件复制到其他节点上	54
		4.4.4 启动 JN 节点	54
		4.4.5 格式化	55
		4.4.6 复制元数据到 node2 节点上	55
		4.4.7 格式化 ZKFC	55
		4.4.8 启动集群	56
		4.4.9 通过浏览器查看集群状态	56
		4.4.10 高可用测试	57
	4.5	小结	58

第 2 篇 Hadoop 核心技术

第 5 章 Hadoop 的分布式协调服务——ZooKeeper — 60

	5.1	ZooKeeper 的核心概念	60
		5.1.1 Session 会话机制	60
		5.1.2 数据节点、版本与 Watcher 的关联	61
		5.1.3 ACL 策略	61
	5.2	ZooKeeper 的安装与运行	61
	5.3	ZooKeeper 服务器端的常用命令	63
	5.4	客户端连接 ZooKeeper 的相关操作	64
		5.4.1 查看 ZooKeeper 常用命令	64
		5.4.2 connect 命令与 ls 命令	65
		5.4.3 create 命令——创建节点	65
		5.4.4 get 命令——获取数据与信息	66
		5.4.5 set 命令——修改节点内容	66
		5.4.6 delete 命令——删除节点	67
	5.5	使用 Java API 访问 ZooKeeper	67
		5.5.1 环境准备与创建会话实例	68

5.5.2 节点创建实例	69
5.5.3 Java API 访问 ZooKeeper 实例	70
5.6 小结	73

第6章 分布式离线计算框架——MapReduce ... 74

6.1 MapReduce 概述	74
6.1.1 MapReduce 的特点	74
6.1.2 MapReduce 的应用场景	75
6.2 MapReduce 执行过程	76
6.2.1 单词统计实例	76
6.2.2 MapReduce 执行过程	77
6.2.3 MapReduce 的文件切片 Split	77
6.2.4 Map 过程和 Reduce 过程	78
6.2.5 Shuffle 过程	78
6.3 MapReduce 实例	79
6.3.1 WordCount 本地测试实例	79
6.3.2 ETL 本地测试实例	84
6.4 温度排序实例	86
6.4.1 时间和温度的封装类 MyKey.Java	87
6.4.2 Map 任务 MyMapper.java	88
6.4.3 数据分组类 MyGroup.Java	89
6.4.4 温度排序类 MySort.java	89
6.4.5 数据分区 MyPartitioner.java	90
6.4.6 Reducer 任务 MyReducer.java	90
6.4.7 主函数 RunJob.java	91
6.5 小结	94

第7章 Hadoop 的集群资源管理系统——YARN ... 95

7.1 为什么要使用 YARN	95
7.2 YARN 的基本架构	96
7.2.1 ResourceManager 进程	96
7.2.2 ApplicationMaster 和 NodeManager	97
7.3 YARN 工作流程	97
7.4 YARN 搭建	98
7.5 小结	100

第8章 Hadoop 的数据仓库框架——Hive ... 101

| 8.1   Hive 的理论基础 | 101 |
| 8.1.1   什么是 Hive | 101 |

- 8.1.2 Hive 和数据库的异同 102
- 8.1.3 Hive 设计的目的与应用 104
- 8.1.4 Hive 的运行架构 104
- 8.1.5 Hive 的执行流程 105
- 8.1.6 Hive 服务 106
- 8.1.7 元数据存储 Metastore 106
- 8.1.8 Embedded 模式 107
- 8.1.9 Local 模式 108
- 8.1.10 Remote 模式 109
- 8.2 Hive 的配置与安装 109
  - 8.2.1 安装 MySQL 110
  - 8.2.2 配置 Hive 112
- 8.3 Hive 表的操作 113
  - 8.3.1 创建 Hive 表 114
  - 8.3.2 导入数据 114
- 8.4 表的分区与分桶 115
  - 8.4.1 表的分区 115
  - 8.4.2 表的分桶 117
- 8.5 内部表与外部表 118
  - 8.5.1 内部表 119
  - 8.5.2 外部表 119
- 8.6 内置函数与自定义函数 121
  - 8.6.1 内置函数实例 121
  - 8.6.2 自定义 UDAF 函数实例 123
- 8.7 通过 Java 访问 Hive 124
- 8.8 Hive 优化 125
  - 8.8.1 MapReduce 优化 126
  - 8.8.2 配置优化 126
- 8.9 小结 127

第 9 章 大数据快速读写——HBase 128

- 9.1 关于 NoSQL 128
  - 9.1.1 什么是 NoSQL 128
  - 9.1.2 NoSQL 数据库的分类 129
  - 9.1.3 NoSQL 数据库的应用 129
  - 9.1.4 关系型数据库与非关系型数据库的区别 130
- 9.2 HBase 基础 130
  - 9.2.1 HBase 简介 130

  9.2.2 HBase 数据模型 131
  9.2.3 HBase 体系架构及组件 132
  9.2.4 HBase 执行原理 134
9.3 HBase 安装 135
9.4 HBase 的 Shell 操作 138
9.5 Java API 访问 HBase 实例 139
  9.5.1 创建表 139
  9.5.2 插入数据 140
  9.5.3 查询数据 141
9.6 小结 142

第 10 章 海量日志采集工具——Flume

10.1 什么是 Flume 143
10.2 Flume 的特点 143
10.3 Flume 架构 144
10.4 Flume 的主要组件 144
  10.4.1 Event、Client 与 Agent——数据传输 145
  10.4.2 Source——Event 接收 145
  10.4.3 Channel——Event 传输 146
  10.4.4 Sink——Event 发送 147
  10.4.5 其他组件 148
10.5 Flume 安装 148
10.6 Flume 应用典型实例 149
  10.6.1 本地数据读取（conf1） 149
  10.6.2 收集至 HDFS 150
  10.6.3 基于日期分区的数据收集 152
10.7 通过 exec 命令实现数据收集 153
  10.7.1 安装工具 153
  10.7.2 编辑配置文件 conf4 155
  10.7.3 运行 Flume 156
  10.7.4 查看生成的文件 156
  10.7.5 查看 HDFS 中的数据 157
10.8 小结 158

第 11 章 Hadoop 和关系型数据库间的数据传输工具——Sqoop

11.1 什么是 Sqoop 159
11.2 Sqoop 工作机制 159
11.3 Sqoop 的安装与配置 161
  11.3.1 下载 Sqoop 161

目录

11.3.2 Sqoop 配置 ·· 162
11.4 Sqoop 数据导入实例 ··· 163
  11.4.1 向 HDFS 中导入数据 ··· 165
  11.4.2 将数据导入 Hive ·· 167
  11.4.3 向 HDFS 中导入查询结果 ·· 170
11.5 Sqoop 数据导出实例 ··· 172
11.6 小结 ·· 173

第 12 章 分布式消息队列——Kafka ··· 174
12.1 什么是 Kafka ·· 174
12.2 Kafka 的架构和主要组件 ··· 174
  12.2.1 消息记录的类别名——Topic ··· 175
  12.2.2 Producer 与 Consumer——数据的生产和消费 ························· 176
  12.2.3 其他组件——Broker、Partition、Offset、Segment ···················· 177
12.3 Kafka 的下载与集群安装 ··· 177
  12.3.1 安装包的下载与解压 ·· 177
  12.3.2 Kafka 的安装配置 ··· 178
12.4 Kafka 应用实例 ··· 181
  12.4.1 Producer 实例 ·· 181
  12.4.2 Consumer 实例 ··· 182
12.5 小结 ·· 184

第 13 章 开源的内存数据库——Redis ··· 185
13.1 Redis 简介 ··· 185
  13.1.1 什么是 Redis ·· 185
  13.1.2 Redis 的特点 ·· 186
13.2 Redis 安装与配置 ··· 186
13.3 客户端登录 ·· 187
  13.3.1 密码为空登录 ·· 187
  13.3.2 设置密码登录 ·· 188
13.4 Redis 的数据类型 ··· 188
  13.4.1 String 类型 ·· 188
  13.4.2 List 类型 ··· 190
  13.4.3 Hash 类型 ··· 191
  13.4.4 Set 类型 ··· 194
13.5 小结 ·· 197

第 14 章 Ambari 和 CDH ··· 198
14.1 Ambari 的安装与集群管理 ··· 198

14.1.1	认识 HDP 与 Ambari	198
14.1.2	Ambari 的搭建	199
14.1.3	配置网卡与修改本机名	199
14.1.4	定义 DNS 服务器与修改 hosts 主机映射关系	200
14.1.5	关闭防火墙并安装 JDK	200
14.1.6	升级 OpenSSL 安全套接层协议版本	201
14.1.7	关闭 SELinux 的强制访问控制	201
14.1.8	SSH 免密码登录	202
14.1.9	同步 NTP	202
14.1.10	关闭 Linux 的 THP 服务	204
14.1.11	配置 UMASK 与 HTTP 服务	204
14.1.12	安装本地源制作相关工具与 Createrepo	205
14.1.13	禁止离线更新与制作本地源	205
14.1.14	安装 Ambari-server 与 MySQL	208
14.1.15	安装 Ambari	210
14.1.16	安装 Agent 与 Ambari 登录安装	211
14.1.17	安装部署问题解决方案	214
14.2	CDH 的安装与集群管理	216
14.2.1	什么是 CDH 和 Cloudera Manager 介绍	216
14.2.2	Cloudera Manager 与 Ambari 对比的优势	216
14.2.3	CDH 安装和网卡配置	217
14.2.4	修改本机名与定义 DNS 服务器	217
14.2.5	修改 hosts 主机映射关系	218
14.2.6	关闭防火墙	218
14.2.7	安装 JDK	219
14.2.8	升级 OpenSSL 安全套接层协议版本	219
14.2.9	禁用 SELinux 的强制访问功能	220
14.2.10	SSH 免密码登录	220
14.2.11	同步 NTP 安装	220
14.2.12	安装 MySQL	222
14.2.13	安装 Cloudera Manager	222
14.2.14	添加 MySQL 驱动包和修改 Agent 配置	223
14.2.15	初始化 CM5 数据库和创建 cloudera-scm 用户	223
14.2.16	准备 Parcels	223
14.2.17	CDH 的安装配置	224
14.3	小结	227

第 15 章 快速且通用的集群计算系统——Spark 228
15.1 Spark 基础知识 228
15.1.1 Spark 的特点 228
15.1.2 Spark 和 Hadoop 的比较 229
15.2 弹性分布式数据集 RDD 230
15.2.1 RDD 的概念 230
15.2.2 RDD 的创建方式 230
15.2.3 RDD 的操作 230
15.2.4 RDD 的执行过程 231
15.3 Spark 作业运行机制 232
15.4 运行在 YARN 上的 Spark 233
15.4.1 在 YARN 上运行 Spark 233
15.4.2 Spark 在 YARN 上的两种部署模式 233
15.5 Spark 集群安装 234
15.5.1 Spark 安装包的下载 234
15.5.2 Spark 安装环境 236
15.5.3 Scala 安装和配置 236
15.5.4 Spark 分布式集群配置 238
15.6 Spark 实例详解 241
15.6.1 网站用户浏览次数最多的 URL 统计 241
15.6.2 用户地域定位实例 243
15.7 小结 246

第 3 篇 Hadoop 项目案例实战

第 16 章 基于电商产品的大数据业务分析系统实战 248
16.1 项目背景、实现目标和项目需求 248
16.2 功能与流程 249
16.2.1 用户信息 250
16.2.2 商品信息 251
16.2.3 购买记录 251
16.3 数据收集 252
16.3.1 Flume 的配置文件 252
16.3.2 启动 Flume 253
16.3.3 查看采集后的文件 253
16.3.4 通过后台命令查看文件 254

16.3.5	查看文件内容	255
16.3.6	上传 user.list 文件	256
16.3.7	上传 brand.list 目录	256

16.4 数据预处理 257
16.5 数据分析——创建外部表 261
16.6 建立模型 264

16.6.1	各年龄段用户消费总额	264
16.6.2	查询各品牌销售总额	265
16.6.3	查询各省份消费总额	266
16.6.4	使用 Sqoop 将数据导入 MySQL 数据库	266

16.7 数据可视化 268
16.8 小结 272

第 17 章 用户画像分析实战 273

17.1 项目背景 273
17.2 项目目标与项目开发过程 274

17.2.1	数据采集	274
17.2.2	数据预处理	275
17.2.3	模型构建	275
17.2.4	数据分析	276

17.3 核心代码解读 277

17.3.1	项目流程介绍	277
17.3.2	核心类的解读	278
17.3.3	core-site.xml 配置文件	279
17.3.4	hdfs-site.xml 配置文件	279
17.3.5	UserProfile.properties 配置文件	280
17.3.6	LoadConfig.java：读取配置信息	280
17.3.7	ReadFile.java：读取文件	281
17.3.8	ReadFromHdfs.java：提取信息	281
17.3.9	UserProfile.java：创建用户画像	282
17.3.10	TextArrayWritable.java：字符串处理工具类	285
17.3.11	MapReduce 任务 1：UserProfileMapReduce.java	285
17.3.12	MapReduce 任务 2：UserProfileMapReduce2.java	289
17.3.13	UserProfilePutInHbaseMap.java：提取用户画像	291
17.3.14	UserProfilePutInHbaseReduce：存储用户画像	292

17.4 项目部署 293
17.5 小结 294

第18章 基于个性化的视频推荐系统实战 ... 295
18.1 项目背景 ... 295
18.2 项目目标与推荐系统简介 ... 295
18.2.1 推荐系统的分类 ... 295
18.2.2 推荐模型的构建流程 ... 296
18.2.3 推荐系统核心算法 ... 297
18.2.4 如何基于Mahout框架完成商品推荐 ... 300
18.2.5 基于Mahout框架的商品推荐实例 ... 300
18.3 推荐系统项目架构 ... 302
18.4 推荐系统模型构建 ... 303
18.5 核心代码 ... 304
18.5.1 公共部分 ... 305
18.5.2 离线部分 ... 307
18.5.3 在线部分 ... 311
18.6 小结 ... 314

第19章 电信离网用户挽留实战 ... 315
19.1 商业理解 ... 315
19.2 数据理解 ... 316
19.2.1 收集数据 ... 316
19.2.2 了解数据 ... 317
19.2.3 保证数据质量 ... 318
19.3 数据整理 ... 318
19.3.1 数据整合 ... 318
19.3.2 数据过滤 ... 319
19.4 数据清洗 ... 319
19.4.1 噪声识别 ... 320
19.4.2 离群值和极端值的定义 ... 321
19.4.3 离群值处理方法 ... 321
19.4.4 数据空值处理示例 ... 323
19.5 数据转换 ... 324
19.5.1 变量转换 ... 324
19.5.2 压缩分类水平数 ... 324
19.5.3 连续数据离散化 ... 325
19.5.4 变换哑变量 ... 326
19.5.5 数据标准化 ... 326
19.5.6 数据压缩 ... 326
19.6 建模 ... 327

- 19.6.1 决策树算法概述 ... 327
- 19.6.2 决策树的训练步骤 ... 327
- 19.6.3 训练决策树 ... 328
- 19.6.4 C4.5 算法 ... 329
- 19.6.5 决策树剪枝 ... 332
- 19.7 评估 ... 335
  - 19.7.1 混淆矩阵 ... 335
  - 19.7.2 ROC 曲线 ... 336
- 19.8 部署 ... 338
- 19.9 用户离网案例代码详解 ... 339
  - 19.9.1 数据准备 ... 339
  - 19.9.2 相关性分析 ... 341
  - 19.9.3 最终建模 ... 342
  - 19.9.4 模型评估 ... 343
- 19.10 小结 ... 346

第1篇
Hadoop 基础知识

▶▶ 第1章 初识 Hadoop

▶▶ 第2章 Hadoop 的安装与配置

▶▶ 第3章 Hadoop 分布式文件系统

▶▶ 第4章 基于 Hadoop 3 的 HDFS 高可用

第 1 章　初识 Hadoop

随着互联网的高速发展，越来越多的用户在日常使用网络的过程中产生了数量庞大的结构化数据，同时在日常生活中也产生了大量的非结构化数据，如视频、音频和图像等。因此，对大量数据的有效存储管理和计算分析成为了信息行业迫切需要解决的问题。大数据就是基于数据爆炸的现状产生的。

Hadoop 的前身由 Doug Cutting 创建，起源于开源的网络搜索引擎 Apache Nutch，本章将从大数据的技术、特点和存储计算模式为起点，初步探究大数据的雏形。

以下是本章主要涉及的知识点。
- 了解大数据的特点及在各行业中的应用。
- 了解大数据技术，掌握大数据与云计算和物联网的关系。
- 了解 Hadoop 的起源、发展和意义。

1.1　大数据初探

本章首先介绍大数据的基本概念，理解这些概念是进一步学习和掌握大数据的基础。了解概念后，才能从大数据和云计算中找到学习的技巧。

在当前的技术领域内，大家提的比较多的当属大数据了，那么到底什么是大数据呢？关于大数据的定义目前有很多种，其实"大数据"就是收集各种数据，经过分析后用来做有意义的事，其中包括对数据进行采集、管理、存储、搜索、共享、分析和可视化。

关于数据的采集、存储和分析较容易理解，因为当数据量足够大的时候，很难存储，如 FaceBook 每天生成 500TB 的数据，如何存储这些数据就成了一个问题。有时我们需要存储大量的数据并进行分析，将分析结果用于运营决策，给决策者提供运营参考，而传统的技术无法实现大批量数据的存储和计算，毕竟单台机器的存储和计算性能都是有限的。

1.1.1　大数据技术

那么，大数据技术又是什么呢？从本质上来说，大数据技术是发现大规模数据中的规律，通过对数据的分析实现对运营层决策的支持。在此处需要注意大数据技术与其他学科之间的关系，Excel 也可以做数据分析，那么为什么还要用到大数据技术呢？

主要原因是，大数据技术面对的是大规模的数据，每一天都会有大批量的数据生成，如何存储与计算这批数据，就是大数据技术要解决的问题。

1.1.2 大数据技术框架

大数据技术框架主要包含 6 个部分，分别是数据收集、数据存储、资源管理、计算框架、数据分析和数据展示，每部分包括的具体技术如图 1.1 所示。

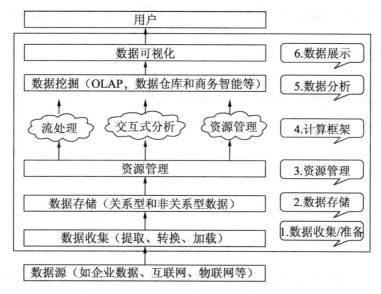

图 1.1 大数据技术框架图

1.1.3 大数据的特点

大数据的特点可以用"4v"来表示，分别为 volume、variety、velocity 和 value，下面具体介绍。

- 海量性（volume）：大数据的数据量很大，每天我们的行为都会产生大批量数据。
- 多样性（variety）：大数据的类型多种多样，比如视频、音频和图片都属于数据。
- 高速性（velocity）：大数据要求处理速度快，比如淘宝"双十一"需要实时显示交易数据。
- 价值性（value）：大数据产生的价值密度低，意思是说大部分数据没有参考意义，少部分数据会形成高价值，比如私家汽车安装的摄像头，大部分情况下是用不到的，但是一旦出现"碰瓷"等现象就会很有价值。

1.1.4 大数据在各个行业中的应用

大数据的本质是发现数据规律，实现商业价值。在生活中有很多大数据应用的场景，包括金融、经济、医疗和体育行业等。

例如在金融行业中，支付宝平台通过大数据进行消费者信用评分，金融机构利用大数据进行金融产品的精准营销。在医疗行业中通过分析病人特征和疗效数据，找到特定病人的最佳治疗方案；还可以在病人档案方面应用高级分析，确定某类疾病的易感人群。在体育行业中可以通过分析数据来制定战术、进行运动员能力评估，定制最佳训练方案。

1.1.5 大数据计算模式

常见的大数据计算模式分为 4 类，如图 1.2 所示。

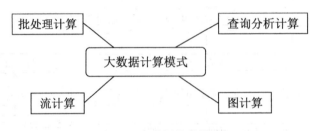

图 1.2 大数据计算模式

- 批处理计算又称为离线计算，是针对大规模历史数据的批量处理，如 MapReduce。
- 流计算是针对流数据的实时计算，可以实时处理产生的数据。商业版的有 IBM InfoSphere Streams 和 IBM StreamBase，开源的有 Storm 和 S4（Simple Scalable Streaming System），还有一部分是企业根据自身需求而定制的，如 Dstream（百度）。
- 图计算是针对大规模图结构数据的处理，常用于社交网络，如 Pregel、GraphX、Giraph（FaceBook）、PowerGraph 和 Hama 等。
- 查询分析计算是针对大规模数据的存储管理和查询分析，如 Hive、Cassandra 和 Impala 等。

1.1.6 大数据与云计算、物联网的关系

关于云计算的解释有很多种，被人们广为接受的是美国国家标准与技术研究院所定义的，即云计算是一种按网络使用量付费的便捷模式，能进入可配置的计算资源共享池（资源包括网络、服务器、存储、应用软件、服务），使资源被利用。

- 云计算的特点：超大规模、通用性、高拓展性、虚拟化、高可靠性、按需服务、极

其廉价、具有潜在危险性。
- 云计算的模式：公有云、私有云、混合云。
- 云计算服务的分类：Saas、Paas 和 Iaas。

大数据与云计算是一种不可分的、相互依存的关系。首先，云计算是计算资源的底层，它的主要作用是支撑上层大数据的处理任务。而大数据的主要处理任务则是提升实时交互式查询效率和分析数据的能力。

物联网，其实就是物物相连的互联网。这其中包含两个意思，一个是在互联网基础上的延伸和扩展，起到核心作用的仍然是互联网；另一个是不管用户端延伸到任何物品上，最终都实现物物相连。

在物联网应用中有 3 项关键技术：传感器技术、RFID 标签和嵌入式系统技术。

物联网产生大数据，大数据助力物联网。随着物联网的发展，产生数据的终端由 PC 转向了包括 PC、智能手机和平板电脑等在内的多样化终端，因此物联网推动了大数据技术的发展。

大数据、云计算和物联网三者息息相关，是互相关联、相互作用的。物联网是大数据的来源（设备数据），大数据技术为物联网数据的分析提供了强有力的支撑；物联网还为云计算提供了广阔的应用空间，而云计算为物联网提供了海量数据存储能力；云计算还为大数据提供了技术基础，而大数据能为云计算所产生的运营数据提供分析和决策依据。三者的关系如图 1.3 所示。

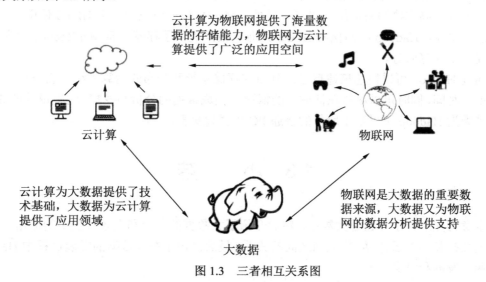

图 1.3　三者相互关系图

1.2　Hadoop 简介

Apache Hadoop 本身是一个框架，它可以用简单的编程模型在计算机集群中对大型数据集进行分布式处理。它可以被设计成单个机器或成千上万台机器的集群，实现提供计算和存储服务。

然而，不同于依赖硬件实现的高可用性，Hadoop 本身被设计为能够检测和处理应用层的错误，因此在计算机集群的整体层面上就提供了高可用服务。

1.2.1 Hadoop 应用现状

随着大数据的快速发展，目前 Hadoop 已经应用在了很多大大小小的互联网企业中。对于国内来说，百度的日志分析、阿里的内部云，以及淘宝都在使用 Hadoop；在国际上，英特尔、微软和 Oracle 等也都有了自己基于 Hadoop 的产品。

1.2.2 Hadoop 简介与意义

Apache 开源软件基金会开发了运行在大规模普通服务器上，用于大数据存储、计算、分析的分布式存储系统和分布式运算框架——Hadoop。Hadoop 的两大核心如下。

- HDFS（Hadoop Distributed File System，分布式存储系统）：是 Hadoop 中的核心组件之一，除了可以保存海量数据，还具有高可靠性、高扩展性和高吞吐率的特点。
- MapReduce：属于分布式计算框架，一般用于对海量数据的计算，它的特点是易于编程、高容错和高扩展等优点。另外，MapReduce 可以独立于 HDFS 使用。

总结来说，Hadoop 中的核心 HDFS 为海量数据提供了存储，而 MapReduce 则为海量数据提供了计算服务。

通过 Hadoop 可以快速搭建自己的分布式存储系统和分布式运算系统，它可以缩短处理数据的时间，同时可以尽量在低成本的情况下完成数据的分析与挖掘。这里说的低成本，主要是因为 Hadoop 可以基于廉价的普通 PC 机搭建集群。

1.3 小　　结

本章首先解释了什么是大数据，然后介绍了大数据的特点和在各行业中的应用，并说明了大数据和当下云计算、物联网之间的关系，最后讲述了在大数据的发展过程中 Hadoop 的起源、发展和意义。

第 2 章 Hadoop 的安装与配置

Hadoop 可以基于 Windows 安装，也可以基于 Linux 安装，本章我们以企业中用的比较多的 Linux 系统为例，讲解 Hadoop 的安装，为了便于读者练习、要先安装虚拟机。本章将深入讲解基于 Linux 的 Hadoop 安装与配置。

以下是本章主要涉及的知识点。
- 学会创建虚拟机。
- 通过本章示例学会安装 Linux 系统和 JDK。
- 通过示例学会 Hadoop 的安装及分布式安装。

2.1 虚拟机的创建

安装虚拟机的步骤如下：

（1）单击主页中"创建新的虚拟机"选项，如图 2.1 所示。

图 2.1 安装第 1 步：创建新的虚拟机

（2）在弹出的界面中选择"自定义（高级）"单选按钮，然后单击"下一步"按钮，如图 2.2 所示。

（3）在进入的界面中单击"下一步"按钮，如图 2.3 所示。

（4）在进入的界面中选择"稍后安装操作系统"单选按钮，然后单击"下一步"按钮，如图 2.4 所示。

（5）在进入的界面中选择"Linux（L）"单选按钮，"版本"选择"CentOS 64 位"，然后单击"下一步"按钮，如图 2.5 所示。

图 2.2　安装第 2 步

图 2.3　安装第 3 步

（6）在进入的界面中将"虚拟机名称"改为 masternode，"位置"改为 C:\masternode 单击"下一步"按钮，如图 2.6 所示。

图 2.4　安装第 4 步　　　　　　　　　图 2.5　安装第 5 步

（7）在进入的界面中，在"处理器数量"下拉列表框中选择 1，在每个处理器的核心数量下拉列表框中选择 1，然后单击"下一步"按钮，如图 2.7 所示。

（8）在进入的界面中根据计算机配置选择内存大小，如果计算机内存是 4GB，这里就选择 1GB，如果是 8GB 以上，就选择 2GB，这里我们选择 2GB，然后单击"下一步"按钮，如图 2.8 所示。

（9）接着会出现创建虚拟机的信息，这里不做任何修改，采取默认设置，直接单击"下一步"按钮，再单击"完成"按钮，如图 2.9 所示。

（10）之后就会出现如下结果，表示虚拟机安装完成，如图 2.10 所示。

第 2 章　Hadoop 的安装与配置

图 2.6　安装第 6 步

图 2.7　安装第 7 步

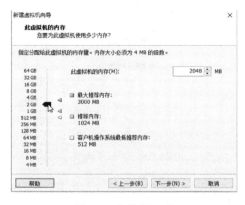

图 2.8　安装第 8 步

图 2.9　安装第 9 步

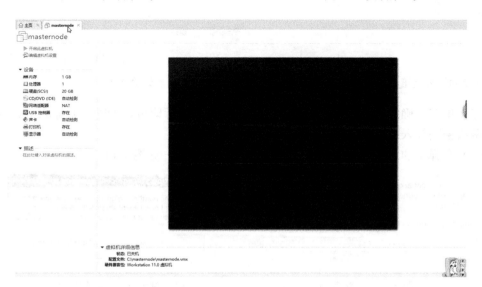

图 2.10　安装完成

2.2 安装 Linux 系统

安装完虚拟机后，接来下就可以基于虚拟机安装 CentOS 系统了。CentOS 系统安装步骤如下。

（1）把 CentOS mini 版本放到光驱中，右击 CD/DVD 选项，如图 2.11 所示。

（2）在弹出的界面中选中"使用 ISO 映像文件"单选按钮，然后单击"浏览"按钮，指向 CentOS 映像文件，如图 2.12 所示。

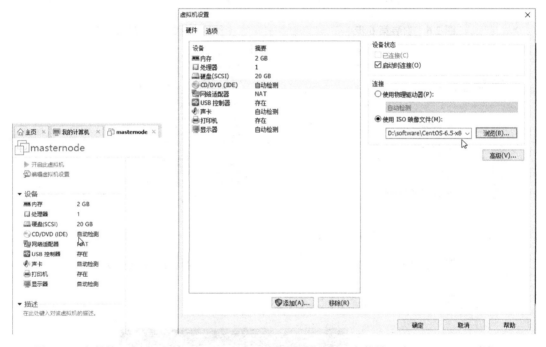

图 2.11　安装第 1 步　　　　　　　　图 2.12　安装第 2 步

（3）单击"确定"按钮，然后按照图示依次选择运行即可，期间需要输入 root 密码，注意两次密码输入要一致，如图 2.13 所示。

（4）在弹出的界面中单击 Write changes to disk 按钮，如图 2.14 所示。

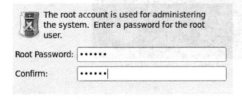
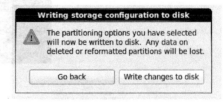

图 2.13　设置密码　　　　　　　　图 2.14　确认创建系统

2.3 配置网络信息

2.2 节中我们已成功创建了 CentOS 系统，但是由于系统还没有配置网络，所以无法访问外网，也无法进行内网机器之间的通信。为了后续搭建集群和访问外网，这里需要进行网络信息的配置，主要分为以下几个步骤。

1. 修改配置信息

```
vi /etc/sysconfig/network-scripts/ifcfg-eth0
```

在 Linux 相应目录下执行以上命令。

```
[root@masternode~]#vi /etc/sysconfig/network-scripts/ifcfg-eth0
```

编辑文件内容如下：

```
DEVICE=eth0
TYPE=Ethernet
ONBOOT=yes
NM_CONTROLLED=yes
BOOTPROTO=static
IPADDR=192.168.245.150
NATMASK=255.255.255.0
GATEWAY=192.168.245.2
```

2. 重启网络服务

```
service network restart
[root@mastermode~]# service network restart
```

当所有结果都显示为 OK 时，才能成功。

```
[root@masternode~]# service network restart
Shutting down loopback interface:                    [ok]
Bringing up loopback interface:                      [ok]
Bringing up interface eth0:Determining if ip address 192.168.245.150 is alread
y in use for device eth0...
```

3. 检验：ping网关（两个）

```
ping www.baidu.com
[root@masternode~]# ping www.baidu.com
Ping:unknown host www.baidu.com
```

4. 测试：ping网关

（1）设置 DNS 服务，输入以下命令：

```
vi /etc/resolv.conf
nameserver 114.114.114.114
```

修改 DNS 后在访问外网时会进行域名的解析，可以 ping 通外网。

（2）访问百度（命令：ping www.baidu.com）：

```
[root@master ~]# ping www.baidu.com
PING www.a.shifen.com (111.13.100.91) 56(84) bytes of data.
64 bytes from 111.13.100.91: icmp_seq=1 ttl=128 time=6.47 ms
64 bytes from 111.13.100.91: icmp_seq=2 ttl=128 time=77.7 ms
64 bytes from 111.13.100.91: icmp_seq=3 ttl=128 time=18.4 ms
64 bytes from 111.13.100.91: icmp_seq=4 ttl=128 time=111 ms
64 bytes from 111.13.100.91: icmp_seq=5 ttl=128 time=124 ms
64 bytes from 111.13.100.91: icmp_seq=6 ttl=128 time=12.6 ms
```

这样就说明能 ping 通网络了。

2.4 克隆服务器

有时为了方便使用，减少重复配置，可以直接将配置好的节点进行克隆，克隆节点时必须在被克隆的节点处于关机状态下。下面我们开始克隆创建出来的 namenode，步骤如图 2.15、图 2.16 和图 2.17 所示。

图 2.15　克隆第 1 步

创建的虚拟机名称为 slavenode，位于 E:\slavenode 文件夹中。

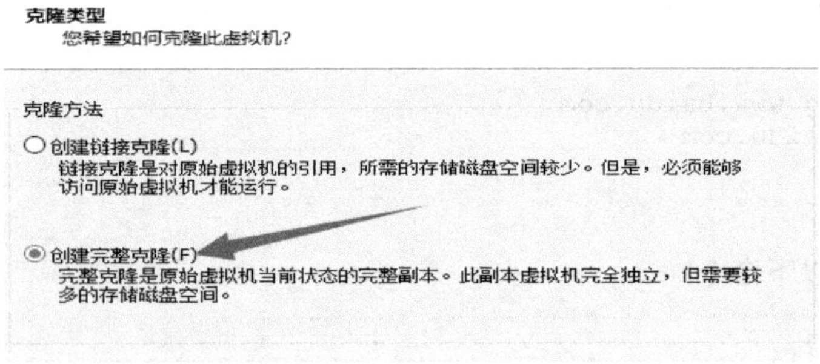

图 2.16　克隆第 2 步 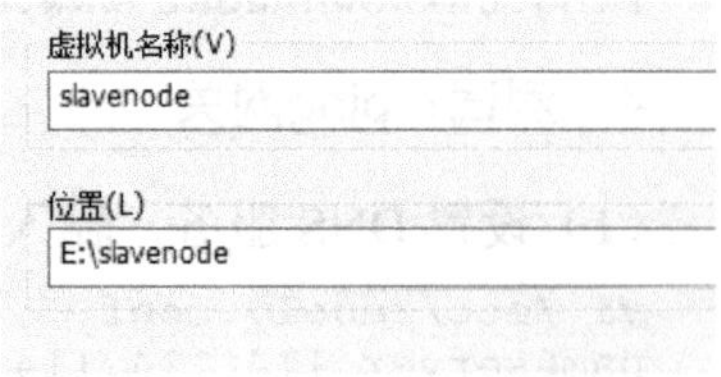　　　　　　　　图 2.17　克隆第 3 步

克隆完成之后，由于克隆的信息与被克隆机器完全一致，所以需要将克隆的机器重新进行网络等信息的设置，设置步骤如下。

（1）删除/etc/sysconfig/network-scripts/ifcfg-eth0 文件中的系统物理地址 HWADDR 和 UUID，并配置 slavenode 的 IP 地址。命令如下：

```
vi /etc/sysconfig/network-scripts/ifcfg-eth0
DEVICE=eth0
TYPE=Ethernet
ONBOOT=yes
NM_CONTROLLED=yes
BOOTPROTO=static
IPADDR=192.168.109.201
NATMASK=255.255.255.0
GATEWAY=192.168.109.2
```

（2）删除 70-persistent-net.rules 文件。命令如下：

```
[root@master~]# rm -rf /etc/udev/rules.d/70-persistent-net.rules
```

（3）通过修改/etc/sysconfig/network 来修改主机名。命令如下：

```
vi/etc/sysconfig/network
```

并将 HOSTNAME 改为 slavenode。

```
NETWORKING=yes
HOSTNAME=slave
```

（4）重新启动系统：

```
Init 6
```

（5）修改 hosts 文件，在文件最后增加一行，格式为：IP 地址 主机名，这样做的目的是将 IP 地址和机器名相映射。这样在 master 和 slave 相互通信时，可以直接使用主机名，也可以使用 IP 地址。

```
vi /etc/hosts
192.168.239.6   master
192.168.239.7   slave
```

修改完 master 主机上的/etc/hosts 文件后，就可以尝试在 master 上通过 ping 的方式与 slave 节点通信。

```
[root@master~]# ping slave
```

2.5　SSH 免密码登录

在 2.4 节中已经创建好了两台机器，并且相互之间可以进行通信，这时就需要进行文件的相互传输，如从 master 节点传到 slave 节点，这样就需要用到 scp（拷贝）等命令来完成。但是在多台服务器之间操作文件传输时总是需要输入密码，每次都输入密码很麻烦，而且有些安全系统高的机器，密码相当难记，而 SSH 免密码登录无疑能大大地提高工作效率。

总的来说，服务器 A 如果要免密码登录到服务器 B 时，需要在服务器 A 上生成密钥对，将生成的公钥上传到服务器 B 上，并把公钥追加到服务器 B 的 authorized_keys 信任文件中。具体操作步骤如下。

1. 在服务器A上执行命令ssh-keygen –t rsa –P' '创建密钥对

```
[root@master~]# ssh-keygen -t rsa -p ''
Genenrating public/private rsa key pair.
Enter file in which to save the key(/root/.ssh/id_rsa):
Created directory'/root/.ssh'.
Your identification has been saved in /root/.ssh/id_rsa.
Your public key has been saved in /root/.ssh/id_rsa.pub
The key fingerprint is:
C8:7d:b6:4a:b8:7a:46:66:87:d4:b2:30:55:2b:1b:cc  root@master
The key's randomart  image is:
+--[RSA 2048]----+1
1     ..         1
1     o. .       1
1     .E..       1
1     o.o*.      1
1     +++S o     1
1      *..o .    1
1     +... .     1
1     oo  .      1
1     .+. .      1
+-----------------+
```

2. 传递公钥

执行完上述命令后，在执行命令的.ssh 文件夹下会生成一个扩展名为.pub 的文件。下面通过 scp 命令将.pub 文件复制到 slave 节点上，假设这里的 slave 节点的 IP 是 192.168.109.201，命令如下：

```
scp .ssh/id_rsa.pub root@192.168.109.201:~
```

将公钥交给 slave，将 ssh/id_rsa.pub 文件复制到 slave 服务器上。

```
[root@master~]#scp .ssh/id_rsa.pub  root@192.168.109.201:~
The  authenticity of host '192.168.109.201(192.168.109.201)'can't be
established.
RSA key fingerprint is 6f:d7:06:f3:47:cf:36:If:9b:ee:3a:60:4d:5C:2d:3c
Are you sure you want to continue connecting (yes/no)?yes
Warning:Permanently added '192.168.109.201'(RSA)to the list of known hosts
root@192.168.109.201's password:
```

需要注意的是，第一次将文件传到 slave 节点时是需要输入密码的，一旦免密码配置完成，后续再换文件时就不需要再输入密码了。

3. 关闭防火墙

为了使两台机器之间进行通信，还需要将每个节点的防火墙都关闭。关闭防火墙有两种方法，一种是永久生效；另一种是立即生效，重启后无效。这两条命令我们可以都运行一遍，这样就代表防火墙立即关闭，并且机器重启后仍然处于关闭状态。

（1）即时生效，重启后失效：

```
service iptables stop
```

（2）重启后永久生效：

```
chkconfig iptables off
```

2.6　安装和配置 JDK

前面已经实现了系统安装、网络设置及防火墙关闭，到目前为止，多台计算机之间就可以进行通信了。

Hadoop 是基于 Java 语言环境的，所以还需要设置 Java 运行环境，在这里我们使用 JDK 7，读者可以从官方网站自行下载。

2.6.1　上传安装包

安装 JDK 的第一步是上传安装包，步骤如图 2.18、图 2.19 和图 2.20 所示，依次操作即可。

图 2.18　上传第 1 步

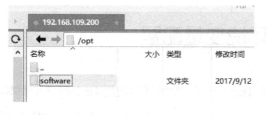

图 2.19　上传第 2 步

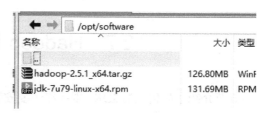

图 2.20　上传第 3 步

2.6.2 安装 JDK

安装 JDK，可以用 rpm 命令。安装命令如下：
`rpm -ivh jdk-7u79-linux-x64.rpm`

在后台运行安装命令后，结果如下，则代表安装成功。

```
[root@ master software]# rpm -ivh jdk -7u79-linux-x64.rpm
Preparing...
1:jdk
Unpacking JAR files...
        Rt.jar...
        Jsse.jar...
        Charsets.jar...
        Tools.jar...
        Localedata.jar...
        Jfxrt.jar...
```

2.6.3 配置环境变量

环境变量是一个具有特定名字的对象，它包含一个或者多个应用程序会使用到的信息。通过使用环境变量，可以很容易地修改涉及的一个或多个应用程序的配置信息。Linux 是一个多用户、多任务的操作系统，通常每个用户默认的环境都是相同的，这个默认环境实际上就是一组环境变量的定义。

用户可以对自己的运行环境进行定制，其方法就是修改相应的系统环境变量。这里我们可以通过修改 home 目录下的隐藏文件 .bash_profile 来修改环境变量。我们设置 JAVA_HOME 指向 JDK 的根目录。

```
export JAVA_HOME=/usr/java/jdk1.7.0_79
export PATH=$PATH:$JAVA_HOME/bin
```

设置完后，记得在 .bash_profile 的所在目录下执行 source .bash_profile，使环境变量立即生效。

2.7　Hadoop 环境变量配置

2.6 节介绍了如何安装 JDK，现在终于要安装 Hadoop 了，此处我们使用 Hadoop 2.x 版本。

另外，仅仅配置环境就经历了这么多步骤，对于没有任何编程基础或者对 Linux 命令陌生的读者，可能感受到了 Hadoop 的复杂性，所以说大数据（Hadoop）入门的门槛较高，但一旦熟练掌握了环境搭建，后续上手就会比较快了，所以让我们继续前行。

2.7.1 解压缩 Hadoop 压缩包

将 Hadopp 压缩包上传到服务器上以后，就需要进行解压缩了，解压代码如下：
```
tar -zxvf hadoop-2.5.1_x64.tar.gz
```

2.7.2 配置 Hadoop 的 bin 和 sbin 文件夹到环境变量中

解压完 Hadoop 后，就可以修改.bash_profile 配置文件，将 Hadoop 的环境信息写到配置文件中，如下代码所示。在这里需要注意，要将 Hadoop 下的 bin 文件夹和 sbin 文件夹都写入配置文件中，其中 sbin 文件夹中是管理命令，如启动和关闭集群等。

关于 Java 环境变量配置已经在 2.6.3 节中介绍过了。本节在 Hadoop 环境变量配置后，配置文件中的完整配置代码如下：
```
export JAVA_HOME=/usr/java/jdk1.7.0_79
export HADOOP_HOME=/opt/software/hadoop-2.5.1
export PATH=$PATH:$JAVA_HOME/bin:$HADOOP_HOME/bin
```

2.7.3 修改/etc/hadoop/hadoop-env.sh

接着需要在 etc/hadoop/hadoop-env.sh 中配置 JAVA_HOME，否则调用 start-dfs.sh 启动时会报错（Error: JAVA_HOME is not set and could not be found）。可以通过以下代码完成 JAVA_HOME 的配置：
```
export JAVA_HOME=/usr/java/jdk1.7.0_79
```

2.8 Hadoop 分布式安装

Hadoop 安装可以是单节点、伪分布式和完全分布式。这里我们着重介绍伪分布式和完全分布式。伪分布式是在一台机器上模拟分布式，主要用于测试；而完全分布式是由两个及两个以上的节点组建的集群，是真正的分布式。下面介绍伪分布式和完全分布式的安装过程。

2.8.1 伪分布式安装

（1）进入 Hadoop 的配置文件目录。
```
cd /opt/software/hadoop-2.5.1/etc/hadoop
```

（2）修改 core-site.xml 文件。

core-site.xml 文件主要配置了访问 Hadoop 集群的主要信息，其中 master 代表主机名称，也可以使用 IP 替换，9000 代表端口。外部通过配置的 hdfs://master:9000 信息，就可以找到 Hadoop 集群。

```
<configuration>
    <property>
        <name>fs.defaultFS</name>
        <value>hdfs://master:9000</value>
    </property>
</configuration>
```

（3）修改 hdfs-site.xml 配置。

hdfs-site.xml 配置文件中配置了 HDFS 的相关信息，其中 dfs.replication 代表副本数，这里设置为 1。

```
<configuration>
    <property>
        <name>dfs.replication</name>
        <value>1</value>
    </property>
</configuration>
```

（4）格式化 HDFS。

格式化 HDFS 的作用是初始化集群，基本配置完成后，就可以通过 hdfs namenode -format 命令初始化集群了。

```
[root@master hadoop -2.5.1]#hdfs namenode -format
```

格式化完成后，会输出以下信息：

```
8/04/27 16:20:20 INFO common.Storage:Storage directory /tmp/hadoop-root/dfs/name has been successfully formatted.
18/04/27 16:20:20 INFO namenode.FSImageFormatProtobuf:Saving image file /tmp/hadoop-root/dfs/name/current/fsimage.ckpt_0000000000000000000 using no compression
18/04/27 16:20:20 INFO namenode.FSImageFormatProtobuf:Image file /tmp/hadoop-root/dfs/name/current/fsimage.ckpt_0000000000000000000 of size 321 bytes saved in 0 seconds.
18/04/27 16:20:20 INFO namenode.NNStorageRetentionManager: Going to retain 1 images with txid >=0
18/04/27 16:20:20 INFO util.ExitUtil: Exiting with status 0
18/04/27 16:20:20 INFO namenode.NameNode: SHUTDOWN_MSG:
/************************************************************
SHUTDOWN_MSG: Shutting down NameNode at master/172.31.228.188
************************************************************/
```

从信息中可以看到 /name has been successfully formatted，代表格式化成功。

（5）启动 HDFS。

```
[root@master hadoop-2.5.1]#start-dfs.sh
```

然后访问网页 http://192.168.109.200:50070/，查看是否安装成功，如图 2.21 所示。出现下图显示的页面则代表伪分布式集群搭建成功。

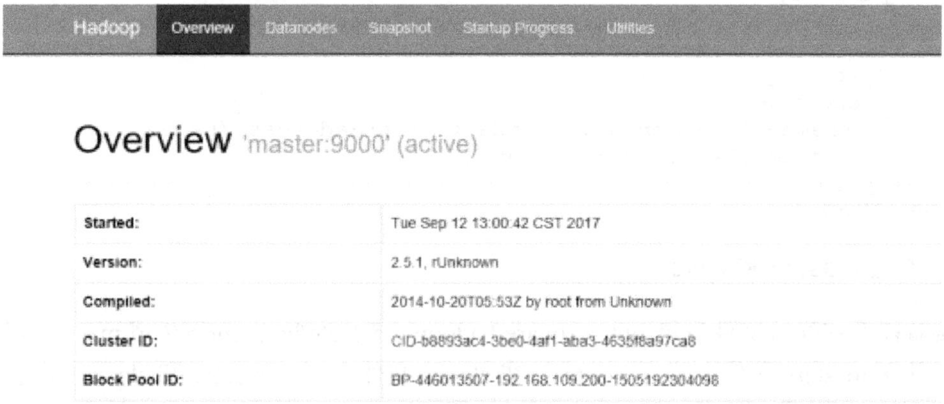

图 2.21　访问网页

2.8.2　完全分布式安装

2.8.1 节介绍的伪分布式是基于单个节点，而完全分布式是基于两个或两个以上节点完成 Hadoop 集群搭建。下面基于两个节点完成，一个节点的名字是 master，另一个节点的名字是 slave。关于搭建伪分布式和完全分布式，主要区别体现在 core-site.xml 和 hdfs-site.xml 的配置不一样，完全分布式会包含更多信息，下面会逐步说明。

1. 修改core-site.xml文件

core-site.xml 文件中，hadoop.tmp.dir 是 Hadoop 文件系统依赖的基础配置，默认存放在 /tmp/{$user}下。但是存放在/tmp 下是不安全的，因为系统重启后文件有可能被删除，所以会指向另外的路径。

```
<configuration>
    <property>
        <name>fs.defaultFS</name>
        <value>hdfs://master:9000</value>
    </property>
    <property>
        <name>hadoop.tmp.dir</name>
        <value>/opt/software/hadoop-2.5.1</value>
    </property>
</configuration>
```

2. 修改hdfs-site.xml文件

这里主要配置了 Secondary NameNode 的信息，其中 slave 是从节点机器名。读者目前可以先"照猫画虎"把系统搭起来，关于 Secondary NameNode 的介绍，后续会详细讲解。

```
<configuration>
    <property>
        <name>dfs.namenode.secondary.http-address</name>
        <value>slave:50090</value>
    </property>
    <property>
        <name>dfs.namenode.secondary.https-address</name>
        <value>slave:50091</value>
    </property>
</configuration>
```

3．配置 masters 和 slaves

接着需要在配置文件目录/opt/software/hadoop-2.5.1/etc/hadoop/下生成 masters 和 slaves 文件，并在 masters 文件中写入 master，在 slaves 文件中写入 master 和 slave，其中 slaves 文件存放的是 datanode，也就是数据节点，如图 2.22 和图 2.23 所示。

```
[root@master hadoop]# cd/opt/software/hadoop-2.5.1/etc/hadoop/
[root@master hadoop]# touch masters
[root@master hadoop]# touch slaves
[root@master hadoop]# vi masters
```

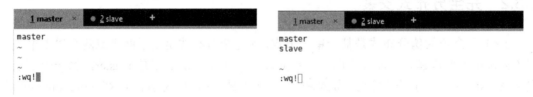

图 2.22　命令在 masters 文件中加入主节点　　图 2.23　slaves 文件中可以添加 master 和 slave 节点

需要注意的是，这里的 master 和 slave 是节点名称，需要与/etc/hosts 中的配置相映射。

4．相关文件的复制

在完全分布式的环境中，master 和 slave 节点上的文件需要一致，因此这里需要将 master 节点中的文件复制到 slave 节点中，主要包括以下文件：

- Hadoop 整个文件夹，如/opt/software/hadoop-2.5.1 下面的所有文件。
- 系统配置文件，如.bash_profile 文件，其中包含各类环境变量的配置。
- /etc/hosts 文件。

操作步骤如下：

（1）复制 Hadoop 整个文件夹。

复制文件夹需要使用 scp -r 指令：

```
scp -r /opt/software/hadoop-2.5.1 root@192.168.109.201:/opt/software
```

（2）复制.bash_profile 到 salve 的 Home 目录。

```
scp .bash_profile root@192.168.109.201:~
```

(3) 复制/etc/hosts 文件，命令如下：

```
scp /etc/hosts root@192.168.109.201:/etc/hosts
```

5. 格式化HDFS

接下来通过格式化 HDFS 实现集群的初始化。

```
[root@master hadoop-2.5.1]# hdfs namenode -format
```

6. 启动HDFS集群

启动命令如下：

```
[root@master hadoop -2.5.1]#sbin/start-dfs.sh
```

7. 在浏览器访问http://ip:50070/，进行测试

测试命令如下：

```
http://192.168.109.200:50070/
```

2.9 小　　结

本章主要介绍了基于 Linux 的 Hadoop 安装与配置，包括虚拟机创建、Linux 系统和 JDK 安装、Hadoop 安装及 Hadoop 分布式安装，并通过实例向读者展示了安装的步骤与技巧。此外，本章还介绍了克隆服务器和 SSH 免密码登录等内容。

第 3 章 Hadoop 分布式文件系统

随着互联网的发展，日常生活和工作中的数据量越来越大，文件和数据被越来越多地存储到系统管理的磁盘中，单台机器已经不能满足大量的文件存储需求，迫切需要一种允许多机器上的多用户通过网络分享文件和存储空间的文件管理系统，这就是分布式文件系统。

分布式文件管理系统有很多，如 DFS 和 HDFS，而 HDFS 适用于一次写入、多次查询的情况。本章我们将详细介绍 DFS 和 HDFS。

本章主要涉及如下知识点。
- DFS 基础知识。
- HDFS 和 DFS 的关系，HDFS 的设计和优缺点，以及 HDFS 的读写文件流程。
- 学会基于 Shell 的操作和基于 Java API 操作 HDFS。

3.1 DFS 介绍

由于一台机器的存储容量有限，一旦数据量达到足够的级别，就需要将数据存放在多台机器上，这就是分布式文件系统，又称之为 DFS（Distributed File System）。

DFS 是 HDFS 的基础，本节将简单讲解一下什么是 DFS 及 DFS 的结构，随后引出 Hadoop 的核心组件 HDFS。

3.1.1 什么是 DFS

分布式文件系统 DFS 是基于 Master/Slave 模式，通常一个分布式文件系统提供多个供用户访问的服务器，一般都会提供备份和容错的功能。分布式文件系统管理的物理资源不一定直接连接在本地节点上，而是通过计算机网络与节点相连，而非文件系统管理的物理存储资源一定直接连在本地节点上。

3.1.2 DFS 的结构

分布式文件系统在物理结构上是由计算机集群中的多个节点构成的，如图 3.1 所示。

这些节点分为两类，一类叫"主节点"（Master Node），也被称为"名称节点"（NameNode）；另一类叫"从节点"（SlaveNode），也被称为"数据节点"（DataNode）。

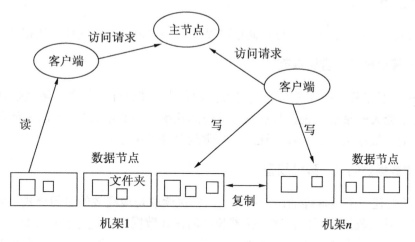

图 3.1　分布式文件系统构图

3.2　HDFS 介绍

前面讲到的 DFS 是统称的分布式文件系统，在 Hadoop 中实现的分布式文件系统被称之为 HDFS，本节将会介绍 HDFS 的基本概念、执行原理及文件的读写流程。

3.2.1　HDFS 的概念及体系结构

HDFS 是 Hadoop 自带的分布式文件系统，即 Hadoop Distributed File System。HDFS 是一个使用 Java 语言实现的分布式、可横向扩展的文件系统。

HDFS 包括一个名称节点（NameNode）和若干个数据节点（DataNode），属于主/从（Master/Slave）关系的结构模型。其中，名称节点负责管理文件系统的命名空间及客户端对文件的访问，也就是中心服务器。

而集群中的数据节点一般是一个节点运行一个数据节点进程，其中每个数据节点上的数据实际上是保存在本地的 Linux 文件系统中，并在名称节点的统一调动下，负责处理文件系统客户端的读/写请求，或删除、创建和复制数据块等操作。

3.2.2　HDFS 的设计

HDFS 的设计主要是为了实现存储大量数据、成本低廉和容错率高、数据一致性，以及顺序访问数据这4个目标。

1．大数据集

HDFS 适合存储大量文件，总存储量可以达到 PB/EB，单个文件一般在几百兆。

2．基于廉价硬件，容错率高

Hadoop 并不需要运行在昂贵且高可靠的硬件上，其设计运行在商用廉价硬件的集群上，因此对于庞大的集群来说，节点发生故障的几率还是非常高的。HDFS 遇到上述故障时被设计成能够继续运行且可以不让用户察觉到明显的中断。

3．流式数据访问（一致性模型）

HDFS 的构建思路是这样的：一次写入、多次读取是最高效的访问模式。数据集通常由数据源生成或从数据源复制而来，接着长时间在此数据集上进行各种分析。

每次分析都将涉及该数据集的大部分数据甚至全部数据，因此读取整个数据集的时间延迟比读取第一条记录的时间延迟更重要。

4．顺序访问数据

HDFS 适用于处理批量数据，而不适合随机定位访问。

3.2.3　HDFS 的优点和缺点

1．HDFS的优点

- 高容错性：数据自动保存多个副本，副本丢失后自动恢复。
- 适合批处理：移动计算而非数据，数据位置暴露给计算机框架。
- 适合大数据处理：GB、TB，甚至 PB 级数据，百万规模以上的文件数量，10k+节点。
- 可构建在廉价机器上：通过副本提高可靠性，提供了容错和恢复机制。

2．HDFS的缺点

- 不适合低延时数据访问：寻址时间长，适合读取大文件，低延迟与高吞吐率。
- 不适合小文件存取：占用 NameNode 大量内存，寻找时间超过读取时间。
- 并发写入、文件随机修改：一个文件只能有一个写入者，仅支持 append（日志），不允许修改文件。

3.2.4　HDFS 的执行原理

从客户端传入文件读写请求时，NameNode（HDFS 的集群管理节点）首先接受客户端的读写服务请求，并根据它保存的 Metadata 元数据，包括元数据的镜像文件（fsimage 和操作日志 edits 信息）和 DataNode（数据存储）通信并进行资源协调，Secondary NameNode 进行 edits 和 fsimage 的合并，同时 DataNode 之间进行数据复制。

如果要存储一个大文件，首先要将文件分割成块，分别放到不同的节点，每块文件都有 3 个副本备份，并且有一个专门记录文件块存放情况的元数据文件以备查询，如图 3.2 和图 3.3 所示。

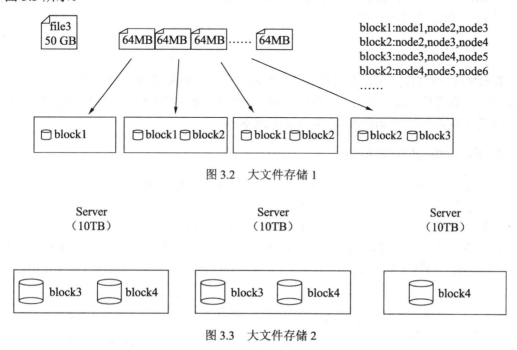

图 3.2　大文件存储 1

图 3.3　大文件存储 2

3.2.5　HDFS 的核心概念

关于 HDFS 有以下核心概念，理解这些概念对于更好地了解 HDFS 的原理有很大帮助。

1．数据块（block）

每个磁盘都有默认的数据块大小，这是磁盘进行数据读/写的最小单位。HDFS 也有块的概念，在 HDFS 1.x 中默认数据块大小为 64MB，在 HDFS 2.x 中默认数据块大小为 128MB。

与单一磁盘上的文件系统相似，HDFS 上的文件也被划分成块大小的多个分块（chunk），作为独立的存储单元。但与面向单一的文件磁盘系统不同的是，HDFS 中小于一个块大小的文件不会占据整个块的空间（例如一个 1MB 的文件存储在一个 128MB 的块中时，文件只会使用 1MB 的磁盘空间，而不是 128MB）。

2. NameNode

NameNode 为 HDFS 集群的管理节点,一个集群通常只有一台活动的 NameNode,它存放了 HDFS 的元数据且一个集群只有一份元数据。NameNode 的主要功能是接受客户端的读写服务,NameNode 保存的 Metadata 信息包括文件 ownership、文件的 permissions,以及文件包括哪些 Block、Block 保存在哪个 DataNode 等信息。这些信息在启动后会加载到内存中。

3. DataNode

DataNode 中文件的储存方式是按大小分成若干个 Block,存储到不同的节点上,Block 大小和副本数通过 Client 端上传文件时设置,文件上传成功后副本数可以变更,BlockSize 不可变更。默认情况下每个 Block 都有 3 个副本。

4. SecondaryNameNode

SecondaryNameNode(简称 SNN),它的主要工作是帮助 NameNode 合并 edits,减少 NameNode 启动时间。SNN 执行合并时机如下:
- 根据配置文件设置的时间间隔 fs.checkpoint.period,默认 3600 秒。
- 根据配置文件设置 edits log 大小 fs.checkpoint.size,规定 edits 文件的最大值默认是 64MB,如图 3.4 所示。

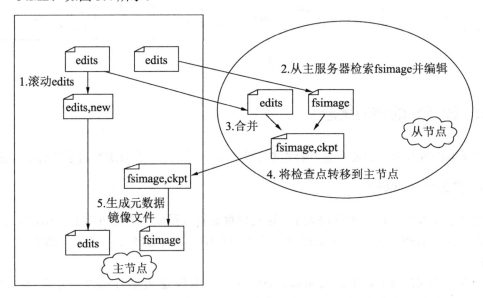

图 3.4　配置文件设置

5．元数据

元数据保存在 NameNode 的内存中，以便快速查询，主要包括 fsimage 和 edits。
- fsimage：元数据镜像文件（保存文件系统的目录树）。
- edits：元数据操作日志（针对目录树的修改操作）被写入共享存储系统中，比如 NFS、JournalNode，内存中保存一份最新的元数据镜像（fsimage+edits）。

3.2.6 HDFS 读文件流程

前面介绍了 HDFS 的核心概念，接下来介绍 HDFS 读写文件的流程。对于存储在 HDFS 上的文件，我们可以通过客户端发送读文件请求，主要步骤如下：

（1）客户端通过调用 FileSystem 对象的 open()方法打开要读取的文件，对于 HDFS 来说，这个对象是 DistributedFileSystem 的一个实例。

（2）DistributedFileSystem 通过使用远程过程调用（RPC）来调用 NameNode，以确定文件起始块的位置。

（3）对于每个块，NameNode 返回到存有该块副本的 DataNode 地址。此外，这些 DataNode 根据它们与客户端的距离来排序。如果该客户端本身就是一个 DataNode，那么该客户端将会从包含有相应数据块副本的本地 DataNode 读取数据。DistributedFileSystem 类返回一个 FSDataInputStream 对象给客户端并读取数据，FSDataInputStream 转而封装 DFSInputStream 对象，该对象管理着 DataNode 和 NameNode 的 I/O。接着，客户端对这个输入流调用 read()方法。

（4）存储着文件起始几个块的 DataNode 地址的 DFSInputStream，接着会连接距离最近的文件中第一个块所在的 DataNode。通过对数据流的反复调用 read()方法，实现将数据从 DataNode 传输到客户端。

（5）当快到达块的末端时，DFSInputStream 会关闭与该 DataNode 的连接，然后寻找下一个块最佳的 DataNode。

（6）当客户端从流中读取数据时，块是按照打开的 DFSInputStream 与 DataNode 新建连接的顺序进行读取的。它也会根据需要询问 NameNode 从而检索下一批数据块的 DataNode 的位置。一旦客户端完成读取，就对 FSDataInputStream 调用 close()方法，如图 3.5 所示。

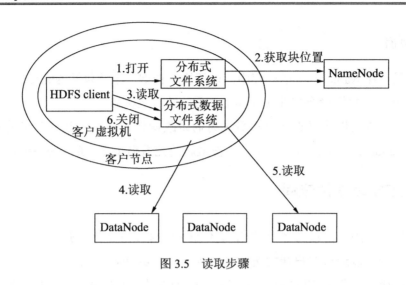

图 3.5 读取步骤

3.2.7 HDFS 写文件流程

对于存储在 HDFS 上的文件也可以写入内容，可以通过客户端发送写文件的请求，主要步骤如下：

（1）客户端调用 DistributedFileSystem 对象的 create()方法新建文件。

（2）DistributedFileSystem 会对 NameNode 创建一个 RPC 调用，在文件系统的命名空间中创建一个新文件，需要注意的是，此刻该文件中还没有相应的数据块。

（3）NameNode 通过执行不同的检查来确保这个文件不存在而且客户端有新建该文件的权限。如果这些检查都通过了，NameNode 就会为创建新文件写下一条记录；反之，如果文件创建失败，则向客户端抛出一个 IOException 异常。

（4）随后 DistributedFileSystem 向客户端返回一个 FSDataOutputStream 对象，这样客户端就可以写入数据了。和读取事件类似，FSDataOutputStream 封装一个 DFSOutputStream 对象，该对象会负责处理 DataNode 和 NameNode 之间的通信。在客户端写入数据的时候，DFSOutputStream 将它分成一个个的数据包，并且写入内部队列，被称之为"数据队列"（data queue）。

（5）DataStream 处理数据队列，它的任务是选出适合用来存储数据副本的一组 DataNode，并据此要求 NameNode 分配新的数据块。这一组 DataNode 会构成一条管线，DataStream 会将数据包流式传输到管线中的第一个 DataNode，然后依次存储并发送给下一个 DataNode。

（6）DFSOutPutStream 也维护着一个内部数据包队列来等待 DataNode 的收到确认回执，称为"确认队列"（ask queue）。收到管道中所有 DataNode 确认信息后，该数据包才会从确认队列删除。

（7）客户端完成数据的写入后，会对数据流调用 close() 方法，如图 3.6 所示。

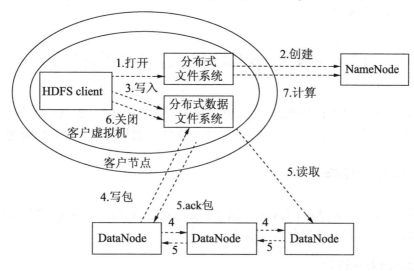

图 3.6　数据流调用步骤

3.2.8　Block 的副本放置策略

HDFS 中的文件作为独立的存储单元，被划分为块（block）大小的多个分块（chunk），在 Hadoop 2.x 中默认值为 128MB。当 HDFS 中存储小于一个块大小的文件时不会占据整个块的空间，也就是说，1MB 的文件存储时只占用 1MB 的空间而不是 128MB。HDFS 的容错性也要求数据自动保存多个副本，副本的放置策略如图 3.7 所示。

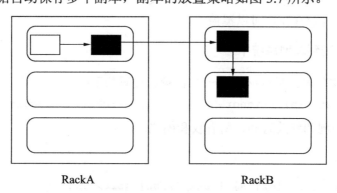

图 3.7　副本的放置策略图示

- 第 1 个副本：放置在上传文件的 DN；如果是集群外提交，则随机挑选一台磁盘不太满、CPU 不太忙的节点。
- 第 2 个副本：放置在与第 1 个副本不同机架的节点上。

- 第 3 个副本：放置在与第 2 个副本相同机架的节点上。
- 更多副本：随机节点。

3.3 Hadoop 中 HDFS 的常用命令

我们已经知道 HDFS 是分布式存储，可以存放大批量的文件，如果要对文件进行操作，可以通过下面的命令来完成，如读取文件、上传文件、删除文件和建立目录等。

3.3.1 对文件的操作

HDFS 的命令都在 Hadoop 的 bin 目录下，如果已经设置好 Hadoop 的环境变量，可以直接输入 HDFS 命令行，常见的相关命令如下。

1．列出HDFS下的文件

```
hdfs dfs -ls [-d] [-h] [-R] [<path> ...]
```

- -d：显示目录。
- -h：以易读的方式显示文件的大小。
- -R：递归列出目录的内容。

2．上传文件

```
hdfs dfs -put [-f] [-p] <localsrc> ... <dst>
```

- -p：保留访问和修改时间、所有权和模式。
- -f：如果文件已存在，可以覆盖。

3．文件被复制到本地系统中

当复制多个文件时，目标必须是一个目录。代码如下：

```
hdfs dfs -get [-p] [-ignoreCrc] [-crc] <src> ... <localdst>
```

- -p：保留访问和修改时间、所有权和模式。

4．删除文档

```
hdfs dfs -rm [-f] [-r|-R] [-skipTrash] <src> ...
```

- -f：如果该文件不存在，则不显示诊断消息或修改退出状态以反映错误。
- -r：递归删除目录的内容。

5．查看文件

```
hdfs dfs -cat [-ignoreCrc] <src> ...
```

6. 建立目录

```
hdfs dfs -mkdir [-p] < paths>
```

7. 复制文件

```
hdfs dfs -copyFromLocal [-f] [-p] <localsrc> ... <dst>
```

- -p：保留访问和修改时间、所有权和模式。
- -f：如果文件已存在，可以覆盖。

3.3.2 管理与更新

3.3.1 节中的命令是对文件的操作，接下来介绍关于 HDFS 的常见情景和管理步骤。

1. 执行基本信息

查看 HDFS 的基本统计信息。代码如下：

```
hdfs dfsadmin -report [-live] [-dead] [-decommissioning]
```

2. 退出安全模式

NameNode 在启动时会自动进入安全模式。安全模式是 NameNode 的一种方式，在这个阶段，文件系统不允许有任何修改，系统显示 Name node in safe mode，说明系统正处于安全模式，这时只需要等待十几秒即可。也可以通过下面的命令退出安全模式。代码如下：

```
hdfs dfsadmin -safemode leave
```

3.4 HDFS 的应用

前面已经说过，HDFS 是一个分布式文件系统，可以对海量数据进行存储并对文件进行操作，并且前面内容中还介绍了一些 HDFS 中的常用命令，本节主要讲解如何应用 HDFS。HDFS 提供了两种访问方式，分别是基于 Shell 和 Java API，下面分别介绍使用 Shell 和 Java API 对 HDFS 进行访问。

3.4.1 基于 Shell 的操作

以下介绍一些我们在 Shell 中操作 HDFS 时经常用到的命令。

1. 创建目录命令

HDFS 创建目录的命令是 mkdir，命令格式如下：

```
hdfs dfs -mkdir 文件夹名
```

例如，在 user 目录下创建 wen 目录。代码如下：

```
hdfs dfs -mkdir /user/wen
```

2．上传文件到HDFS

上传文件时，文件首先复制到 DataNode 上，只有所有的 DataNode 都成功接收完数据，文件上传才是成功的。命令格式如下：

```
hdfs dfs -put filename newfilename
```

例如，通过"-put 文件 1 文件 2"命令将 test1 文件上传到 HDFS 上并重命名为 test2。代码如下：

```
hdfs dfs -put test1 test2
```

3．列出HDFS上的文件

采用-ls 命令列出 HDFS 上的文件，需要注意的是，在 HDFS 中没有"当前工作目录"这个概念。命令格式如下：

```
hdfs dfs -ls
```

例如，列出 HDFS 特定目录下的所有文件。代码如下：

```
hdfs dfs -ls /wen
```

4．查看HDFS下某个文件的内容

通过"-cat 文件名"命令查看 HDFS 下某个文件的内容。命令格式如下：

```
hdfs dfs -cat 文件名
```

例如，查看 HDFS 上 wen 目录下 test.txt 中的内容。代码如下：

```
hdfs dfs -cat /wen/test.txt
```

5．将HDFS中的文件复制到本地系统中

通过"-get 文件 1 文件 2"命令将 HDFS 中某目录下的文件复制到本地系统的某文件中，并对该文件重新命名。命令格式如下：

```
hdfs dfs -get 文件名 新文件名
```

例如，将 HDFS 中的 in 文件复制到本地系统并重命名为 IN1。代码如下：

```
hdfs dfs -get in IN1
```

-get 命令与-put 命令一样，既可以操作目录，也可以操作文件。

6．删除HDFS下的文档

通过"-rmr 文件"命令删除 HDFS 下的文件。命令格式如下：

```
hdfs dfs -rmr 文件
```
例如，删除 HDFS 下的 out 文档。代码如下：
```
hdfs dfs -rmr out
```
-rmr 删除文档命令相当于 delete 的递归版本。

7. 格式化HDFS

通过-format 命令实现 HDFS 格式化。命令格式如下：
```
hdfs namenode -format
```

8. 启动HDFS

通过运行 start-dfs.sh，就可以启动 HDFS 了。命令格式如下：
```
start-dfs.sh
```

9. 关闭HDFS

当需要退出 HDFS 时，通过 stop-dfs.sh 就可以关闭 HDFS。命令格式如下：
```
stop-dfs.sh
```

3.4.2 基于 Java API 的操作

本节将介绍通过 Java API 来访问 HDFS，首先介绍 HDFS 中的文件操作主要涉及的几个类。

- Configuration 类：该类的对象封装了客户端或者服务器的配置。
- FileSystem 类：该类的对象是一个文件系统对象，可以用该对象的一些方法对文件进行操作。FileSystem fs = FileSystem.get(conf);通过 FileSystem 的静态方法 get 获得该对象。
- FSDataInputStream 和 FSDataOutputStream：这两个类是 HDFS 中的输入/输出流，分别通过 FileSystem 的 open 方法和 create 方法获得。

接下来通过实例介绍如何利用 Java API 进行文件夹的创建、文件列表显示、文件上传和文件下载操作，为了使读者对实例具有完整性的理解，我们把主函数也列了出来。实例代码如下：

```
import java.io.IOException;
import org.apache.hadoop.conf.Configuration;
import org.apache.hadoop.fs.FileStatus;
import org.apache.hadoop.fs.FileSystem;
import org.apache.hadoop.fs.Path;
public class HdfsDemo {
public static void main(String[] args) {
    createFolder();
    //uploadFile();
```

```
        //downloadFile();
//listFile(new Path("/"));
}
```

在主函数中，分别调用了 createFolder()、uploadFile()、downloadFile()和 listFile(new Path("/")函数来实现文件夹的创建、文件上传、文件下载和递归显示文件夹功能。

3.4.3 创建文件夹

如果想要将文件放入不同的文件夹中，则可以有针对性地动态创建文件夹。关于文件夹的创建比较简单。代码如下：

```
public static void createFolder() {
    // 定义一个配置对象
    Configuration conf = new Configuration();
    try {
        // 通过配置信息得到文件系统的对象
        FileSystem fs = FileSystem.get(conf);
        //在指定的路径下创建文件夹
        Path path = new Path("/yunpan");
        fs.mkdirs(path);
    } catch (IOException e) {
        e.printStackTrace();
    }
}
```

3.4.4 递归显示文件

如果要显示文件夹中的文件，需要用到递归算法，因为文件夹中可能有文件，也可能有文件夹。代码如下：

```
public static void listFile(Path path) {
    Configuration conf = new Configuration();
    try {
        FileSystem fs = FileSystem.get(conf);
        //传入路径，表示显示某个路径下的文件夹列表
        //将给定路径下所有的文件元数据放到一个 FileStatus 的数组中
//FileStatus 对象封装了文件和目录的元数据，包括文件长度、块大小、权限等信息
        FileStatus[] fileStatusArray = fs.listStatus(path);
        for (int i = 0; i < fileStatusArray.length; i++) {
            FileStatus fileStatus = fileStatusArray[i];
    //首先检测当前是否是文件夹，如果"是"则进行递归
            if (fileStatus.isDirectory()) {
                System.out.println("当前路径是: " + fileStatus.getPath());
                listFile(fileStatus.getPath());
            } else {
                System.out.println("当前路径是: " + fileStatus.getPath());
            }
        }
```

```
    } catch (IOException e) {
e.printStackTrace();
    }
}
```

3.4.5 文件上传

前面创建了文件夹及文件的递归显示，接着来看一下文件的上传。关于文件上传的代码如下：

```
public static void uploadFile() {
    Configuration conf = new Configuration();
    try {
        FileSystem fs = FileSystem.get(conf);
        //定义文件的路径和上传的路径
        Path src = new Path("e://upload.doc");
        Path dest = new Path("/yunpan/upload.doc ");
        //从本地上传文件到服务器上
        fs.copyFromLocalFile(src, dest);
    } catch (IOException e) {
        // TODO Auto-generated catch block
        e.printStackTrace();
    }
}
```

关于文件的上传，首先需要定义上传源 src，这里的 src 是指本地文件路径，即将要上传的目的地路径是 dest，也就是 HDFS 上的路径。一旦设定了文件上传的数据源和目的路径之后，就可以调用 FileSystem 的 copyFromLocalFile()方法来实现文件的上传了。

3.4.6 文件下载

与文件上传所对应的就是文件下载。文件下载代码和文件上传类似，只是函数略有区别。代码如下：

```
public static void downloadFile() {
    Configuration conf = new Configuration();
    try {
        FileSystem fs = FileSystem.get(conf);
        //定义下载文件的路径和本地下载路径
        Path src = new Path(
                "/yunpan/download.doc");
        Path dest = new Path("e://download.doc");
        //从服务器下载文件到本地
        fs.copyToLocalFile(src, dest);
    } catch (IOException e) {
        // TODO Auto-generated catch block
        e.printStackTrace();
    }
}
```

关于文件的下载，首先需要定义下载源 src，这里的 src 是指 HDFS 上的路径，将要下载的目的地为 dest，也就是本地系统的路径。设定了文件下载的数据源和目的路径之后，就可以调用 FileSystem 的 copyToLocalFile()方法实现文件的下载了。

3.5 小　　结

本章首先介绍了 DFS，即分布式文件系统，接着介绍了 Hadoop 的分布式文件系统即 HDFS 的核心概念、读写文件的流程及基于 Shell 和 Java API 对 HDFS 的操作。

第 4 章　基于 Hadoop 3 的 HDFS 高可用

前面已经介绍了 HDFS 的作用是存储数据。本章主要讨论如何基于完全分布式实现大批量数据存储，同时为了保障集群的运行正常，将 NameNode 配置为高可用。本章我们基于 2018 年 4 月份发布的 Hadoop 3.1.0 版本讲解 HDFS 完全分布式环境及高可用环境搭建。

本章主要涉及知识点如下：
- Hadoop 3 新特性。
- HDFS 高可用及其实现原理。
- HDFS 分布式环境搭建。
- HDFS 高可用环境搭建。

4.1　Hadoop 3.x 的发展

Hadoop 版本包括 Hadoop 1.x、Hadoop 2.x 和 Hadoop 3.x。在编写本书时，Hadoop 的最新版本是 2018 年 4 月正式发布的 Hadoop 3.1.0。本章我们将以 Hadoop 3.1.0 为例，完成 HDFS 高可用的搭建。在介绍 Hadoop 3.x 之前，先介绍一下 Hadoop 2.x 产生的背景。

4.1.1　Hadoop 3 新特性

Hadoop 3.1.0 GA 版本于 2017 年 12 月份正式发布。Hadoop 3 相较于 Hadoop 2 有一些新特性，包括基于 JDK 1.8、HDFS 可擦除编码、MR Native Task 优化、基于 Cgroup 的内存隔离和 IO Disk 隔离，以及支持更改分配容器的资源 Container resizing 等。

Hadoop 3 的新特性介绍如下。

1. classpath isolation

防止不同版本的 JAR 包发生冲突。

2. Shell重写

启动脚本和 Hadoop 2.x 不同。

3．支持HDFS中的擦除编码

主要用于做数据恢复，这一特性使 HDFS 的存储节省了一半空间，同时还不降低可靠性。擦除编码目前主要针对的是大数据块。

擦除编码的工作原理是把存储系统接收到的大块数据进行切割并且编码，接着再对切割之后的数据进行再次切割并编码，持续重复这个操作，直到数据切割到合适的数据块大小为止，这样数据块就分散成多个数据块，再进行冗余校验，把不重复的数据块和编码写到存储系统中。

4．MapReduce任务级本地优化

提高 MR 的执行速度，Hadoop 3 为 MapReduce 增加了基于 C/C++ 的 map output collector。

5．MapReduce内存参数自动推断

Hadoop 2.x 中通过配置 mapreduce.{map,reduce}.memory.mb 和 mapreduce.{map. reduce}. Java.opts 来配置所使用的内存，如果设置不合理，则会使内存资源严重浪费，而在 Hadoop 3.x 中则不需要再配置。

6．端口区别

Hadoop 2 和 Hadoop 3 的端口区别如表 4.1 所示。

表 4.1　Hadoop 2 和Hadoop 3 的端口区别

分　　类	应　　用	Hadoop 2.x端口	Hadoop 3 端口
NN ports	NameNode	8020	9820
	NN Http UI	50070	9870
	NN Https UI	50470	9871
SNN ports	SNN Http	50091	9869
	SNN Http UI	50090	9868
DN ports	DN IPC	50020	9867
	DN	50010	9866
	DN Http UI	50075	9864
	DN Https UI	50475	9865

4.1.2　Hadoop 3 HDFS 集群架构

如图 4.1 所示，HDFS 集群中包括 NameNode、DataNode 和 Secondary NameNode，具体介绍如下。

- NameNode：接受客户端的读写服务，比如文件的上传和下载，保存元数据，包括文件大小、文件创建时间、文件的拥有者、权限、路径和文件名。元数据存放在内存中，不会和磁盘发生交互。
- DataNode：简称 DN，与 NameNode 对应，主要用来存储数据内容，本地磁盘目录存储数据块（Block），以文件形式分别存储在不同的 DataNode 节点上，同时存储 Block 的元数据信息文件。
- Secondary NameNode：前面提到 NameoNode 的元数据存储在内存中，为了保证数据不丢失，需要将数据保存起来，这里涉及的文件包括 fsimage 和 edits。fsimage 是整个元数据文件，在集群刚开始搭建时是空的，对元数据增删改的操作放到 edits 文件中。Secondary NameNode 完成数据的合并操作，每隔 3600 秒更新一次。

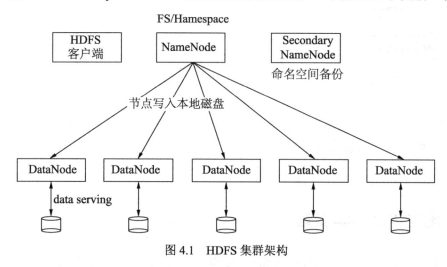

图 4.1　HDFS 集群架构

4.2　Hadoop 3 HDFS 完全分布式搭建

在搭建 Hadoop3 完全分布式之前需要先准备 3 台机器作为 3 个节点，这里假设分别为 node1、node2 和 node3，其中 node1 为 NameNode（简写为 NN），node2 为 Secondary NameNode（简写为 SN），同时，把 node1、node2、node3 都设置为 DataNode（简写为 DN），如表 4.2 所示。

表 4.2　环境搭建节点分配

node1	node2	node3
NN	SN	—
DN	DN	DN

假设要在 node1 上启动集群，则需要基于 node1 免密码登录到 node2 和 node3 后开启

相应的服务。接下来基于表 4.2 介绍完全分布式搭建，步骤如下所述。

4.2.1 安装 JDK

请读者参考第 2 章虚拟机和 Linux 的安装。

在这里需要注意的是，Hadoop 3.1.0 需要用到 JDK 1.8，读者可以从官网下载。这里我们以 JDK-8u171-linux-x64 为例，安装命令如下：

```
[root@node51 software]# rpm -ivh jdk-8u171-linux-x64.rpm
Preparing...                ########################################### [100%]
   1:jdk1.8                 ########################################### [100%]
Unpacking JAR files...
        tools.jar...
        plugin.jar...
        javaws.jar...
        deploy.jar...
        rt.jar...
        jsse.jar...
        charsets.jar...
        localedata.jar...
```

安装完成后，通过 java –version 命令查看 Java 的版本号，如果出现 java version 1.8.0_171，则表示 Java 安装成功。

```
[root@node51 software]# java -version
java version "1.8.0_171"
```

4.2.2 配置 JDK 环境变量

打开 bash_profile，并在 bash_profile 中配置 JAVA_HOME，配置代码如下：

```
export JAVA_HOME=/usr/java/jdk1.8.0_171-amd64
```

同时将$JAVA_HOME/bin 加入 PATH。

配置完成后，通过 echo $JAVA_HOME 命令可查看是否配置成功，如果出现如下提示，则代表环境变量配置成功。

```
[root@node51 ~]# echo $JAVA_HOME
/usr/java/jdk1.8.0_171-amd64
```

4.2.3 配置免密码登录

根据需要，在/etc/sysconfig/network 文件中配置主机名，比如我们选择了一台机器将主机名设置为 node1，命令如下：

```
vi /etc/sysconfig/network
```

打开 network 文件，接着在 network 文件中写入以下内容：

```
NETWORKING=yes
HOSTNAME=node1
```

重新登录系统后，命令行显示如下，则代表主机修改成功：

```
[root@node1 ~]#
```

4.2.4 配置 IP 和主机名字映射关系

接着需要将 IP 和主机名进行映射，这样在后续的配置文件中可以直接输入主机名即可，配置方式是打开/etc/hosts 文件，并写入以下内容：

```
192.168.19.10 node1
192.168.19.20 node2
192.168.12.30 node3
```

需要注意的是，读者需要根据自己的 IP 进行修改，其中 node1、node2 和 node3 为节点名，比如后续在配置文件中写入 node1 时，就类似写入了 192.168.19.10。

4.2.5 SSH 免密码登录设置

本节以 node1 为中心，设置到其他机器的 SSH 免密码登录方式。

设置 SSH 免密码登录主要的步骤是生成密钥和复制公钥。由于前面已经说过以 node1 为 NameNode，具体步骤如下。

（1）登录 node1，在 node1 上输入 ssh-keygen 命令，然后一直按回车键操作。代码如下：

```
[root@node1 ~]# ssh-keygen
Generating public/private rsa key pair.
Enter file in which to save the key (/root/.ssh/id_rsa):
Enter passphrase (empty for no passphrase):
Enter same passphrase again:
Your identification has been saved in /root/.ssh/id_rsa.
Your public key has been saved in /root/.ssh/id_rsa.pub.
The key fingerprint is:
49:ce:b3:fa:73:d4:b6:cf:ef:fb:62:b6:86:c0:29:91 root@node1
The key's randomart image is:
+--[ RSA 2048]----+
|                 |
|                 |
|        ..       |
|       +E.       |
|       So o      |
|       .o= o     |
|       .o o o    |
|       .. . o.=  |
|       ...o =+** |
+-----------------+
```

（2）查看生成的私钥和公钥。

执行完第 1 步操作之后，会在执行命令的同级目录下生成.ssh 文件，注意这个.ssh 为隐藏文件夹。进入.ssh 文件后，执行 ls 命令列出文件，其中的 id_rsa 为私钥文件，id_rsa.pub 为公钥文件。

```
[root@node1 .ssh]# ls
authorized_keys  id_rsa  id_rsa.pub  known_hosts
```

（3）把公钥文件复制到 node1、node2 和 node3 节点上，命令如下：

```
ssh-copy-id -i ./id_rsa.pub root@node1
ssh-copy-id -i ./id_rsa.pub root@node2
ssh-copy-id -i ./id_rsa.pub root@node3
```

这里需要注意的是，node1 本身也要做免密码登录设置。

完成上述 3 步之后，就可以检测 SSH 免密码登录是否成功了。测试命令如下，如果没有出现输入密码提示，则代表设置成功了。

```
[root@node1 .ssh]# ssh node2
Last login: Wed May  9 16:49:02 2018 from 192.168.19.1
```

4.2.6　配置 Hadoop 3.1.0

首先需下载 Hadoop 3.1.0 GA 版本，我们从官网下载 Hadoop 3.1.0 版本，官网如图 4.2 所示，目前最新的版本是 3.1.0，本节以 Hadoop 3.1.0 为例完成完全分布式的安装。

图 4.2　官网下载 Hadoop 3.1.0 版本

解压缩 Hadoop。下载到本机后，就可以进行解压缩了，解压缩命令如下：

```
[root@node1 software]# tar -zxvf hadoop-3.1.0.tar.gz
```

1．部署及配置

Hadoop 的配置涉及以下几个文件，分别是：hadoop-env.sh、core-site.xml、hdfs-site.xml

和workers。其中，hadoop-env.sh是Hadoop运行环境变量配置；core-site.xml是Hadoop公共属性的配置；hdfs-site.xml是关于HDFS的属性配置；workers是DataNode分布配置。下面我们分别配置这几个文件。

（1）hadoop-env.sh文件

在/etc/hadoop/hadoop-env.sh中配置运行环境变量，在默认情况下，这个文件是没有任何配置的。我们需要配置JAVA_HOME、HDFS_NAMENODE_USER和HDFS_DATANODE_USER等，HDFS_SECONDARYNAMENODE_USER配置代码如下：

```
export JAVA_HOME=/usr/java/jdk1.8.0_171-amd64
export HDFS_NAMENODE_USER=root
export HDFS_DATANODE_USER=root
export HDFS_SECONDARYNAMENODE_USER=root
```

其中，JAVA_HOME=/usr/java/jdk1.8.0_171-amd64是指定JDK的位置，HDFS_NAMENODE_USER=root是指定操作NameNode进程的用户是root。同理，HDFS_DATANODE_USER和HDFS_SECONDARYNAMENODE_USER分别指定了操作DataNode和Secondary NameNode的用户，在这里我们设置为root用户，具体应用时，读者根据情况进行设置即可。

在这里需要注意的是，HDFS_NAMENODE_USER、HDFS_DATANODE_USER和HDFS_SECONDARYNAMENODE_USER是Hadoop 3.x为了提升安全性而引入的。

（2）core-site.xml文件

core-site.xml中主要配置Hadoop的公共属性，配置代码如下：

```
<?xml version="1.0" encoding="UTF-8"?>
<?xml-stylesheet type="text/xsl" href="configuration.xsl"?>

<configuration>
<property>
<name>fs.defaultFS</name>
<value>hdfs://node1:9820</value>
</property>
<property>
<name>hadoop.tmp.dir</name>
<value>/opt/hadoopdata</value>
</property>
</configuration>
```

其中，fs.defaultFS是指定NameNode所在的节点，在这里配置为node1；9820是默认端口；hdfs:是协议；hadoop.tmp.dir是配置元数据所存放的配置，这里配置为/opt/hadoopdata，后续如果需要查看fsiamge和edits文件，可以到这个目录下查找。

（3）hdfs-site.xml文件

hdfs-site.xml文件中主要是HDFS属性配置，配置代码如下：

```
<?xml version="1.0" encoding="UTF-8"?>
<?xml-stylesheet type="text/xsl" href="configuration.xsl"?>
<configuration>
<property>
```

```
<name>dfs.namenode.secondary.http-address</name>
<value>node2:9868</value>
</property>
</configuration>
```

其中，dfs.namenode.secondary.http-address 属性是配置 Secondary NameNode 的节点，在这里配置为 node2。端口为 9868。

关于这些配置，读者可以从官网上查找，网址为 http://hadoop.apache.org/docs/r3.1.0/，其中的左下角有个 Configuration 项，其中包括 core-default.xml 等配置文件。

（4）workers 文件

在 workers 中配 DataNode 节点，在其中写入：

```
node1
node2
node3
```

2. 将Hadoop复制到其他节点

前面我们在 node1 上完成了 Hadoop 中 NameNode、Secondary NameNode 和 DataNode 的配置，现在需要将 Hadoop 及这些配置复制到其他几个节点的特定目录下，复制命令如下：

```
[root@node1 software]# scp -r ./hadoop-3.1.0 root@node2:/opt/software/
[root@node1 software]# scp -r ./hadoop-3.1.0 root@node3:/opt/software/
```

这样Hadoop和配置文件就都复制到了其他节点上。

3. 格式化

第一次安装 Hadoop 需要进行格式化，以后就不需要了。格式化命令在 hadoop/bin 下面，执行如下命令：

```
[root@node1 ~]# hdfs namenode -format
```

格式化后会创建一个空白的 fsimage 文件，可以在 opt/hadoopdata/dfs/name/current 中找到 fsimage 文件，注意此时没有 edits 文件。

4. 启动命令

进入 hadoop/sbin 下面运行 start-dfs.sh，启动 HDFS 集群，启动命令如下：

```
[root@node1 current]# start-dfs.sh
Starting namenodes on [node1]
Starting datanodes
node2: WARNING: /opt/software/hadoop-3.1.0/logs does not exist. Creating.
node3: WARNING: /opt/software/hadoop-3.1.0/logs does not exist. Creating.
Starting secondary namenodes [node2]
2018-05-0918:29:57,190WARNutil.NativeCodeLoaUnabltoloadnative-hadooplib
raryforyourplatform...usingbuiltin-java classes where applicable
```

这时，可以在不同节点中通过 jps 命令查看不同的进程。

node1 节点：

```
[root@node1 current]# jps
2010 NameNode
2428 Jps
2127 DataNode
```

node2 节点：

```
[root@node2 ~]# jps
1523 Jps
1429 SecondaryNameNode
1367 DataNode
```

node3 节点：

```
[root@node3 ~]# jps
1473 Jps
1400 DataNode
```

由此可见，node1 是 NameNode 也是 DataNode，node2 是 Secondary NameNode、DataNode，node3 是 DataNode。启动正常。

5．打开浏览器查看HDFS监听页面

在浏览器中输入 http://ip:9870，比如这里输入 http://192.168.19.10:9870/，出现以下界面则表示 Hadoop 完全分布式搭建成功，如图 4.3 所示。

| Hadoop | Overview | Datanodes | Datanode Volume Failures | Snapshot | Startup Progress | Utilities |

Overview 'node1:9820' (active)

Started:	Wed May 09 18:29:29 +0800 2018
Version:	3.1.0, r16b70619a24cdcf5d3b0fcf4b58ca77238ccbe6d
Compiled:	Fri Mar 30 08:00:00 +0800 2018 by centos from branch-3.1.0
Cluster ID:	CID-ff0428d6-783d-4b0c-803d-71b41ebb0559
Block Pool ID:	BP-2045811995-192.168.19.10-1525861547422

图 4.3　Node1 状态为 active

选择 Datanodes 选项，可以看到 DataNode 的利用率和 DataNode 的节点状态，如图 4.4 和图 4.5 所示。

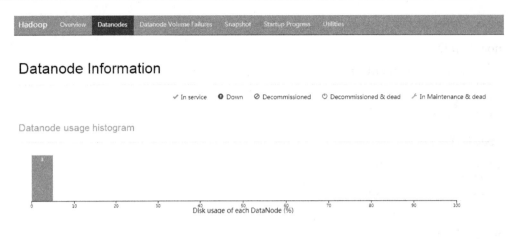

图 4.4　Node1 的 Datanodes（数据节点）信息

图 4.5　Node1 有 3 个 Datanode

6．上传文件

在前面已经讲过了如何将文件上传到 HDFS 上，这里我们将 test.txt 上传到 HDFS 的根目录"/"下面，命令如下：

```
[root@node1 ~]# hdfs dfs -put test.txt /
```

如果上传成功，则可以在Utilities选项的下拉菜单中选择Browse the file system命令进行查看，如图4.6所示。

图 4.6　Node1 的 Utilities 选项

在输入框中输入"/"，单击 Go!按钮，则可以看到刚才的文件已经上传到了 HDFS 上，如图 4.7 所示。

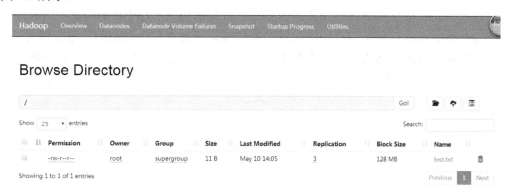

图 4.7　Node1 文件上传成功页面

4.3　什么是 HDFS 高可用

NameNode 存在单点失效的问题。如果 NameNode 失效了，那么所有的客户端——包括 MapReduce 作业均无法读、写文件，因为 NameNode 是唯一存储元数据与文件到数据块映射的地方。在这种情况下，Hadoop 系统无法提供服务，为了减少由计算机硬件和软件易错性所带来的损失而导致 NameNode 节点失效的问题，可以通过搭建 HDFS 高可用集群来实现 NameNode 的高可用性。

HDFS 高可用是配置了一对活动-备用（active-standby）NameNode。当活动 NameNode 失效，备用 NameNode 就会接管它的任务并开始服务于来自客户端的请求，不会有任何明显的中断。

在这样的情况下，要想从一个失效的 NameNode 中恢复，系统管理员需要启动一个拥有头文件系统元数据副本的新 NameNode，并配置 DataNode 和客户端以便使用这个新的 NameNode。新的 NameNode 直到满足以下情形后才能响应服务：

- 将命名空间的映像导入内存中。
- 重做编辑日志。
- 接收到足够多的来自 DataNode 的数据块报告并退出安全模式。对于一个大型并拥有大量文件和数据块的集群，NameNode 的冷启动需要 30 分钟甚至更长时间。
- 系统恢复时间太长，也会影响到日常维护。

4.3.1　HDFS 高可用实现原理

在高可用的实现中，主要配置了活动-备用（active-standby）NameNode。当活动

NameNode 失效，备用 NameNode 就会接管它的任务并响应来自客户端的服务请求，不会有任何明显中断。实现这一目标需要在架构上作如下修改。

在活动 NameNode 失效之后，备用 NameNode 能够快速（几十秒的时间）实现任务接管，因为最新的状态存储在闪存中：包括最新的编辑日志条目和最新的数据块映射信息。实际观察到的失效时间略长一点（需要 1 分钟左右），这是因为系统需要保守地确定活动 NameNode 是否真的失效了，如图 4.8 所示。

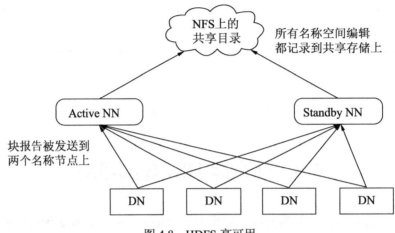

图 4.8　HDFS 高可用

通过双 NameNode 消除单点故障，其协调工作的要点如下。

元数据管理方式需要改变，内存中各自保存一份元数据，edits 日志只能有一份，只有 Active 状态的 NameNode 节点可以做写操作，两个 NameNode 都可以读取 edits，共享的 edits 放在一个共享存储中进行管理，通常使用 Journal Node 实现。

当备用 NameNode 接管工作之后，它将通读共享编辑日志直至末尾，以实现与活动 NameNode 的状态同步，并继续读取由活动 NameNode 写入的新条目。

需要一个状态管理功能模块，实现一个 ZKFailover，常驻在每一个 NameNode 所在的节点上。每一个 ZKFailover 负责监控自己所在的 NameNode 节点，利用 ZK 进行状态标识。当需要进行状态切换时，由 ZKFailover 来负责切换，切换时需要防止分裂混乱现象的发生。

DataNode 需要同时向两个 NameNode 发送数据块处理报告，因为数据块的映射信息存储在 NameNode 的内存中，而非磁盘上。

客户端需要使用特定的机制来处理 NameNode 的失效问题，这一机制对用户是透明的。

Secondary NameNode 的角色被备用 NameNode 所包含，备用 NameNode 为活动的 NameNode 命名空间设置周期性检查点。

4.3.2　HDFS 高可用实现

HDFS 高可用的架构图如图 4.9 所示。

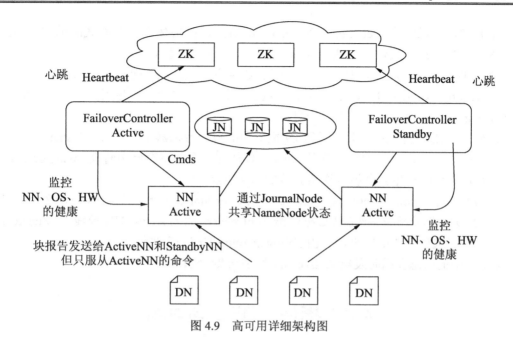

图 4.9 高可用详细架构图

在高可用架构图中的节点包括 NameNode、FailoverController、JournalNode、DataNode 和 ZooKeeper，下面分别介绍每个节点。

1. NameNode（NN）节点

NameNode 节点主要有两种状态：active 和 standby。

2. FailoverController（ZKFC）节点

FailoverController 节点用于监控和控制 NameNode 的状态切换，当一个集群中的 Active NameNode"挂掉"后，会把 Standby NameNode 状态切换成 active。

3. JournalNode（JN）节点

JournalNode 节点共享 edits 日志文件，因为 edits 文件一旦丢失，就会导致元数据丢失，数据也就丢失了，这里的 JN 往往是一个集群，来保障 edits 不会丢失。

4. DataNode（DN）

DataNode 节点定时和 NameNode 进行通信，接受 NameNode 的指令，同时 DataNode 之间还会相互通信，执行数据块复制任务。

5. ZooKeeper（ZK）

选择其中一个备用的 NameNode 为 active。在这里需要注意的是，在搭建 HA（HDFS

高可用）时，可以没有 Secondary NameNode 节点，在搭建完全分布式时的 Secondary NameNode 合并 edits 文件的功能由 Standby 的 NameNode 替代了。

前面我们已经提到，HDFS 高可用实现方式是通过配置了一对活动 - 备用（active-standby）NameNode。如果主 NameNode 发生故障，则切换到备用的 NameNode 上，通过双 NameNode 消除单点故障，同时：

- Active NN 会把 edits 文件写到 JN 集群中的每一台机器上，active 和 standby 的 NN 都可以读取 edits 文件，共享的 edits 文件放在一个共享存储中管理。fsimage 文件会在格式化时产生并会被推送给其他的 NN 节点。
- FailoverController Active 和 FailoverController Standby 分别对 Active NN 和 Standby NN 进行健康检查，把心跳数据传给 ZooKeeper。如果 Active NN 掉线了，ZooKeeper 会进行选举，选举出来后，由 FailoverController 进行切换。
- DN 中的信息会同时发给 Active NN 和 Standby NN。

4.4　搭建 HDFS 高可用

前面已经了解了高可用的架构和各组件间的协作关系，下面开始搭建 HDFS 高可用，在这个过程中，我们还是以 node1、node2 和 node3 来搭建高可用环境，每个节点所分配的作用如表 4.3 所示。

表 4.3　节点的分配

node1	node2	node3
NN	NN	—
ZKFC	—	ZKFC
JN	JN	JN
ZK	ZK	ZK
DN	DN	DN

由表 4.3 中可以看出，NameNode（NN）分别配置在 node1 和 node2 上，ZKFC 配置在 node1 和 node3 上，JournalNode（JN）配置在 node1、node2 和 node3 上，ZooKeeper（ZK）配置在 node1、node2 和 node3 上，DataNode（DN）配置在 node1、node2 和 node3 上。接下来开始高可用的搭建，首先来配置 ZooKeeper。

4.4.1　配置 ZooKeeper

首先从官网 http://zookeeper.apache.org/ 下载 ZooKeeper，这里以 ZooKeeper-3.4.6 为例，并上传到 node1 上。

解压缩 ZooKeeper：

```
[root@node1 software]# tar -zxvf ZooKeeper-3.4.6.tar.gz
```

在 ZooKeeper-3.4.6 的 conf 文件夹下创建 zoo.cfg 文件并写入以下配置：

```
tickTime=2000
clientPort=2181
initLimit=5
syncLimit=2
dataDir=/opt/ZooKeeper/data
server.1=node1:2888:3888
server.2=node2:2888:3888
server.3=node3:2888:3888
```

其中，dataDir 代表存放 ZooKeeper 数据文件的目录，server.1、server.2 和 server.3 中的 1、2、3 分别对应第 1 个节点、第 2 个节点和第 3 个节点，其中 2888 和 3888 代表接收和发送数据的端口。

创建 myid，为了保障 server.1、server.2 和 server.3，可以分别和第 1 个节点（node1）、第 2 个节点（node2）和第 3 个节点（node3）对应起来，还需要在 dataDir 所对应的文件夹中创建 myid 文件。具体操作如下。

在 node1 节点上创建/opt/ZooKeeper/data 文件夹，并在其中创建 myid 文件。

```
[root@node1 conf]# mkdir -p /opt/ZooKeeper/data
[root@node1 conf]# cd /opt/ZooKeeper/data
[root@node1 data]# vi myid
```

接着在 myid 文件中写入 1，这样就和 server.1 对应起来了。通过 more myid 命令可以查看 myid 文件中的内容。

```
[root@node1 data]# more myid
1
```

在 node2 和 node3 中的/opt/ZooKeeper/data 下分别创建 myid 文件，并分别写入 2 和 3，此处与 node1 操作类似，写入后分别通过 more 查看文件内容。

```
[root@node2 data]# more myid
2
[root@node3 data]# more myid
3
```

复制 node1 上的 ZooKeeper 到 node2 和 node3 节点上。

```
[root@node1 software]# scp -r ZooKeeper-3.4.6 root@node2:/opt/software/
[root@node1 software]# scp -r ZooKeeper-3.4.6 root@node3:/opt/software/
```

启动 ZooKeeper。进入每个节点的 ZooKeeper 的 bin 目录并启动 ZooKeeper，启动命令如下：

```
[root@node3 bin]# zkServer.sh start
JMX enabled by default
Using config: /opt/software/ZooKeeper-3.4.6/bin/../conf/zoo.cfg
Starting ZooKeeper ... STARTED
```

检查 ZooKeeper 是否配置成功，在每个节点输入 jps 命令查看进程，会发现有个

QuorumPeerMain，这就是 ZooKeeper 的进程，命令如下：

```
[root@node1 bin]# jps
1322 Jps
1292 QuorumPeerMain
```

这样就代表 ZooKeeper 集群已经配置成功。关于 ZooKeeper 的介绍读者可以参考相应章节。

4.4.2 配置 Hadoop 配置文件

Hadoop 配置文件一共有 4 个，分别是：

```
hadoop-env.sh
core-site.xml
hdfs-site.xml
workers
```

hadoop-env.sh 的配置，是关于 Hadoop 运行环境的配置，在配置高可用时，分别定义了 JAVA_HOME 及运行 NameNode、DataNode、ZKFC 和 JournalNode 进程的用户：

```
export JAVA_HOME=/usr/java/jdk1.8.0_171-amd64
export HDFS_NAMENODE_USER=root
export HDFS_DATANODE_USER=root
export HDFS_ZKFC_USER=root
export HDFS_JOURNALNODE_USER=root
```

core-site.xml 文件的配置如下：

```
<configuration>
<property>
<name>fs.defaultFS</name>
<value>hdfs://mycluster</value>
</property>
<property>
<name>hadoop.tmp.dir</name>
<value>/opt/hadoopdata</value>
</property>
<property>
<name>hadoop.http.staticuser.user</name>
<value>root</value>
</property>
<property>
<name>ha.ZooKeeper.quorum</name>
<value>node1:2181,node2:2181,node3:2181</value>
</property>
</configuration>
```

其中，hadoop.http.staticuser.user 是定义在网页界面访问数据时使用的用户名，ha.ZooKeeper.quorum 是定义 ZooKeeper 集群。

hdfs-site.xml 文件的配置如下：

```xml
<configuration>
<property>
        <name>dfs.nameservices</name>
    <value>mycluster</value>
</property>
<property>
    <name>dfs.ha.namenodes.mycluster</name>
    <value>nn1,nn2</value>
</property>
<property>
    <name>dfs.namenode.rpc-address.mycluster.nn1</name>
    <value>node1:8020</value>
</property>
<property>
    <name>dfs.namenode.rpc-address.mycluster.nn2</name>
    <value>node2:8020</value>
</property>
<property>
    <name>dfs.namenode.http-address.mycluster.nn1</name>
    <value>node1:9870</value>
</property>
<property>
    <name>dfs.namenode.http-address.mycluster.nn2</name>
    <value>node2:9870</value>
</property>
<property>
    <name>dfs.namenode.shared.edits.dir</name>
    <value>qjournal://node1:8485;node2:8485;node3:8485/wen</value>
</property>
<property>
    <name>dfs.client.failover.proxy.provider.mycluster</name>
    <value>
    org.apache.hadoop.hdfs.server.namenode.ha.ConfiguredFailoverProxyProvider
    </value>
</property>
<property>
     <name>dfs.ha.fencing.methods</name>
     <value>sshfence</value>
</property>
<property>
     <name>dfs.ha.fencing.ssh.private-key-files</name>
     <value>/root/.ssh/id_rsa</value>
</property>
<property>
    <name>dfs.journalnode.edits.dir</name>
    <value>/opt/journalnode/data</value>
</property>
<property>
    <name>dfs.ha.automatic-failover.enabled</name>
    <value>true</value>
```

```
</property>
</configuration>
```

其中，dfs.nameservices 是配置 HDFS 集群 ID，dfs.ha.namenodes.mycluster 是配置 NameNode 的 ID 号，dfs.namenode.rpc-address.mycluster.nn1 是定义 NameNode 的主机名和 RPC 协议的端口，dfs.namenode.http-address.mycluster.nn1 是定义 NameNode 的主机名和 HTTP 协议的端口。

dfs.namenode.shared.edits.dir 是定义共享 edits 的 URL，dfs.client.failover.proxy.provider.mycluster 是定义返回 active namenode 的类。

dfs.ha.fencing.methods 是定义 NameNode 切换时的隔离方法，主要是为了防止"脑裂"问题，dfs.ha.fencing.ssh.private-key-files 是定义隔离方法的密钥，dfs.journalnode.edits.dir 是保存 edits 文件的目录，dfs.ha.automatic-failover.enabled 用于定义开启自动切换。

workers 文件的配置如下：

```
node1
node2
node3
```

workers 主要配置 DataNode 节点，在 workers 配置文件中写入以上内容，代表 node1、node2 和 node3 是 DataNode 节点。

4.4.3 将配置文件复制到其他节点上

将所有的配置文件复制到其他节点上，复制命令如下：

```
[root@node1 hadoop]# scp hadoop-env.sh root@node2:/opt/software/hadoop-3.1.0/etc/hadoop/
hadoop-env.sh100%    16KB    16.2KB/s    00:00
[root@node1 hadoop]# scp core-site.xml root@node2:/opt/software/hadoop-3.1.0/etc/hadoop/
core-site.xml100%    2107    2.1KB/s    00:00
[root@node1 hadoop]# scp hdfs-site.xml root@node2:/opt/software/hadoop-3.1.0/etc/hadoop/
hdfs-site.xm100%    1131    1.1KB/s    00:00
[root@node1 hadoop]# scp workers root@node2:/opt/software/hadoop-3.1.0/etc/hadoop/
workers100%    18    0.0KB/s    00:00
[root@node1 hadoop]# scp hadoop-env.sh root@node3:/opt/software/hadoop-3.1.0/etc/hadoop/
hadoop-env.sh100%    16KB    16.2KB/s    00:00
[root@node1 hadoop]# scp hdfs-site.xml root@node3:/opt/software/hadoop-3.1.0/etc/hadoop/
hdfs-site.xml100%    1131    1.1KB/s    00:00
[root@node1 hadoop]# scp core-site.xml root@node3:/opt/software/hadoop-3.1.0/etc/hadoop/
core-site.xml 100%    2107    2.1KB/s    00:00
[root@node1 hadoop]# scp workers root@node3:/opt/software/hadoop-3.1.0/etc/hadoop/
workers100%    18    0.0KB/s    00:00
```

4.4.4 启动 JN 节点

启动 JN 节点，命令如下：

```
hdfs --daemon start journalnode
```

在 node1、node2 和 node3 上分别启动 JournalNode，然后通过 jps 命令进行查看，这

里以 node1 为例。代码如下：

```
[root@node1 opt]# hdfs --daemon start journalnode
[root@node1 opt]# jps
1531 Jps
1292 QuorumPeerMain
1487 JournalNode
```

4.4.5 格式化

在 node1 上执行格式化操作，执行命令如下：

```
hdfs namenode -format
```

4.4.6 复制元数据到 node2 节点上

由于我们需要把 node1 和 node2 设置为两个 NameNode，所以搭建高可用时要求 node1 和 node2 上的元数据是一样的，因此需要将 node1 上的元数据复制到 node2 上。在这里需要注意的是，执行格式化命令后，会在/opt/hadoopdata 目录下生成元数据，执行命令如下：

```
[root@node1 opt]# scp -r ./hadoopdata root@node2:/opt/
fsimage_0000000000000000000100%   389     0.4KB/s   00:00
seen_txid 100%    2     0.0KB/s   00:00
VERSION 100%  216    0.2KB/s   00:00
fsimage_0000000000000000000.md5 100%   62    0.1KB/s   00:00
```

4.4.7 格式化 ZKFC

ZKFC（ZooKeeper Failover Controller）是在 HDFS 高可用前提下，基于 ZooKeeper 的自动切换原理触发 NameNode 切换的一个进程。

在 NameNode 节点上启动的 ZKFC 进程内部，运行着如下 3 个对象服务。

- HealthMonitor：定期检查 NameNode 是否不可用或是否进入了一个不健康的状态，并及时通知 ZooKeeper Failover Controller。
- ActiveStandbyElector：控制和监控 NameNode 在 ZooKeeper 上的状态。
- ZKFailoverController：协调 HealthMonitor 和 ActiveStandbyElector 对象并处理它们通知的 event 变化事件，完成自动切换的过程。

通 hdfs zkfc –formatZK 命令格式化 ZKFC（ZooKeeper Failover Controller），执行命令如下：

```
[root@node1 opt]# hdfs zkfc -formatZK
2018-05-11 17:22:26,467 INFO tools.DFSZKFailoverController: STARTUP_MSG:
STARTUP_MSG: Starting DFSZKFailoverController
```

```
STARTUP_MSG:   host = node1/192.168.19.10
STARTUP_MSG:   args = [-formatZK]
STARTUP_MSG:   version = 3.1.0
```

启动 ZKFC：

```
[root@node1 opt]# hdfs --daemon start zkfc
```

4.4.8 启动集群

执行 start-dfs.sh 命令启动集群：

```
[root@node1 opt]# start-dfs.sh
Starting namenodes on [node1 node2]
Starting datanodes
Starting journal nodes [node2 node3 node1]
```

从上面的提示中可以看出 node1 和 node2 作为 HDFS 高可用的 NameNode，node1、node2 和 node3 作为用来保证 NameNode 间数据共享的 JournalNode。

4.4.9 通过浏览器查看集群状态

在浏览器中分别打开 http://192.168.19.10:9870/ 和 http://192.168.19.20:9870/，会看到下面两个界面，可以看出 node2 为 active 状态，node1 是 standby 状态，如图 4.10 和图 4.11 所示。

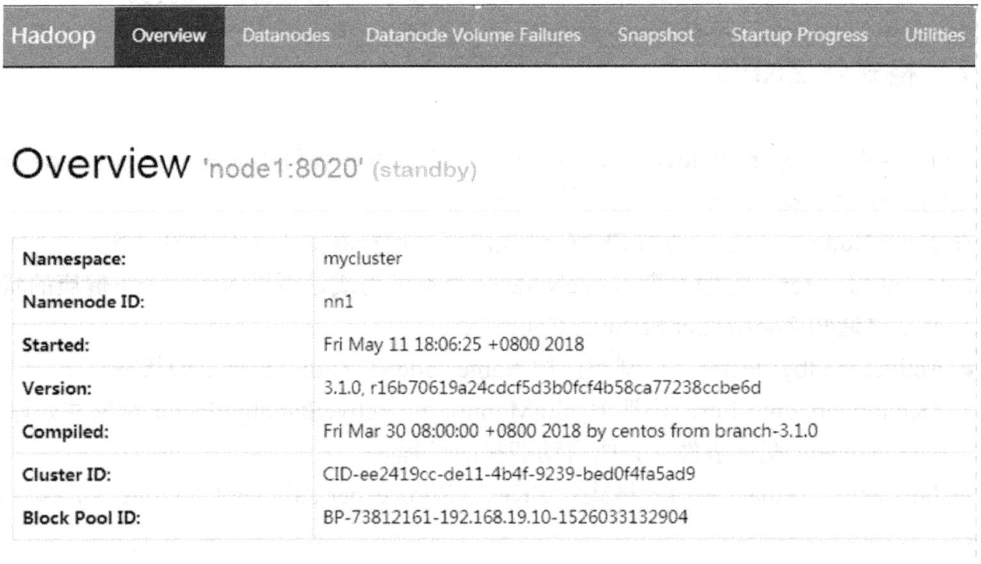

图 4.10 node1 界面

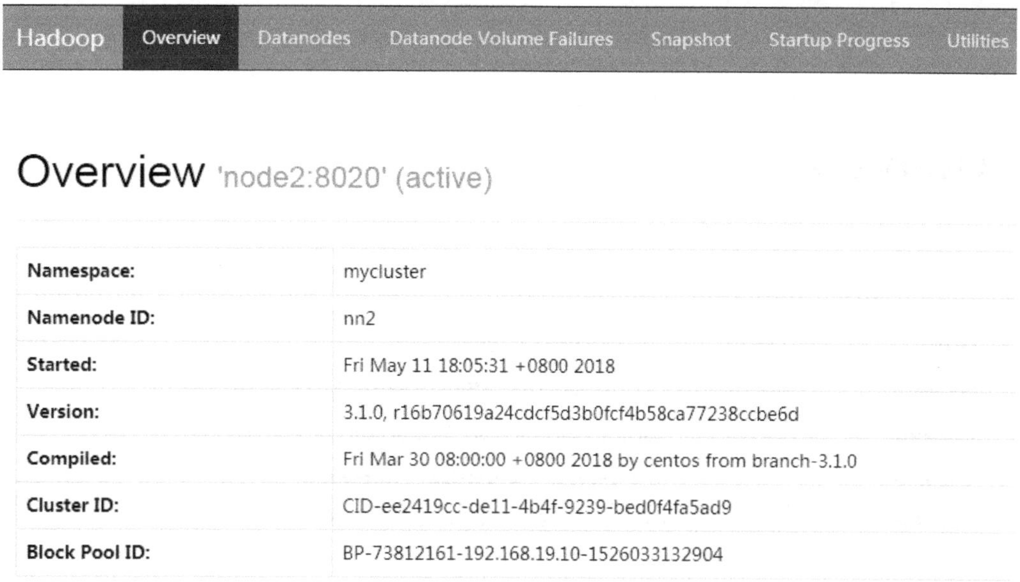

图 4.11　node2 界面

4.4.10　高可用测试

目前 node1 是 standby（备用）状态，node2 是 active（活跃）状态，现在我们先查看 NameNode 进程的 ID 号，随后杀死在 node2 中处于 active 状态的 NameNode 进程，测试当此节点失效时，备用 NameNode（即 node1）能否接管已失效节点的任务并开始服务于来自客户端的请求，命令如下：

```
[root@node2 dfs]# jps
6016 NameNode
1270 QuorumPeerMain
6086 DataNode
6297 DFSZKFailoverController
5914 JournalNode
6395 Jps
[root@node2 dfs]# kill -9 6016
```

接着我们打开 node1，发现 node1 的状态已经变成了 active 状态，如图 4.12 所示。

这样，关于 Hadoop 3 的高可用就配置成功了。关于高可用的配置，读者可以参考官网 http://hadoop.apache.org/docs/r3.1.0/hadoop-project-dist/hadoop-hdfs/HDFSHighAvailabilityWithQJM.html。

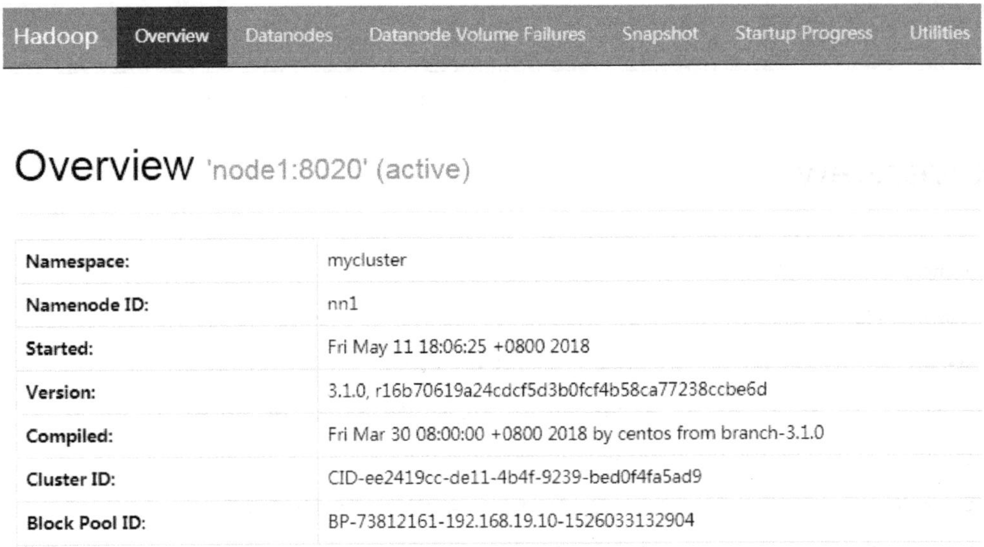

图 4.12　Hadoop 3 的高可用配置成功

4.5　小　　结

本章主要讨论了 Hadoop 3.x 的新特性及如何基于 Hadoop 3 搭建完全分布式和 NameNode 高可用的环境，搭建过程比较烦琐，读者需要多做几遍，遇到问题可以多看看 log 日志，同时需要注意操作的前后顺序。

第 2 篇
Hadoop 核心技术

- 第 5 章　Hadoop 的分布式协调服务——ZooKeeper
- 第 6 章　分布式离线计算框架——MapReduce
- 第 7 章　Hadoop 的集群资源管理系统——YARN
- 第 8 章　Hadoop 的数据仓库框架——Hive
- 第 9 章　大数据快速读写——HBase
- 第 10 章　海量日志采集工具——Flume
- 第 11 章　Hadoop 和关系型数据库间的数据传输工具——Sqoop
- 第 12 章　分布式消息队列——Kafka
- 第 13 章　开源的内存数据库——Redis
- 第 14 章　Ambari 和 CDH
- 第 15 章　快速且通用的集群计算系统——Spark

第 5 章　Hadoop 的分布式协调服务——ZooKeeper

ZooKeeper 在分布式应用中提供了诸如统一命名服务、配置管理和分布式锁的基础，成为高效、稳健的分布式协调服务。另外，在分布式数据一致的情况下，ZooKeeper 采用了一种被称为 ZAB（ZooKeeper Automic Broadcast）的一致性协议。

本章主要涉及如下知识点。
- 掌握 ZooKeeper 的基本概念。
- 通过示例学会 ZooKeeper 的安装与运行。
- 服务器端常用命令，以及如何通过 java API 访问 ZooKeeper。

5.1　ZooKeeper 的核心概念

在分布式系统构建的集群中，每一台机器都有自己的角色定位。其中最典型的是 Master/Slave 模式，在这种模式中，所有写操作的机器都可以称为 Master 机器；所有通过异步复制方式获取最新数据并提供读服务的机器都可以称为 Slave 机器。

在 ZooKeeper 中，不同于以往的是引入了全新的 Leader、Follower 和 Observer 三种角色概念，即 ZooKeeper 会通过选举选定一台被称为 Leader 的机器，这台服务器将为客户端提供读写服务。

除 Leader 外，其他机器包括 Follower 和 Observer 都能够提供读服务，唯一不同的是，Leader 选举过程和写操作的"过半写功能"策略——Observer 都是不参与的。所以在不影响写性能的情况下，Observer 可以提升集群的读性能。下面我们来介绍一下 ZooKeeper 中的核心概念。

5.1.1　Session 会话机制

在 ZooKeeper 中，当客户端与服务器端成功建立连接后，Session 会话随之建立，同时会生成一个全局唯一的会话 ID（Session ID）。在 ZooKeeper 中，一个客户端连接是指客户端和服务器之间的一个 TCP 长连接。

ZooKeeper 对外的服务端口默认是 2181，当客户端启动时，新建立的 TCP 连接也将第一次启动，它能通过心跳检测与服务器保持有效会话，同时还会向 ZooKeeper 发送请求并接收响应，另外还能够接收来自服务器的 Watch 事件通知。

Session 的 SessionTimeout 值用来设置一个客户端会话的超时时间。当出现故障而又想要保存之前创建的会话时，只需在 SessionTimeout 规定的时间内重新连接上集群的任意一台服务器即可。

5.1.2 数据节点、版本与 Watcher 的关联

在 ZooKeeper 中，"节点"是指数据模型中的数据单元，也叫数据节点——Znode。数据模型是以树（Znode Tree）的格式进行存储，并通过斜杠（/）来分割路径，分割后的每一个 Znode 都会保存自己的数据内容，同时还会保存一系列属性值，如/zoo/path。

ZooKeeper 的每一个 Znode 都会对应一个叫做 Stat 的数据结构。而 Stat 记录了 version（当前 Znode 的版本）、cversion（当前 Znode 子节点的版本）和 aversion（当前 Znode 的 ACL 版本）当前的 3 个数据版本。

当用户在 ZooKeeper 中注册一些 Watcher（时间监听器）后，在一些特定事件触发的情况下，ZooKeeper 将会把事件通知发送到感兴趣的客户端上，这是 ZooKeeper 分布式协调服务的重要特性。

5.1.3 ACL 策略

ZooKeeper 的权限控制系统类似于 UNIX 文件系统，它采用的是 ACL（Access Control Lists）策略。

ZooKeeper 定义了如下 5 种权限。
- CREATE：创建子节点的权限。
- READ：获取节点数据和子节点的权限。
- WRITE：更新节点数据的权限。
- DELETE：删除子节点的权限。
- ADMIN：设置节点 ACL 的权限。

5.2 ZooKeeper 的安装与运行

下面我们将详细介绍 ZooKeeper 的安装步骤。由于 ZooKeeper 需要 JDK，因此在安装 ZooKeeper 前首先要确认已经安装了 JDK。安装 ZooKeeper 的步骤如下。

（1）从 Apache 的关于 ZooKeeper 的发布页面下载 ZooKeeper，下载页面是：

```
http://hadoop.apache.org/ZooKeeper/release.html
```
读者可以下载稳定版本的 ZooKeeper，这里我们下载使用的版本是 ZooKeeper-3.4.6。

（2）接着将文件上传到 Linux 系统中的文件夹下，比如上传到/opt/software 目录下，并进行解压缩，解压缩命令是 tar-zxvf ZooKeeper-3.4.6，然后将解压缩生成的文件夹重命名为 ZooKeeper。

（3）创建配置文件 zoo.cfg，在/opt/software/ZooKeeper/conf 下有 zoo_sample.cfg 文件，复制文件并重命名为 zoo.cfg，在其中加入以下代码：

```
tickTime=2000
dataDir=/opt/ZooKeeper/data
clientPort=2181
initLimit=10
syncLimit=5
server.1=192.168.106.10:2888:3888
server.2=192.168.106.20:2888:3888
server.3=192.168.106.30:2888:3888
```

在该配置文件中，tickTime 属性指定了 ZooKeeper 中的基本时间单元；dataDir 属性指定了 ZooKeeper 存储持久数据的本地文件系统设置；clientPort 属性指定了 ZooKeeper 用于监听客户端连接的端口。

initLimit 参数设定了所有跟随者与 Leader 节点进行连接并同步的时间范围。如果在设定的时间段内，半数以上的跟随者未能完成同步，Leader 节点便会宣布放弃 Leader 地位，然后进行另外一次 Leader 选举。

如果这种情况经常发生，则表明设定的值太小。syncLimit 参数设定了允许一个 Follower 与 Leader 进行同步的时间。其中 192.168.106.10、192.168.106.20 和 192.168.106.30，分别代表 3 个节点的 IP 地址。

（4）复制 ZooKeeper 文件夹到其他节点上。

假定还有另外两个节点，分别是 slave0 和 slave1，一旦完成上面的配置，我们就可以将整个 ZooKeeper 文件发送到 slave0 和 slaver1 的/opt/software/下。

（5）创建 my.id 文件。

在 3 个节点的 opt/ZooKeeper/data 文件夹下分别创建 myid 文件，并且写入数字 1、2、3，与配置文件中保持一致。比如根据上面的配置文件，我们需要做的是在 192.168.106.10 节点的 myid 文件中写入 1；在 192.168.106.20 节点的 myid 文件中写入 2；在 192.168.106.30 节点的 myid 文件中写入 3。

（6）修改环境变量。

在.bash_profile 中配置环境变量以保证可以在任意目录下启动 ZooKeeper，环境变量中的 ZooKeeper 路径如下：

```
ZOOKEEPER_HOME=/opt/software/ZooKeeper
PATH=$PATH:$HOME/bin:$JAVA_HOME/bin:$CATALINA_HOME/bin:$HADOOP_HOME/
bin:$HADOOP_HOME/sbin:$ZOOKEEPER_HOME/bin
```

(7）启动 ZooKeeper 服务。

完成以上配置后就可以启动 ZooKeeper 服务了，在所有节点中输入 zkServer.sh start，保证所有节点都启动 ZooKeeper。

在 ZooKeeper 启动完成后，可以通过 zkServer.sh status 查看状态，当出现一个 Leader 两个 Follower 时，则代表启动成功。

5.3 ZooKeeper 服务器端的常用命令

5.2 节中介绍了 ZooKeeper 的安装与配置。本节我们将介绍 ZooKeeper 服务器端最常见的几种命令，包括启动服务、查看服务状态、停止服务和重启服务。

1. 启动ZooKeeper服务

通过 zkServer.sh start 命令可以启动 ZooKeeper 服务，如果启动成功，则出现 STARTED 状态。

```
[root@master bin]# ./zkServer.sh start
Using config: /opt/software/ZooKeeper/bin/../conf/zoo.cfg
Starting ZooKeeper ... STARTED
```

其中，Using config 是 zoo.cfg 配置文件所在的路径。

2. 查看ZooKeeper服务状态

通过 zkServer.sh status 命令可以查看 ZooKeeper 的服务状态，比如 Leader 或者 Follower 状态。

```
[root@master bin]# ./zkServer.sh status
JMX enabled by default
Using config: /opt/software/ZooKeeper/bin/../conf/zoo.cfg
Mode: follower
```

3. 停止ZooKeeper服务

通过 zkServer.sh stop 命令结束 ZooKeeper 服务，如果结束服务成功，则出现提示 STOPPED 状态：

```
[root@master bin]# ./zkServer.sh stop
JMX enabled by default
Using config: /opt/software/ZooKeeper/bin/../conf/zoo.cfg
Stopping ZooKeeper ... STOPPED
```

4. 重启ZooKeeper服务

命令如下：
zkServer.sh restart

5.4　客户端连接 ZooKeeper 的相关操作

5.3 节我们介绍了 ZooKeeper 服务的启动和状态查看等命令，现在就可以通过客户端连接到 ZooKeeper 了。连接方式可以使用 ZooKeeper/bin 下面的 zkCli.sh 命令，连接格式是 zkCli.sh -server IP:port。比如，现在要连接本机的 2181 端口。代码如下：

```
$./zkCli.sh -server 127.0.0.1:2181
```

连接成功后，可以看到以下结果：

```
Connecting to 127.0.0.1:2181
2018-04-13 15:11:49,775 [myid:] - INFO  [main:Environment@100] - Client environment:ZooKeeper.version=3.4.6-1569965, built on 02/20/2014 09:09 GMT
2018-04-13 15:11:49,779 [myid:] - INFO  [main:Environment@100] - Client environment:host.name=master
2018-04-13 15:11:49,779 [myid:] - INFO  [main:Environment@100] - Client environment:java.version=1.7.0_79
2018-04-13 15:11:49,781 [myid:] - INFO  [main:Environment@100] - Client environment:java.vendor=Oracle Corporation
2018-04-13 15:11:49,781 [myid:] - INFO  [main:Environment@100] - Client environment:java.home=/usr/java/jdk1.7.0_79/jre
2018-04-13 15:11:49,782 [myid:] - INFO  [main:Environment@100] - Client environment:java.class.path=/opt/software/ZooKeeper/ZooKeeper
[zk: 127.0.0.1:2181(CONNECTED) 0]
```

5.4.1　查看 ZooKeeper 常用命令

一旦连接到 ZooKeeper 上就可以通过相关命令进行操作了，可以通过 help 命令查看有哪些命令，具体如下：

```
[zk: 127.0.0.1:2181(CONNECTED) 0] help
ZooKeeper -server host:port cmd args
    connect host:port
    get path [watch]
    ls path [watch]
    set path data [version]
    rmr path
    delquota [-n|-b] path
    quit
    printwatches on|off
    create [-s] [-e] path data acl
    stat path [watch]
    close
    ls2 path [watch]
    history
    listquota path
    setAcl path acl
    getAcl path
    sync path
```

```
redo cmdno
addauth scheme auth
delete path [version]
setquota -n|-b val path
```

通过这些命令，我们就可以实现对 ZooKeeper 的操作了。这里的操作都是对 ZooKeeper 的节点进行操作，比如列出节点、创建节点和删除节点，可以把 ZooKeeper 理解为以下结构，如图 5.1 所示。

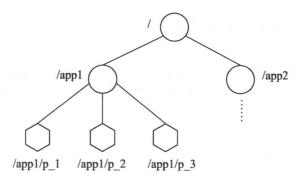

图 5.1　ZooKeeper 结构

接下来我们学习一些常见的命令。

5.4.2　connect 命令与 ls 命令

connect 命令用于连接 ZooKeeper 服务器端，比如通过 connect 127.0.0.1:2181，可以连接到本机的 2181 端口。

ls 命令格式：ls path，表示列出 path 下的文件。

ls 命令用于获取路径下的节点信息，需要注意的是该路径为绝对路径。

比如：ls / 可以列出根目录下的所有文件，从下面的运行结果中可以看出根目录下有一个 ZooKeeper 节点。

```
[zk: 127.0.0.1:2181(CONNECTED) 4] ls /
[ZooKeeper]
```

如果想要列出 ZooKeeper 下的信息，需要用到绝对路径 ls /ZooKeeper。

```
[zk: 127.0.0.1:2181(CONNECTED) 5] ls /ZooKeeper
[quota]
```

5.4.3　create 命令——创建节点

create 命令格式如下：

```
create [-s] [-e] path data acl.
```

其中，-s 和-e 参数分别指定节点为持久节点或临时节点，在不指定的情况下，则表示是持久节点； acl 用来进行权限控制。如果是临时节点，会话关闭后节点也就不存在了。比如 create /mynode1 content1 命令用于创建 mynode1 节点，并且给 mynode1 赋值为 content1：

```
[zk: 127.0.0.1:2181(CONNECTED) 6] create /mynode1 content1
Created /mynode1
```

5.4.4　get 命令——获取数据与信息

get 命令用于获取 ZooKeeper 节点的数据内容和相关信息，比如我们要取得 mynode1 的内容，执行 get /mynode1 命令后，运行结果是节点的数据内容 content1 和此节点的相关匹配信息。

```
[zk: 127.0.0.1:2181(CONNECTED) 8] get /mynode1
content1
cZxid = 0x100000009
ctime = Fri Apr 13 15:57:29 CST 2018
mZxid = 0x100000009
mtime = Fri Apr 13 15:57:29 CST 2018
pZxid = 0x100000009
cversion = 0
dataVersion = 0
aclVersion = 0
ephemeralOwner = 0x0
dataLength = 8
numChildren = 0
```

5.4.5　set 命令——修改节点内容

set 命令格式如下：

```
set path data
```

set 命令用于修改节点内容，其中 data 参数是需要更新的新内容，在这里第 1 次把 mynode1 的内容修改为 content2，第 2 次修改为 content3。具体代码如下：

```
[zk: 127.0.0.1:2181(CONNECTED) 9] set /mynode1 content2
cZxid = 0x100000009
ctime = Fri Apr 13 15:57:29 CST 2018
mZxid = 0x10000000a
mtime = Fri Apr 13 16:15:39 CST 2018
pZxid = 0x100000009
cversion = 0
dataVersion = 1
aclVersion = 0
ephemeralOwner = 0x0
dataLength = 8
numChildren = 0
```

```
[zk: 127.0.0.1:2181(CONNECTED) 10] set  /mynode1 content3
cZxid = 0x100000009
ctime = Fri Apr 13 15:57:29 CST 2018
mZxid = 0x10000000b
mtime = Fri Apr 13 16:15:48 CST 2018
pZxid = 0x100000009
cversion = 0
dataVersion = 2
aclVersion = 0
ephemeralOwner = 0x0
dataLength = 8
numChildren = 0
```

通过运行结果会发现,在运行 set /mynode1 content2 后,dataVersion=1,运行完 set /mynode1 content3 后,dataVersion=2。

修改完后,通过 get 命令查看结果,发现内容已经变成了 Content 3,具体如下:

```
[zk: 127.0.0.1:2181(CONNECTED) 13] get /mynode1
content3
cZxid = 0x100000009
ctime = Fri Apr 13 15:57:29 CST 2018
mZxid = 0x10000000b
mtime = Fri Apr 13 16:15:48 CST 2018
pZxid = 0x100000009
cversion = 0
dataVersion = 2
aclVersion = 0
ephemeralOwner = 0x0
dataLength = 8
numChildren = 0
```

5.4.6 delete 命令——删除节点

delete 命令格式如下:

```
delete path [version]
```

通过 delete 命令可以删除 ZooKeeper 上的指定节点。

```
[zk: 127.0.0.1:2181(CONNECTED) 14] delete /mynode1
[zk: 127.0.0.1:2181(CONNECTED) 15] ls /
[ZooKeeper]
```

可以看到,运行完 delete 命令后,mynode1 节点已经不存在了。在这里需要注意的是,如果节点存在子节点,需要先将子节点删除。

5.5 使用 Java API 访问 ZooKeeper

5.4 节中介绍了可以通过 zkCli.sh 命令访问 ZooKeeper,本节将介绍如何通过 Java API 来实现对 ZooKeeper 的访问。

5.5.1 环境准备与创建会话实例

在 Java 调用 ZooKeeper 时，需要导入 ZooKeeper-3.X.X.jar，比如 ZooKeeper-3.4.9.jar，创建会话是为了访问 ZooKeeper，首先需要创建一个 ZooKeeper 实例来连接 ZooKeeper 服务器。格式如下：

`ZooKeeper(String connectString, int sessionTimeout, Watcher watcher)`

- connectString：代表连接到的 ZooKeeper，如 47.104.214.68:2181。
- sessionTimeout：代表会话过期时间，单位是毫秒。
- watcher：通过注册 Watcher，实现对节点状态变化的监听，在节点被创建、删除及内容更新时，客户端都会收到通知。

首先来看一个创建会话的实例：

`ZooKeeper ZooKeeper = new ZooKeeper("47.104.214.68:2181", 3000,watcher);`

ZooKeeper 的构造方法还有以下几个：

```
ZooKeeper(String connectString,int sessionTimeout,Watcher watcher,boolean canBeReadOnly)
ZooKeeper(String connectString,int sessionTimeout,Watcher watcher,long sessionId,boolean canBeReadOnly)
ZooKeeper(String connectString,int sessionTimeout,Watcher watcher,long sessionId,byte[] sessionPasswd, boolean canBeReadOnly)
```

除了前面介绍的几个参数，这里的 canBeReadOnly 用于表示当前会话是否支持 read-only，sessionId 和 sessionPasswd 分别代表会话的 ID 和密钥，这两个参数可以唯一确定一个会话。可以通过 ZooKeeper 实例调用 getSesssionId()和 getSessionPasswd()方法获得会话的 ID 和密钥。

再说一下第 3 个参数 watcher，watcher 是 org.apache.ZooKeeper.Watcher 类型的对象，当客户端和服务器建立好连接之后会调用 Watcher 中的 process 方法，process 方法会接受一个 WatchedEvent 类型的参数，用于标识发生了哪种事件。以下是一个 process 内容实例，先让读者有一个初步的认识，后续还会通过实例具体说明。

```
public void process(WatchedEvent watchedEvent) {
    //判断是否已连接
    if (Event.KeeperState.SyncConnected==watchedEvent.getState()) {
    // 第一次与 ZooKeeper 服务器建立好连接
        if(Event.EventType.None==watchedEvent.getType()  && watchedEvent.getPath()==null) {
        //子节点变化事件
        } else if(EventType.NodeCreated == event.getType()) {
        
        }
        // 继续监听其他事件类型
    }

}
```

其中，WatchedEvent 主要包括以下两方面信息：
- watchedEvent.getState()取得与服务器连接状态，通过调用 watchedEvent.getState()得到和 ZooKeeper 服务器的连接状态信息。这里的状态信息主要包括 SyncConnected、Disconnected、ConnectedReadOnly 和 AuthFailed 等几类。
- watchedEvent.getType()取得节点事件类型，可以通过 watchedEvent.getType()方法获取具体的事件类型。事件类型的取值包括 None、NodeCreated、NodeDeleted、NodeDataChanged 和 NodeChildrenChanged。

需要注意的是，Watcher 通知是一次性的，也就是说一旦触发一次通知后，这个 Watcher 就无效了，所以客户端需再次注册 Watcher，也就是在 process 里再次注册 Watcher，否则无法再次监听到节点的状态变化。

5.5.2 节点创建实例

前面介绍了连接 ZooKeeper 实例和 Watcher，下面就可以实现对节点的创建、内容修改和删除等操作了。

节点创建代码如下：

```
ZooKeeper.create("/mynode","mycontent1".getBytes(), Ids.OPEN_ACL_UNSAFE,
CreateMode.PERSISTENT);
```

第 1 个参数是要创建的节点路径；第 2 个参数是创建节点的数据值；第 3 个参数是这个节点的访问权限，我们这里指定该节点可以被任何人访问；第 4 个参数是创建节点的类型，这里的 PERSISTENT 是指会话重启后，节点仍然存在。

具体实例如下：

```
import org.apache.ZooKeeper.CreateMode;
import org.apache.ZooKeeper.KeeperException;
import org.apache.ZooKeeper.WatchedEvent;
import org.apache.ZooKeeper.Watcher;
import org.apache.ZooKeeper.Watcher.Event.EventType;
import org.apache.ZooKeeper.Watcher.Event.KeeperState;
import org.apache.ZooKeeper.ZooDefs.Ids;
import org.apache.ZooKeeper.ZooKeeper;

public class ZooKeeperDemo implements Watcher {
    private static final int SESSION_TIMEOUT = 30000;
    public static ZooKeeper ZooKeeper;
    public static void main(String args[]){
        String path = "/mynode";
        try {
```

```
                ZooKeeper = new ZooKeeper("47.104.214.68:2181", SESSION_
                TIMEOUT, new ZooKeeperDemo());
                //注册
ZooKeeper.exists(path, true);
                //创建节点
                ZooKeeper.create(path, "mycontent1".getBytes(), Ids.OPEN_ACL_
                UNSAFE, CreateMode.PERSISTENT);
                Thread.sleep(3000);
            } catch (Exception e) {
                // TODO Auto-generated catch block
                e.printStackTrace();
            }

        }

        @Override
        public void process(WatchedEvent event) {

            if (KeeperState.SyncConnected == event.getState()) {

                if (EventType.NodeCreated == event.getType()) {
                    //当节点创建成功时进行回调,此处进行提示打印
                    System.out.println("Node created success. ");
                    try {
                        ZooKeeper.exists(event.getPath(), true);
                    } catch (KeeperException e) {

                        e.printStackTrace();
                    } catch (InterruptedException e) {

                        e.printStackTrace();
                    }
                }
            }
        }
    }
```

运行以上代码,节点会被创建。同时,由于注册了 Watcher,会回调 Watcher 的 process() 方法,并且打印出 Node created success。

5.5.3 Java API 访问 ZooKeeper 实例

在上一节中讲解了节点的创建,关于节点的其他相关操作都与之类似,下面的实例中包括节点创建、取值、修改和删除等操作,以及 Watcher 的方法回调,具体代码如下:

```
import org.apache.ZooKeeper.CreateMode;
import org.apache.ZooKeeper.KeeperException;
import org.apache.ZooKeeper.WatchedEvent;
import org.apache.ZooKeeper.Watcher;
import org.apache.ZooKeeper.Watcher.Event.EventType;
import org.apache.ZooKeeper.Watcher.Event.KeeperState;
```

```java
import org.apache.ZooKeeper.ZooDefs.Ids;
import org.apache.ZooKeeper.ZooKeeper;

public class ZooKeeperDemo implements Watcher {

    private static final int SESSION_TIMEOUT = 30000;

    public static ZooKeeper ZooKeeper;

    public static void main(String args[]){

        String path = "/zknode10";

        try {

            ZooKeeper = new ZooKeeper("47.104.214.68:2181", SESSION_TIMEOUT, new ZooKeeperDemo());

            ZooKeeper.exists(path, true);

            //创建节点
            ZooKeeper.create(path, "mycontent1".getBytes(), Ids.OPEN_ACL_UNSAFE, CreateMode.PERSISTENT);
            Thread.sleep(3000);

            //得到节点内容
            byte[] bytes1 = ZooKeeper.getData(path, null, null);

            String result1 = new String(bytes1);

            System.out.println("result:"+result1);

            //设置节点内容
            ZooKeeper.setData(path, "testSetData000".getBytes(), -1);

            //再次得到节点内容
            byte[] bytes2 = ZooKeeper.getData(path, null, null);
            String result2 = new String(bytes2);

            System.out.println("result after modified:"+result2);

            Thread.sleep(3000);

            //删除节点
            ZooKeeper.delete(path, -1);

            ZooKeeper.close();

        } catch (Exception e) {
            // TODO Auto-generated catch block
            e.printStackTrace();
        }
```

```java
        }
        @Override
        public void process(WatchedEvent event) {

            if (KeeperState.SyncConnected == event.getState()) {

                if (EventType.NodeCreated == event.getType()) {
                    //当节点创建成功时进行回调,此处进行提示打印
                    System.out.println("Node created success. ");
                    try {
                        ZooKeeper.exists(event.getPath(), true);
                    } catch (KeeperException e) {

                        e.printStackTrace();
                    } catch (InterruptedException e) {

                        e.printStackTrace();
                    }
                }else if (EventType.NodeDeleted == event.getType()) {
                    try {
                        ZooKeeper.exists(event.getPath(), true);
                    } catch (KeeperException e) {

                        e.printStackTrace();
                    } catch (InterruptedException e) {

                        e.printStackTrace();
                    }
                    System.out.println("Node deleted success. " );

                }else if (EventType.NodeDataChanged == event.getType()) {
                    try {
                        ZooKeeper.exists(event.getPath(), true);
                    } catch (KeeperException e) {

                        e.printStackTrace();
                    } catch (InterruptedException e) {

                        e.printStackTrace();
                    }
                    System.out.println("Node  changed success. " );

                }
            }
        }
    }
}
```

运行以上程序,结果如下:

```
Node created success.
result:mycontent
Node changed success.
result after modified:mycontent after modified
Node deleted success.
```

5.6 小　　结

本章首先介绍了 ZooKeeper 的如 Session、数据节点（Znode）、版本、Watcher 和 ACL 等核心概念；然后详细讲解了 ZooKeeper 的安装步骤，以及 ZooKeeper 中服务器端和客户端的常用命令；最后利用实例介绍了如何通过 Java API 访问 ZooKeeper。

第 6 章　分布式离线计算框架
——MapReduce

Hadoop 中有两个重要的组件：一个是 HDFS，另一个是 MapReduce，HDFS 用来存储大批量的数据，而 MapReduce 则是通过计算来发现数据中有价值的内容。本章我们主要介绍 MapReduc 中的以下几方面内容：
- MapReduce 的应用场景、工作机制和编程模型。
- MapReduce 的执行原理。
- WordCount 本地测试实例、ETL 本地测试实例，以及温度排序实例。

6.1　MapReduce 概述

Hadoop 作为开源组织下最重要的项目之一，自推出后得到了全球学术界和工业界的广泛关注、推广和普及。它是开源项目 Lucene（搜索索引程序库）和 Nutch（搜索引擎）的创始人 Doug Cutting 于 2004 年推出的。当时 Doug Cutting 发现 MapReduce 正是其所需要解决大规模 Web 数据处理的重要技术，因而模仿 Google MapReduce，基于 Java 设计开发了一个称为 Hadoop 的开源 MapReduce 并行计算框架和系统。

6.1.1　MapReduce 的特点

前面我们已经讲到 Hadoop 的 HDFS 用于存储数据，MapReduce 用来计算数据。接着来介绍一下 MapReduce 的特点。MapReduce 适合处理离线的海量数据，这里的"离线"可以理解为存在本地，非实时处理。离线计算往往需要一段时间，如几分钟或者几个小时，根据业务数据和业务复杂度有所区别。MapReduce 往往处理大批量数据，比如 PB 级别或者 ZB 级别。MapReduce 有以下特点，如图 6.1 所示。
- 易于编程：如果要编写分布式程序，只需要实现一些简单接口，与编写普通程序类似，避免了复杂的过程。同时，编写的这个分布式程序可以部署到大批量廉价的普通机器上运行。

- 具有良好的扩展性：是指当一台机器的计算资源不能满足存储或者计算的时候，可以通过增加机器来扩展存储和计算能力。
- 具有高容错性：MapReduce 设计的初衷是可以使程序部署运行在廉价的机器上，廉价的机器坏的概率相对较高，这就要求其具有良好的容错性。当一台机器"挂掉"以后，相应数据的存储和计算能力会被移植到另外一台机器上，从而实现容错性。

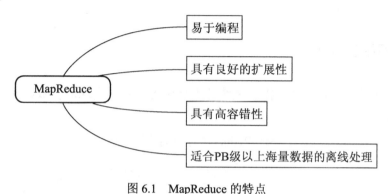

图 6.1　MapReduce 的特点

6.1.2　MapReduce 的应用场景

MapReduce 的应用场景主要表现在从大规模数据中进行计算，不要求即时返回结果的场景，比如以下典型应用：
- 单词统计。
- 简单的数据统计，比如网站 PV 和 UV 统计。
- 搜索引擎建立索引。
- 搜索引擎中，统计最流行的 K 个搜索词。
- 统计搜索词频率，帮助优化搜索词提示。
- 复杂数据分析算法实现。

前面提到，Hadoop 的 MapReduce 是来自于 Google 的 MapReduce，其实 Google 公司很早就将"搜索引擎建立索引"应用到了搜索中。

前面介绍了 MapReduce 的优点和适用场景，下面介绍 MapReduce 不适用的方面。
- 实时计算，MapReduce 不合适在毫秒级或者秒级内返回结果。
- 流式计算，MapReduce 的输入数据集是静态的，不能动态变化，所以不适合流式计算。
- DAG 计算，如果多个应用程序存在依赖关系，并且后一个应用程序的输入为前一个的输出，在这种情况下也不适合 MapReduce。

6.2 MapReduce 执行过程

前面我们了解了 MapReduce 的基本概念，接下来介绍 MapReduce 的执行过程。MapReduce 的执行过程比较复杂，我们先从一个 WordCount 实例着手，从总体上理解 MapReduce 的执行过程。

6.2.1 单词统计实例

假设有一个非常大的文件，需求是统计文件中每个单词出现的次数，MapReduce 的执行过程如图 6.2 所示。

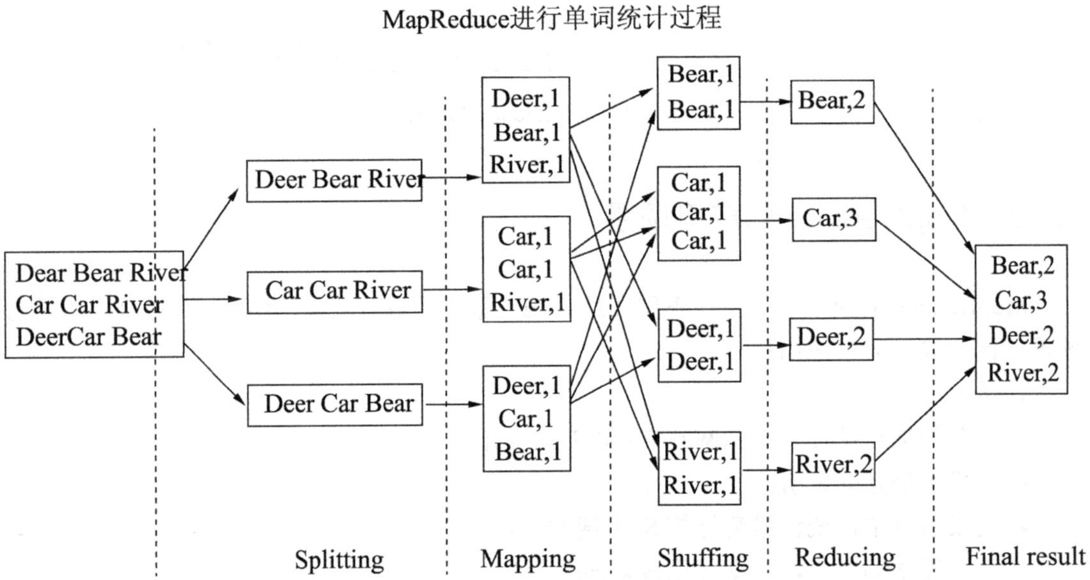

图 6.2 MapReduce 的执行过程 1

图 6.2 中主要分为 Split、Map、Shuffle 和 Reduce 阶段，每个阶段在 WordCount 中的作用如下：

- Split 阶段，首先大文件被切分成多份，假设这里被切分成了 3 份，每一行代表一份。
- Map 阶段，解析出每个单词，并在后边记上数字 1。
- Shuffle 阶段，将每一份中的单词分组到一起，并默认按照字母进行排序。
- Reduce 阶段，将相同的单词进行累加。
- 输出结果。

6.2.2 MapReduce 执行过程

从 WordCount 实例中，可以基于单词统计大概了解 MapReduce 的过程，接下来我们从理论层面来介绍 MapReduce 的执行过程，如图 6.3 所示。

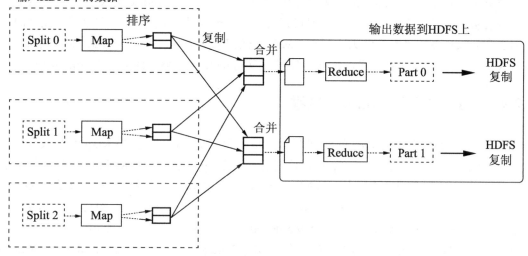

图 6.3 MapReduce 的执行过程 2

具体执行过程如下：

（1）数据会被切割成数据片段。

（2）数据片段以 key 和 value 的形式被读进来，默认是以行的下标位作为 key，以行的内容作为 value。

（3）数据会传入 Map 中进行处理，处理逻辑由用户自行定义，在 Map 中处理完后还是以 key 和 value 的形式输出。

（4）输出的数据传给了 Shuffle（洗牌），Shuffle 完成对数据的排序和合并等操作，但是 Shuffle 不会对输入的数据进行改动，所以还是 key2 和 value2。

（5）数据随后传给了 Reduce 进行处理，Reduce 处理完后，生成 key3 和 value3。

（6）Reduce 处理完的数据会被写到 HDFS 的某个目录中。

如果读者是第一次看到这个执行过程可能不太好理解，其实这就是 MapReduce 程序自己的处理流程，都是按照这个"套路"运行的。下面对 Split 阶段、Map 和 Reduce 阶段，以及 Shuffle 阶段分别展开介绍。

6.2.3 Map Reduce 的文件切片——Split

Split 的大小默认与 block 对应，也可以由用户任意控制。MapReduce 的 Split 大小计算公式如下：

```
max(min.split,min(max.split,block))
```

其中，max.split=totalSize/numSpilt，totalSize 为文件大小，numSplit 为用户设定的 map task 个数，默认为 1；min.split=InputSplit 的最小值，具体可以在配置文件中配置参数 mapred.min.split.size，不配置时默认为 1B，block 是 HDFS 中块的大小。

举例来说：把一个 258MB 的文件上传到 HDFS 上，假设 block 块大小是 128MB，那么它就会被分成 3 个 block 块，与之对应产生 3 个 Split，所以最终会产生 3 个 map task。而第 3 个 block 块里存的文件大小只有 2MB，它的 block 块大小是 128MB，那么它实际占用多大空间呢？通过以上公式可知其占用的是实际的文件大小，而非一个块的大小。

6.2.4　Map 过程和 Reduce 过程

Map 的实现逻辑和 Reduce 的实现逻辑都是由程序员完成的，其中 Map 的个数和 Split 的个数对应起来，也就是说一个 Split 切片对应一个 Map 任务，关于 Reduce 的默认数是 1，程序员可以自行设置。另外需要注意的是，一个程序可能只有一个 Map 任务却没有 Reduce 任务，也可能是多个 MapReduce 程序串接起来，比如把第一个 MapReduce 的输出结果当作第二个 MapReduce 的输入，第二个 MapReduce 的输出成为第三个 MapReduce 的输入，最终才可以完成一个任务，通过阅读后面的 MapReduce 实例，读者会对 Map 和 Reduce 有进一步的理解。

6.2.5　Shuffle 过程

Shuffle 又叫"洗牌"，它起到连接 Map 任务与 Reduce 任务的作用，在这里需要注意的是，Shuffle 不是一个单独的任务，它是 MapReduce 执行中的步骤，如图 6.4 所示。

从图 6.4 中可以看出，Shuffle 分为两部分，一部分在 Map 端，另一部分在 Reduce 端，Map 处理后的数据会以 key、value 的形式存在缓冲区中（buffer in memory），缓冲区大小为 128MB。当该缓冲区快要溢出时（默认 80%），会将数据写到磁盘中生成文件，也就是溢写操作（spill to disk）。溢写磁盘的过程是由一个线程来完成，溢写之前包括 Partition（分区）和 Sort（排序），Partition 和 Sort 都有默认实现，其中 Partition 分区默认是按照"hash 值%reduce 数量"进行分区的，分区之后的数据会进入不同的 Reduce，而 Sort 是默认按照字母顺序进行排序的。读者可以根据业务需求进行编写，具体可以参考后面的实例。溢写之后会在磁盘上生成多个文件，多个文件会通过 merge 线程完成文件的合并，由多个小文件生成一个大文件。

合并完之后的数据（以 key 和 value 的形式存在）会基于 Partition 被发送到不同的 Reduce 上，如图 6.4 中任务之间的长箭头所示，Reduce 会从不同的 Map 上取得"属于"自己的数据并写入磁盘，完成 merge 操作减少文件数量，并调用 Reduce 程序，最终通过

Output 完成输出。

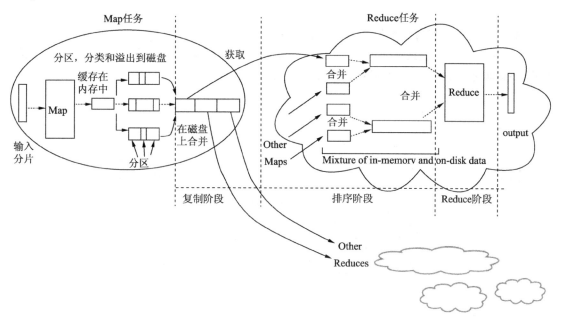

图 6.4 MapReduce 执行中的步骤

6.3 MapReduce 实例

本节中,我们将从实现层面来介绍如何开发 MapReduce 程序,MapReduce 的编程遵循一个特定流程,主要是编写 Map 和 Reduce 函数。

6.3.1 WordCount 本地测试实例

前面我们通过一个 WordCount 实例介绍了 MapReduce 执行过程,在这里用一个 WordCount 的单词统计实例来介绍如何编写 MapReduce 程序。

一个完整的 MapReduce 程序主体主要分为两部分,一个是 Mapper,另一个是 Reducer。

用户自定义的 Mapper.java 类解析 key/value 对值,然后产生一个中间 key/value 对值的集合,把所有具有相同中间 key 值的中间 value 值集合在一起后传递给 Reduce 函数。

用户自定义的 Reducer.java 类接受一个中间 key 的值和相关的一个 value 值的集合。Reduce 函数合并这些 value 值,形成一个较小的 value 值的集合。每次 Reduce 函数调用时只产生 0 或 1 个 value 输出值。通常我们通过一个迭代器把中间的 value 值提供给 Reduce 函数,这样就可以处理无法全部放入内存中的大量 value 值的集合。

在编写代码之前，可以先进行本机环境配置，读者也可以将程序打包后上传到 CentOS 上执行。

本地运行环境配置的编写步骤如下：

（1）在本机配置 Hadoop 的环境变量，右击"我的电脑"，在弹出的快捷菜单中选择"属性>高级系统设置"，如图 6.5 所示。

图 6.5 "高级系统设置"页面

（2）进入系统属性界面之后选择"高级"选项卡，单击"环境变量"按钮，如图 6.6 所示。

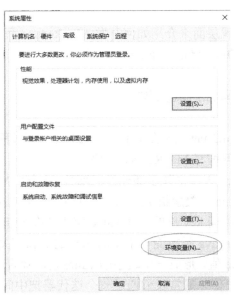

图 6.6 环境变量设置页面

（3）在"环境变量"设置界面中，在"变量值"中输入 HADOOP_HOME，并复制加入 Hadoop 的解压目录，如图 6.7 所示。在 Path 环境变量中增加%HADOOP_HOME%\bin;，如图 6.8 所示。

图 6.7　新建 HADOOP_HOME 变量

图 6.8　增添 HADOOP_HOME 环境变量

（4）将 winutil.exe 文件放入 Hadoop 解压目录下的 bin 文件夹里，如图 6.9 所示。

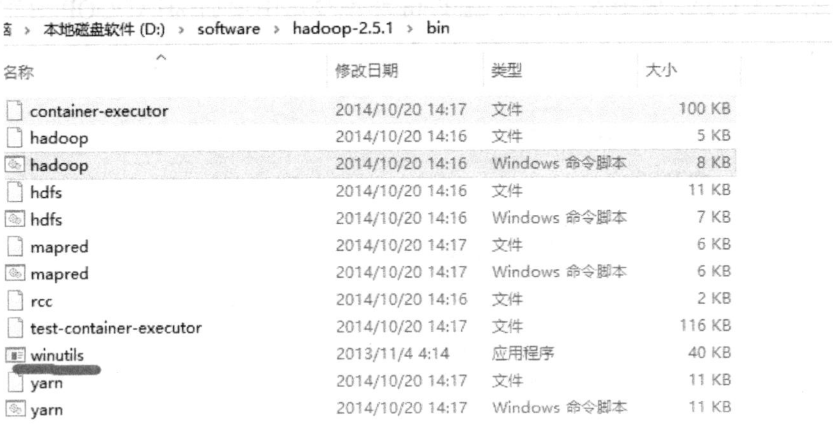

图 6.9　操作页面

（5）新建一个 Java 工程，导入需要的 JAR 包。新建一个 WordCountMapper 类，继承 Mapper，这是一个 Map 过程，对输入文本进行词汇的分割并循环输出给 Reducer。代码如下：

```java
package com.sendto.wordcount;

import java.io.IOException;
import org.apache.hadoop.io.IntWritable;
import org.apache.hadoop.io.LongWritable;
import org.apache.hadoop.io.Text;
import org.apache.hadoop.mapreduce.Mapper;

public class WordCountMapper extends Mapper<LongWritable, Text, Text,
IntWritable >{
    @Override
    protected void map(LongWritable key,Text value,Context context)
        throws IOException,InterruptedException{
        //用空格进行分词
        String [] str=value.toString().split(" ");
        //for 循环输出
        for(int i=0;i<str.length;i++){
            //new Tex, new IntWritable 进行可序列化
            context.write(new Text(str[i]), new IntWritable(1));
        }
    }
}
```

（6）新建 WordCountReducer 类，继承 Reducer；这是一个 Reduce 过程，将从 Map 传入的词汇进行分组合并，并通过文本和单词统计量的方式输出。代码如下：

```java
package com.sendto.wordcount;

import java.io.IOException;
```

```java
import org.apache.hadoop.io.IntWritable;
import org.apache.hadoop.io.Text;
import org.apache.hadoop.mapreduce.Reducer;
//mapper 切分后是 a 1 b 1 a 1 c 1 的形式, 输出 a 2 b 1 c 1 形式
public class WordCountReducer extends Reducer<Text,IntWritable,Text,
IntWritable>{
    //数据分组合并输出
    @Override
    protected void reduce (Text arg0,Iterable<IntWritable> arg1,Context
arg2)
        throws IOException,InterruptedException{
        int sum=0;
        for(IntWritable i:arg1){
            sum=sum+i.get();
        }
        arg2.write(arg0, new IntWritable(sum));
    }
}
```

（7）创建主方法。上面编写了 Mapper 和 Reducer，为了使 Mapper 和 Reducer 正常运行，还需要编写主方法。主方法中需要先设置要连接的 HDFS 和要读取的文件及处理后的文件在 HDFS 中的路径，指明我们所要进行的 Map 和 Reduce 过程的类，然后开始 MapReduce 的离线数据处理。代码如下：

```java
package com.sendto.wordcount;

import java.io.IOException;
import org.apache.hadoop.conf.Configuration;
import org.apache.hadoop.fs.Path;
import org.apache.hadoop.io.IntWritable;
import org.apache.hadoop.mapreduce.Job;
import org.apache.hadoop.mapreduce.lib.input.FileInputFormat;
import org.apache.hadoop.mapreduce.lib.output.FileOutputFormat;
import com.sun.jersey.core.impl.provider.entity.XMLJAXBElementProvider.Text;

public class MRRunJob {
    public static void main (String []args){
        Configuration conf=new Configuration();
        //NameNode 入口 IP
        conf.set("fs.defaultFS", "hdfs://192.168.8.30:9000");
        Job job =null;
            try {
                //任务名字
                job = Job.getInstance(conf, "mywc");
            } catch (IOException e1) {
                e1.printStackTrace();
            }
    //主方法
            job.setJarByClass(MRRunJob.class);
                //Map 方法名
            job.setMapperClass(WordCountMapper.class);
            //Reducer 方法名
```

```java
        job.setReducerClass(WordCountReducer.class);
            //Map 输出的 key 类型
        job.setOutputKeyClass(Text.class);
            //Map 输出的 value 类型
        job.setOutputValueClass(IntWritable.class);
        try {
        //读取的文件位置
    FileInputFormat.addInputPath(job, new Path("/usr/input/data/wc/"));
            //处理完之后的数据存放位置,注意输出的文件夹如果已经存在会报错
    FileOutputFormat.setOutputPath(job, new Path("/usr/output/data/wc/"));
            boolean f = job.waitForCompletion(true);
        } catch (Exception e) {
            e.printStackTrace();
        }
    }
}
```

6.3.2 ETL 本地测试实例

数据的分析和挖掘能够为管理层提供决策支持,下面介绍一个数据清洗的实例。为了更直观地处理和分析数据,进行数据清洗,我们将数据中的时间格式进行标准化处理,使其更加直观。

(1) 新建一个 Java Project 工程,导入需要的 JAR 包。

(2) 新建一个 ETLMapper 类,继承 Mapper,将成行的数据通过逗号分隔开,提取出数据中的时间,然后对时间进行标准化处理。代码如下:

```java
package com.sendto.etltest;

import java.io.IOException;
import java.text.SimpleDateFormat;
import org.apache.hadoop.io.LongWritable;
import org.apache.hadoop.io.NullWritable;
import org.apache.hadoop.io.Text;
import org.apache.hadoop.mapreduce.Mapper;
//数据示例格式:ShangHai,XinJiang,TIAMAO,4051524318584579,YUANTONG,119.168.
12.132,eu
//LongWritable 读取数据的下标偏移量,text 指可序列化字符串
public class ETLMapper extends Mapper<LongWritable,Text,Text,
NullWritable> {
    @Override
    protected void map(LongWritable key, Text value, Context context)
            throws IOException, InterruptedException {
        //通过逗号分词
        String [] strArray = value.toString().split(",");
        String strContent= "";
        for(int i = 0;i<strArray.length;i++){
            //将第 3 个数据转换成 yyyy-MM-dd HH:mm:ss 时间格式
            if(i==3){
    SimpleDateFormat sdf = new SimpleDateFormat("yyyy-MM-dd HH:mm:ss");
```

```
            //Long.parseLong 将 String 类型转化成 Long 类型，因为元数据中是以秒为单位，
所以要加 000
            String str = sdf.format(Long.parseLong(strArray[i]+"000"));
                        //中间以空格隔开
            strContent = strContent+str+",";
                }else {
                    strContent = strContent+strArray[i]+",";
                }
            }
            context.write(new Text(strContent),NullWritable.get());
        }
    }
```

（3）新建一个 ETLReducer 类，继承 Reducer；将 Mapper 传过来的内容写入磁盘。代码如下：

```
package com.sendto.etltest;

import java.io.IOException;
import org.apache.hadoop.io.NullWritable;
import org.apache.hadoop.io.Text;
import org.apache.hadoop.mapreduce.Reducer;

//输入和输出都是这一行记录
public class ETLReducer extends Reducer<Text,NullWritable,NullWritable,Text>{
    @Override
    protected void reduce(Text arg0, Iterable<NullWritable> arg1,Context arg2)throws IOException, InterruptedException {
        //将 Mapper 传过来的内容写入磁盘
        arg2.write(NullWritable.get(), arg0);
    }
}
```

（4）创建主方法并设置要连接的 HDFS 和要读取的文件及处理后的文件在 HDFS 中的路径，指明 Map 和 Reduce 过程的类。代码如下：

```
package com.sendto.etltest;

import java.io.IOException;
import org.apache.hadoop.conf.Configuration;
import org.apache.hadoop.fs.FileSystem;
import org.apache.hadoop.fs.Path;
import org.apache.hadoop.io.NullWritable;
import org.apache.hadoop.io.Text;
import org.apache.hadoop.mapreduce.Job;
import org.apache.hadoop.mapreduce.lib.input.FileInputFormat;
import org.apache.hadoop.mapreduce.lib.output.FileOutputFormat;

public class MRRunJob {
    public static void main(String[] args) {
        Configuration conf = new Configuration();
        //NameNode 入口 IP
        conf.set("fs.defaultFS", "hdfs://192.168.8.30:9000");
```

```java
        //初始化 fs
        FileSystem fs  = null;
try {
    fs = FileSystem.get(conf);
} catch (IOException e2) {
    // TODO Auto-generated catch block
    e2.printStackTrace();
}
    Job job =null;
    try {
        job = Job.getInstance(conf, "mywc");
    } catch (IOException e1) {
        e1.printStackTrace();
    }
//NameNode 入口 IP
    job.setJarByClass(MRRunJob.class);
    //Map 方法名
    job.setMapperClass(ETLMapper.class);
    //Reducer 方法名
    job.setReducerClass(ETLReducer.class);
    //Map 输出的 key 类型
    job.setOutputKeyClass(Text.class);
    //Map 输出的 value 类型
    job.setOutputValueClass(NullWritable.class);
    try {
    //读取的文件位置
        FileInputFormat.addInputPath(job, new Path("/usr/input/data/etl0/"));
    //处理完之后数据的存放位置
        Path path = new Path("/usr/output/data/etl/");
    //如果这个目录已经存在就删掉
        if(fs.exists(path)){
            fs.delete(path, true);
        }
        FileOutputFormat.setOutputPath(job, path);
        boolean f = job.waitForCompletion(true);
    } catch (Exception e) {
        e.printStackTrace();
    }
  }
}
```

6.4 温度排序实例

任务：假设有多年气温数据，如下：
- 1949-10-01 14:21:02   34
- 1949-10-02 14:01:01   36
- 1950-01-01 11:21:02   32

- 1950-10-03 12:21:02   27
- 1950-10-01 12:21:02   37
- 1950-10-02 12:21:02   41
- 1951-07-01 12:21:02   45
- 1951-07-02 12:21:02   46
- 1951-07-03 12:21:03   47
- 1951-12-01 12:21:02   23

现在需要找到每年每月的 3 个最高温度时刻并进行降序排列，同时由于数据量可能会很大，为了提高效率，将每一年的数据分别由不同的 Reduce 执行，产生不同的文件。把每年的温度数据通过处理，找到每月最高的三个温度的时刻。该实例中涉及的类包括：MyKey、MyGroup、MyMapper、MyPartitioner、MyReducer、MySort 和 RunJob。其中，在 MySort 中进行排序，在 MyGroup 进行分组，下面我们分别看一下这几个类的核心代码。

6.4.1 时间和温度的封装类 MyKey.Java

MyKey 的主要作用是实现年、月、温度的封装，同时实现序列化和反序列化，重写 compareTo()方法。代码如下：

```java
package com.sendto.airsort;

import java.io.DataInput;
import java.io.DataOutput;
import java.io.IOException;

import org.apache.hadoop.io.WritableComparable;
@SuppressWarnings("rawtypes")
public class MyKey implements WritableComparable{    //可序列化；可比较
    private int year;                                //年
    private int month;                               //月
    private double air;                              //温度
    public int getYear() {
        return year;
    }
    public void setYear(int year) {
        this.year = year;
    }
    public int getMonth() {
        return month;
    }
    public void setMonth(int month) {
        this.month = month;
    }
    public double getAir() {
        return air;
    }
    public void setAir(double air) {
```

```
            this.air = air;
        }
        @Override                                              //序列化
        public void write(DataOutput out) throws IOException {
    //通过write方法写入序列化的数据流
            out.writeInt(year);
            out.writeInt(month);
            out.writeDouble(air);
    }
        @Override                                              //反序列化
        public void readFields(DataInput in) throws IOException {
    //通过readFields方法从序列化的数据流中读出进行赋值
            year = in.readInt();
            month = in.readInt();
            air = in.readDouble();
        }
        @Override
    //按照字典顺序进行比较，返回的值是一个int型
        public int compareTo(Object o) {
            return this==o?0:-1;
        }
    }
```

6.4.2 Map 任务 MyMapper.java

通过 Mapper 将数据解析为 key-value 形式，其中 key:1949-10-01 14:31:02 和 value:34，将时间进行分割，提取出年、月信息并封装到 key 中，由 key 作为排序的信息进行排序。代码如下：

```
package com.sendto.airsort;

import java.io.IOException;
import org.apache.hadoop.io.Text;
import org.apache.hadoop.mapreduce.Mapper;
//输入的key和value都是text类型，将温度和年、月分割后，输出的是封装后的MyKey，
温度仍为我们想看到的text类型
public class MyMapper extends Mapper<Text,Text,MyKey,Text> {
    @Override
    protected void map(Text key, Text value,Context context)
            throws IOException, InterruptedException {

//年月日通过-进行分割提取
        String [] strArray = key.toString().split("-");
//将年、月、温度数据封装到myKey中
        MyKey myKey = new MyKey();
        myKey.setYear(Integer.parseInt(strArray[0]));
        myKey.setMonth(Integer.parseInt(strArray[1]));
```

```
            myKey.setAir(Double.parseDouble(value.toString()));
                                    //"\t"相当于 Tab 键
            context.write(myKey, new Text(key.toString()+"\t"+value));
    }
}
```

6.4.3 数据分组类 MyGroup.Java

MyGroup 主要是完成分组,然后把年的数据进行对比,如果在同一年中就返回所在月份。代码如下:

```
package com.sendto.airsort;

import org.apache.hadoop.io.WritableComparable;
import org.apache.hadoop.io.WritableComparator;
//1949-10-29 23:29:12 39
public class MyGroup extends WritableComparator{
    //继承 WritableComparator 类来实现排序
    public MyGroup(){
        super(MyKey.class,true);
    }
    @Override
    public int compare(WritableComparable a, WritableComparable b) {
MyKey myKey1 = (MyKey) a;
        MyKey myKey2 = (MyKey) b;
        //以年做对比,如果在同一年中就返回所在月份,不在同一年就返回比较结果
        int r1 = Integer.compare(myKey1.getYear(),myKey2.getYear());
        //如果年相等
        if(r1==0){
            return  Integer.compare(myKey1.getMonth(), myKey2.getMonth());

        }
        return r1;
    }
}
```

6.4.4 温度排序类 MySort.java

根据年、月、温度进行排序,先以年作对比,如果年相等就比较月,年不相等就返回比较结果;如果月相等就以温度倒序排序,月不相等就返回比较结果。代码如下:

```
package com.sendto.airsort;

import org.apache.hadoop.io.WritableComparable;
import org.apache.hadoop.io.WritableComparator;

public class MySort extends WritableComparator{
    //使用 super()调用序列化的构造函数
    public MySort(){
        super(MyKey.class,true);
```

```
        }
        @Override
        public int compare(WritableComparable a, WritableComparable b) {
            //通过 MyKey 进行排序处理分组合并
            MyKey myKey1 = (MyKey) a;
            MyKey myKey2 = (MyKey) b;
            //以年作对比
            int r1 = Integer.compare(myKey1.getYear(),myKey2.getYear());
            //如果年相等就比较月，年不等就返回年的比较结果
            if(r1==0){
                //以月作对比
                int r2 = Integer.compare(myKey1.getMonth(), myKey2.getMonth());
                //如果月相等就以温度排序，月不相等就返回月的比较结果
                if(r2==0){
                    //月相等就把温度按倒序排序，加-，表示倒序排序
return -Double.compare(myKey1.getAir(), myKey2.getAir());
                }
                return r2;
            }
            return r1;
        }
}
```

6.4.5 数据分区 MyPartitioner.java

通过 MyPartitioner 来控制 Reducer 的数量，在这里是把每一年的数据进行分区，比如 1949、1950、1951 这三年，每一年的数据给一个 Partitioner。代码如下：

```
package com.sendto.airsort;

import org.apache.hadoop.io.Text;
import org.apache.hadoop.mapreduce.Partitioner;
public class MyPartitioner extends Partitioner<MyKey,Text>{
    @Override
//已知最小年份是1949，计算分区的位置，numbers 为 Reduce 的数量
    public int getPartition(MyKey key, Text value, int numPartitions) {
        //假如1949、1950和1951这3年
        //1949-1949=0；得到第一个分区
        //1950-1949=1；得到第二个分区
        //1951-1949=2；得到第三个分区
        return (key.getYear()-1949)%numPartitions;
    }
}
```

6.4.6 Reducer 任务 MyReducer.java

Reducer 的作用是将每月三个最高的温度记录取出。代码如下：

```
package com.sendto.airsort;
```

```java
import java.io.IOException;
import org.apache.hadoop.io.NullWritable;
import org.apache.hadoop.io.Text;
import org.apache.hadoop.mapreduce.Reducer;
public class MyReducer extends Reducer<MyKey,Text,NullWritable,Text> {
    @Override                                      //取出前3条记录
    protected void reduce(MyKey arg0, Iterable<Text> arg1,Context ctx)
            throws IOException, InterruptedException {
        int sum = 0;                               //计数器
        for(Text t : arg1){
            sum++;
            //如果大于3
            if(sum>3){
                //跳出
                break;
            }else{
                //写入记录
                ctx.write(NullWritable.get(), t);
            }
        }
    }
}
```

6.4.7 主函数 RunJob.java

前面的这些类需要通过一个主函数进行连接，设置要连接的 HDFS，以及处理后的文件在 HDFS 中的路径，指明我们所要进行的 Map 和 Reduce 过程的类。代码如下：

```java
package com.sendto.airsort;

import java.io.IOException;
import org.apache.hadoop.conf.Configuration;
import org.apache.hadoop.fs.FileSystem;
import org.apache.hadoop.fs.Path;
import org.apache.hadoop.io.IntWritable;
import org.apache.hadoop.io.Text;
import org.apache.hadoop.mapreduce.Job;
import org.apache.hadoop.mapreduce.lib.input.FileInputFormat;
import org.apache.hadoop.mapreduce.lib.input.KeyValueTextInputFormat;
import org.apache.hadoop.mapreduce.lib.output.FileOutputFormat;

public class RunJob {
    public static void main(String[] args) {
        Configuration conf = new Configuration();
        //基本配置 NameNode 入口 IP
        conf.set("fs.defaultFS", "hdfs://192.168.21.100:9000");

        FileSystem fs=null;
        try {
            fs = FileSystem.get(conf);
        } catch (IOException e2) {
            //自动生成的代码块
```

```java
            e2.printStackTrace();
        }
        
        Job job = null;
        try {
            //任务名称
            job = Job.getInstance(conf, "weather");
        } catch (IOException e1) {
            //自动生成的代码块
            e1.printStackTrace();
        }
        //主方法
        job.setJarByClass(RunJob.class);
        
        //Map 方法名
        job.setMapperClass(MyMapper.class);
//InputFormat 方法名
        job.setInputFormatClass(KeyValueTextInputFormat.class);
        
        //Reducer 方法名
        job.setReducerClass(MyReducer.class);
        
        //Partitioner 方法名
        job.setPartitionerClass(MyPartitioner.class);
        
        //SortComparator 方法名
        job.setSortComparatorClass(MySort.class);
        
        //GroupingComparator 方法名
        job.setGroupingComparatorClass(MyGroup.class);
        
        //Reducer Text 的数量
        job.setNumReduceTasks(3);
        
        //Map 输出的 Key 类型
        job.setOutputKeyClass(MyKey.class);
        
        //Map 输出的 Value 类型
        job.setOutputValueClass(Text.class);
        try {
            //读取文件的位置
    FileInputFormat.addInputPath(job, new Path("/usr/input11"));
            
            //Path path = new Path("/usr/output1");
            
        //处理完的数据的存放位置，注意，输出的文件夹如果已经存在则会删除
            Path path = new Path("/usr/output11");
            
            if(fs.exists(path)){
                
                fs.delete(path, true);
            }
    FileOutputFormat.setOutputPath(job, path);
```

```
            boolean f = job.waitForCompletion(true);
            System.out.println("f:"+f);
        } catch (Exception e) {
            //自动生成的代码块
            e.printStackTrace();
        }
    }
}
```

上面的主函数写好之后,直接运行就可以处理相应的数据了,接着我们把 weather 文件上传到/usr/input11 中,在主函数中会读取这个文件夹中的数据,上传代码如下:

```
[root@master software]# hdfs dfs -put weather /usr/input11
```

查看 weather 文件在/usr/input11 中是否存在。代码如下:

```
[root@master software]# hdfs dfs -ls /usr/input11
```

运行前面编写的 RunJob 主函数,完成后查看输出文件,在 Linux 中输入 hdfs dfs -ls /usr/output11/指令。代码如下:

```
[root@master software]# hafs dfs -ls/usr/output11/
Found 4 items
-rw-r--r--   3 Administrator supergroup      0 2017-09-18 09:43 /usr/output11/_SUCCESS
-rw-r--r--   3 Administrator supergroup     46 2017-09-18 09:43 /usr/output11/part-r-00000
-rw-r--r--   3 Administrator supergroup     92 2017-09-18 09:43 /usr/output11/part-r-00001
-rw-r--r--   3 Administrator supergroup     92 2017-09-18 09:43 /usr/output11/part-r-00002
```

根据上面的查看结果,可以看出来生成了相应的结果文件,命令如下:

hdfs dfs - cat /usr/oupu11/part-r-0000, hdfs dfs - cat /usr/oupu11/part-r-0001, hdfs dfs - cat /usr/oupu11/part- r-0002。

运行这几条命令得到相应结果如下:

```
[root@master software]# hdfs dfs-cat/usr/output11/part-r-00000
1949-10-02 14:01:02   36
1949-10-01 14:21:02   34
[root@master software]# hdfs dfs-cat/usr/output11/part-r-00001
1950-01-01 11:21:02   32
1950-10-02 12:21:02   41
1950-10-01 12:21:02   37
1950-10-03 12:21:02   27
[root@master software]# hdfs dfs-cat/usr/output11/part-r-00002
1951-07-03 12:21:03   47
1951-07-02 12:21:02   46
1951-07-01 12:21:02   45
1951-12-01 12:21:02   23
49 年是 0; /usr/output11/part-r-00000
```

```
50 年是 1：/usr/output11/part-r-00001
51 年是 2：/usr/output11/part-r-00002
```

根据上述结果可以看出，最后实现了每月三个最高温度的时刻按倒序排序，说明程序运行顺利。

6.5 小　　结

本章主要介绍了 MapReduce 出现的背景及其特点，同时基于几个实例讲解了 MapReduce 的具体应用。读者需要理解并掌握这几个实例，并在自己搭建的环境中"跑"起来，根据运行结果再次理解 MapReduce 的执行逻辑。

第 7 章 Hadoop 的集群资源管理系统——YARN

旧版本 MapReduce 中的 JobTracker/TaskTracker 在可扩展性、内存消耗、可靠性和线程模型方面存在很多问题,需要开发者做很多调整来修复。

Hadoop 的开发者对这些问题进行了 Bug 修复,可是由此带来的成本却越来越高,为了从根本上解决旧 MapReduce 存在的问题,同时也为了保障 Hadoop 框架后续能够健康地发展,从 Hadoop 0.23.0 版本开始,Hadoop 的 MapReduce 框架就被动了"大手术",从根本上发生了较大变化。同时新的 Hadoop MapReduce 框架被命名为 MapReduce V2,也叫 YARN(Yet Another Resource Negotiator,另一种资源协调者)。

本章主要涉及的知识点如下:
- 为何要使用 YARN,以及 YARN 的基本架构。
- 掌握 YARN 的环境搭建。

7.1 为什么要使用 YARN

与旧 MapReduce 作比较,YARN 采用了一种分层的集群框架,具有以下几个优势。
- 解决了 NameNode 的单点故障问题,可以通过配置 NameNode 高可用来解决。
- 提出了 HDFS 联邦,通过 HDFS 联邦可以使多个 NameNode 分别管理不同的目录,从而实现访问隔离及横向扩展。
- 将资源管理和应用程序管理分离开,分别由 ResouceManager 和 ApplicationMaster 负责。
- 具有向后兼容的特点,运行在 MR1 上的作业不需要做任何修改就可以运行在 YARN 上。
- YARN 是一个框架管理器,用户可以将各种计算框架移植到 YARN 上,统一由 YARN 进行管理和资源调度。目前支持的计算框架有 MapReduce、Storm、Spark 和 Flink 等。

7.2 YARN 的基本架构

YARN 的核心思想是将功能分开,在 MR1 中,JobTracker 有两个功能,一个是资源管理,另一个是作业调用。在 YARN 中则分别由 ResourceManager 和 ApplicationMaster 进程来实现。其中,ResourceManager 进程完成整个集群的资源管理和调度,而 ApplicationMaster 进程则负责应用程序的相关事务,如任务调度、容错和任务监控等。

系统中所有应用资源调度的最终决定权由 ResourceManager 担当。每个应用的 ApplicationMaster 实际上是框架指定的库,其从 ResourceManager 调度资源并和 NodeManeger 一同执行监控任务,NodeManager 会通过心跳信息向 ResourceManager 汇报自己所在节点的资源使用情况,如图 7.1 所示。

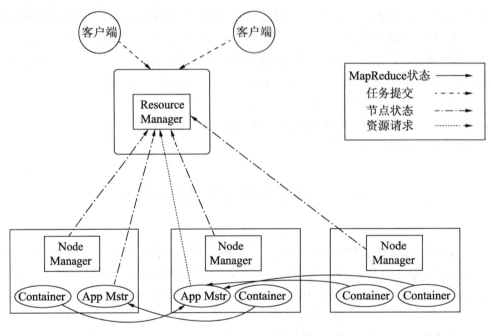

图 7.1 执行和监控过程图

7.2.1 ResourceManager 进程

ResourceManager 进程包含两个主要内容:Scheduler 和 ApplicationManager。

Scheduler 依据容量和队列等类似的约束分配资源到运行的不同应用中。Scheduler 是一个纯调度器,它不监督也不跟踪应用的状态。同样地,它不确保重启由应用失败或硬件失败所造成的失败任务。Scheduler 根据应用所需的资源执行调度，调度内存、CPU、硬盘和网络等资源到 Container 中。

Scheduler 有一个用于不同队列和应用中集群资源划分的插件，目前 MapReduce 调度器如 CapacityScheduler 和 FairScheduler 就是插件的一些实例。

CapacityScheduler 支持层级队列，这使得集群资源的共享更加可预测。

7.2.2 ApplicationMaster 和 NodeManager

ApplicationManager 进程负责接收作业提交，协商首个 Container 执行应用指定的 ApplicationManager 并提供重启失败的 ApplicationManager Container 的服务。

每个机器上的 NodeManager 作为框架代理，负责监控 Container 资源使用的监控并提供类似 ResourceManager 或者 Scheduler 之类提供的报告。

每个应用的 ApplicationMaster 进程负责协调 Scheduler 上合适的资源容器，并跟踪容器状态和监控执行。

7.3　YARN 工作流程

YARN 的工作流程如图 7.2 所示。

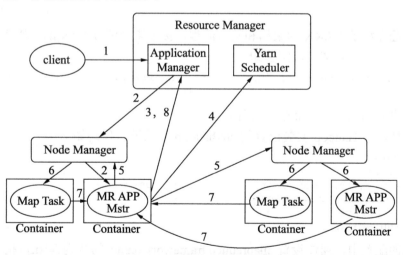

图 7.2　YARN 的工作流程

YARN 的工作流程主要分为以下几个步骤。

（1）用户向 YARN 中的 Resource Manager 提交应用程序，包括用户程序、启动 ApplicationMaster 命令和 ApplicationMaster 程序等。

（2）ResourceManager 为应用程序分配 Container，随后与 Container 所在的 NodeManager 进行通信，并且由 NodeManager 在 Container 中启动对应的 ApplicationMaster。

（3）ApplicationMaster 会在 ResourceManager 中进行注册，这样用户就能够通过 ResourceManager 来查看应用程序的运行情况，然后它会为这个应用程序的各项任务申请资源，同时监控其运行状态直到结束。

（4）ApplicationMaster 采用的是轮询方式，基于 RPC 协议向 ResourceManager 申请和获取所需要的资源。

（5）在 ApplicationMaster 申请到资源后，它会和申请到的 Container 所对应的 NodeManager 进行交互通信，同时要求在该 Container 中启动任务。

（6）NodeManager 为要启动的任务准备好运行环境，并且将启动命令写在一个脚本中，通过该脚本来运行任务。

（7）每个任务基于 RPC 协议向对应的 ApplicationMaster 汇报自己的运行状态与进度，以便让 ApplicationMaster 随时掌握各个任务的运行状态，这样就可以在任务运行失败时重启任务。

（8）在应用程序运行完之后，其对应的 ApplicationMaster 会通过与 ResourceManager 通信来要求注销和关闭自己。

7.4　YARN 搭建

前面已经介绍了 YARN 的架构和基本流程，接下来介绍 YARN 的环境搭建，关于搭建过程，可以参考官方文档，网址如下：

```
http://hadoop.apache.org/docs/r2.5.2/hadoop-yarn/hadoop-yarn-site/YARN.html
```

（1）编辑 mapred-site.xml 配置文件。

配置文件位于 Hadoop 安装位置的 etc/hadoop/目录下。代码如下：

```
<configuration>
    <property>
        <name>mapreduce.framework.name</name>
        <value>yarn</value>
    </property>
</configuration>
```

在上面的配置中，通过设置 mapreduce.framework.name 的值为 yarn，指明通过 yarn 进行资源管理。

（2）伪分布 yarn 的 yarn-site.xml 配置文件内容如下：

```
<configuration>
    <property>
            <name>yarn.nodemanager.aux-services</name>
            <value>mapreduce_shuffle</value>
```

```xml
        </property>
</configuration>
```

(3) 完全分布式的 yarn-site.xml 修改内容如下：

```xml
<configuration>
    <property>
        <name>yarn.nodemanager.aux-services</name>
        <value>mapreduce_shuffle</value>
    </property>
    <property>
        <name>yarn.resourcemanager.resource-tracker.address</name>
        <value>192.168.109.200:8031</value>
    </property>
    <property>
        <name>yarn.resourcemanager.address</name>
        <value>192.168.109.200:8032</value>
    </property>
    <property>
        <name>yarn.resourcemanager.scheduler.address</name>
        <value>192.168.109.200:8034</value>
    </property>
    <property>
        <name>yarn.resourcemanager.webapp.address</name>
        <value>192.168.109.200:8088</value>
    </property>
    <property>
        <name>yarn.log-aggregation-enable</name>
        <value>true</value>
    </property>
    <property>
        <name>yarn.log.server.url</name>
        <value>http://master:19888/jobhistory/logs/</value>
    </property>
</configuration>
```

(4) 将 mapred-site.xml 和 yarn-site.xml 发送给 slave 节点，执行命令如下：

```
cd /opt/software/hadoop-2.5.1/etc/hadoop
scp -r mapred-site.xml root@slave:/opt/software/hadoop-2.5.1/etc/hadoop
scp -r yarn-site.xml root@slave:/opt/software/hadoop-2.5.1/etc/hadoop
```

(5) 启动 HDFS，启动命令和启动后的结果如下：

```
[root@master hadoop]# start-dfs.sh
Starting namenodes on [master]
master:startingnamenode,loggingto/opt/software/hadoop-2.5.1/logs/hadoop-
root-namenode-master.out
slave: starting datanode,logging to/opt/software/hadoop-2.5.1/logs/
hadoop-root-datanode-slave.out
master:startingdatanode,loggingto/opt/software/hadoop-2.5.1/logs/hadoop-
root-datanode-master.out
Starting secondary namenodes [slave]
slave:startingsecondarynamenode,loggingto/opt/software/hadoop-2.5.1/logs/
hadoop-root-secondarynamenode-slave.out
```

接着通过 JPS 查看进程，发现有 NameNode 和 DataNode：

```
[root@master hadoop]# jps
1880 NameNode
1970 DataNode
2190 JPS
```

（6）启动 YARN，启动命令和启动后的结果如下：

```
[root@master hadoop]# start-yarn.sh
starting yarn daemons
startingresourcemanager,loggingto/opt/software/hadoop-2.5.1/logs/yarn-
root-resourcemanager-master.ut
master:startingnodemanager,loggingto/opt/software/hadoop-2.5.1/logs/
yarn-root-nodemanager-master.out
slave: startingnodemanager,loggingto/opt/software/hadoop-2.5.1/logs/
yarn-root-nodemanager-slave.out
```

（7）通过 JPS 查看进程，发现有了 NodeManager 和 ResourceManager 进程。

```
[root@master hadoop]# jps
2613 JPS
1880 NameNode
2325 NodeManager
1970 DataNode
2240 ResourceManager
```

YARN 成功启动后，界面如图 7.3 所示。

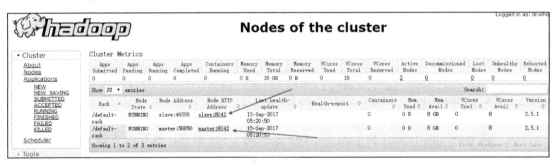

图 7.3　YARN 成功启动

从图 7.3 中可以看到，有 slave 和 master 两个运行中的节点。

7.5　小　　结

本章主要介绍了 Apache YARN。YARN 是 Hadoop 的集群资源管理系统，在 Hadoop 2 中被引入，最初是为了改善 MapReduce 的缺陷，同时 YARN 也具有通用性，同样可以支持其他的分布式计算模式。

第 8 章 Hadoop 的数据仓库框架——Hive

在 Facebook Jeff 团队所构建的信息平台中，最庞大的组成部分是 Apache Hive（https://hive.apache.org/）。Hive 是一个构建在 Hadoop 上的数据仓库框架，是应 Facebook 每天产生的海量网络数据进行管理和（机器）学习的需求而产生和发展的。

本章主要涉及的知识点如下：
- Hive 的应用和运行架构及执行原理。
- 通过示例讲解 Hive 的安装过程。
- 内部表及外部表，以及它们之间的区别。
- MapReduce 优化、配置优化，以及从程序角度进行优化。

8.1 Hive 的理论基础

前面我们已经介绍了 HDFS 和 MapReduce。HDFS 用于存储数据，MapReduce 用于处理、分析数据。对于有一定开发经验，特别是有 Java 基础的程序员，学习 MapReduce 相对比较容易；可是对于习惯使用 SQL 的程序员来说，学习 MapReduce 的成本相对比较大。

那么如何能使习惯使用 SQL 的程序员在最短的时间内基于大数据技术完成数据的分析呢？在这种背景下 Hive 应运而生。

本章我们将介绍什么是 Hive 及 Hive 与传统数据库的比较，并且从 Hive 的运行架构方面掌握 Hive 的核心知识。

8.1.1 什么是 Hive

Hive 本身是数据仓库。那么什么又是数据仓库呢？首先先来了解一下数据仓库的概念及数据仓库的特点。

数据仓库是为了协助分析报告，支持决策，为需要业务智能的企业提供业务流程的改进和指导，从而节省时间和成本，提高质量。它与数据库系统的区别是，数据库系统可以很好地解决事务处理，实现对数据的"增、删、改、查"操作，而数据仓库则是用来做查询分析的数据库，通常不会用来做单条数据的插入、修改和删除。

Hive 作为一个数据仓库工具，非常适合数据的统计分析，它可以将数据文件组成表格并具有完整的类 SQL 查询功能，还可将类 SQL 语句自动转换成 MapReduce 任务来运行。因此，如果使用 Hive，可以大幅提高开发效率。

和传统的数据仓库一样，Hive 主要用来访问和管理数据。与传统数据仓库较大的区别是，Hive 可以处理超大规模的数据，可扩展性和容错性非常强。由于 Hive 有类 SQL 的查询语言，所以学习成本相对比较低。

8.1.2 Hive 和数据库的异同

很多初学者会将 Hive 与关系型数据库混淆，认为 Hive 就是关系型数据库，主要原因是由于 Hive 采用了类 SQL 的查询语言 HQL（Hive Query Language）所致。其实，HQL 的引入仅仅是为了降低学习成本，底层还是 MapReduce。Hive 本身是数据仓库，并不是数据库系统，清楚这一点，有助于理解 Hive 的特性。

Hive 和数据库的主要区别在查询语言、存储位置、数据格式、数据更新、索引、执行、执行延迟、可扩展性和数据规模几方面，详见表 8.1 所示。

表 8.1 Hive和RDBMS的区别

区别	Hive	RDBMS
查询语言	HQL	SQL
数据存储位置	HDFS	Local FS
数据格式判断	查询时判断	插入时判断
执行	MR	Executor
执行延迟	高	低
处理数据规模	大	小

1. 查询语言

前面我们提到，类 SQL 的查询语言 HQL 可以使熟悉 SQL 的开发人员很快上手，可以很方便地使用 Hive 进行开发。那么 SQL 和 HQL 有哪些区别呢？读者可以参考表 8.2 所示。

表 8.2 SQL和HiveQL的比较

特性	SQL	HiveQL
更新	UPDATE、INSERT、DELETE	INSERT
事务	支持	有限支持
索引	支持	支持
延迟	亚秒级	分钟级
函数	数百个内置函数	几十个内置函数
多表插入	不支持	支持
Create table as select	SQL-92 中不支持，但有些数据库支持	支持

(续)

特　　性	SQL	HiveQL
SELECT	SQL-92	支持排序的SORT BY。可限制返回行数量的LIMIT
子查询	在任何子句中支持"相关"的（correlated）或不相关的（noncorrelated）	只能在FROM、WHERE或HAVING子句中（不支持相关子查询）
视图	可更新	用户定义函数
扩展点	用户定义函数	MapReduce脚本

2．数据存储位置

在数据存储位置方面来说，数据库是将数据存储在块设备或本地文件系统中。而 Hive 是将所有数据存储在 HDFS 中，并建立在 Hadoop 之上。

3．数据格式

在 Hive 中，并没有定义特有的数据格式，数据格式是由用户指定，用户在定义数据格式时需要指定 3 个属性，分别是列分隔符，比如通常为空格、\t、\x001，行分隔符，如"\n"，以及读取文件数据的方法。

因为在加载数据的过程中，不需要从用户数据格式到 Hive 本身定义的数据格式中进行转换，所以，在 Hive 加载的过程当中不会对数据本身做任何调整，而只是将数据内容本身复制到了相应的 HDFS 目录中。而在传统的数据库中，由于不同的数据库有不同的存储引擎，各自定义了自己的数据格式，全部的数据都会按照一定的组织结构进行存储，因此数据库在加载数据的过程中会比较耗时。

4．数据更新

前面提到，Hive 本身是针对数据仓库而设计的，同时，由于数据仓库的内容往往是读多写少。所以 Hive 中不支持对数据的修改和增加，所有的数据都是在加载的过程中完成的。而数据库中的数据往往需要经常进行修改、查询、增加等操作。

5．索引

在索引方面，Hive 在加载数据的过程中不会对数据做任何处理，也不会对数据进行扫描处理，所以也没有对数据中的某些键值创建索引。在 Hive 访问数据中满足条件的数据值时，需要扫描全部数据，由此造成的访问延迟较高。由于 HQL 最终会转化成 MapReduce，因此 Hive 可以并行访问数据，即使在没有索引的情况下，对于大批量数据的访问，Hive 仍可以表现出优势。在数据库中，通常会针对某一列或者某几列创建索引，所以对于少批量的满足特定条件的数据访问，数据库具有很高的效率，以及较低的延迟。因此，Hive

数据访问的延迟较高，不适合在线查询数据。

6．执行

关于执行，前面提到 Hive 中大多数查询的执行最终是通过 MapReduce 来实现的。而数据库则具有自己的执行引擎。

7．执行延迟

由于 Hive 在查询数据的时候并没有索引，需要扫描整个表，由此造成的延迟较高，同时，由于 MapReduce 自身具有较高的延迟，也会导致查询延迟。相比较来说，在数据量规模小的情况下数据库的执行延迟较低。只有当数据规模大到超过数据库的处理能力的时候，Hive 的并行计算优势才能体现出来。

8．可扩展性

Hive 与 Hadoop 的可扩展性是一致的，原因是 Hive 本身是建立在 Hadoop 之上的。而数据库由于 ACID 语义的严格限制，扩展行非常有限。

9．处理数据规模

由于 Hive 建立在集群上，同时可以基于 MapReduce 进行并行计算，所以可以支撑大规模的数据，而数据库可以支撑的数据规模相对较小。

8.1.3 Hive 设计的目的与应用

前面已经提到，Hive 的设计目的是让那些熟悉 SQL 但编程技能相对比较薄弱的分析师可以对存放在 HDFS 中的大规模数据集进行查询。目前，Hive 已经是一个非常成功的 Apache 项目，很多组织把它用作一个通用的、可伸缩的数据处理平台。

Hive 主要应用于传统的数据仓库任务 ETL（Extract-Transformation-Loading）和报表生成。其中，报表生成中可以完成大规模数据分析和批处理的任务，进行海量数据离线分析和低成本进行数据分析，可以应用于日志分析，如统计网站一个时间段内的 PV、UV，以及多维度数据分析等。大部分互联网公司使用 Hive 进行日志分析，包括百度、淘宝等。

8.1.4 Hive 的运行架构

Hive 的用户接口主要有 3 个，分别是 CLI（Command Line）、Client 和 WUI。其中 CLI 是最常用的。

在 CLI 启动时，一个 Hive 的副本也会随之启动。Client，顾名思义是 Hive 的客户端，用户会连接至 Hive Server，在启动 Client 模式时，需要指出 Hive Server 在哪个节点上，同时在该节点启动 Hive Server。WUI 则是通过浏览器来访问 Hive。

HiveServer 是 Hive 的一种实现方式，客户端可以对 Hive 中的数据进行相应操作，而不启动 CLI，HiveServer 和 CLI 两者都允许远程客户端使用 Java、Python 等多种编程语言向 Hive 提交请求，并取回结果。

下面我们根据 Hive 的运行架构图解释一下 Hive 的体系结构，如图 8.1 所示。

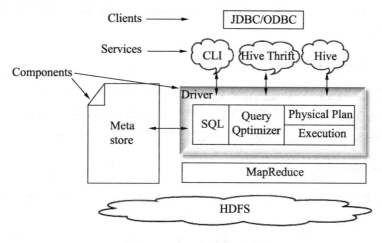

图 8.1 Hive 的运行架构图

在图 8.1 中，Metastore 主要用来存储元数据，Hive 是将元数据存储在数据库中，如 MySQL、derby。在 Hive 中的元数据包括表的名字、表的列和分区及其属性、表的属性（是否为外部表等）、表的数据所在目录等。

解释器、编译器、优化器完成 HQL 查询语句从词法分析、语法分析、编译、优化到查询计划的生成。生成的查询计划存储在 HDFS 中，并在随后用 MapReduce 调用执行。

Hive 的数据存储在 HDFS 中，大部分的查询、计算由 MapReduce 完成。

8.1.5 Hive 的执行流程

在 Hive 上执行查询时，整体流程大致步骤如下：

（1）用户提交查询任务到 Driver。

（2）编译器 Compiler 获得用户的任务计划。

（3）编译器 Compiler 根据用户任务从 Metastore 中得到所需的 Hive 元数据信息。

（4）编译器 Compiler 对任务进行编译，首先将 HQL 转换为抽象语法树，接着把抽象语法树转换成查询语句块，并将查询语句块转化为逻辑的查询计划。

（5）把最终的计划提交到 Driver。

（6）Driver 将计划提交到 Execution Engine，获得元数据信息，接着提交到 JobTracker 或者 Source Manager 运行该任务，该任务会直接从 HDFS 中读取文件并进行相应的操作。

（7）取得并返回执行结果。

8.1.6 Hive 服务

Hive 的 Shell 环境仅仅是 Hive 命令提供的一项服务。我们可以在运行时使用--service 选项指明要使用哪种服务。同时，通过输入 hive--service help 可以获得可用服务列表。下面介绍一些常用服务。

1．CLI（Common Line Interface）服务

前面我们说过，CLI 是 Hive 的命令行接口，也就是 Shell 环境。CLI 启动的时候会同时启动一个 Hive 的副本,这也是默认的服务。我们可以通过 bin/hive 或 bin/hive --service cli 命令来指出 Hive Server 所在的节点，并且在该节点启动 Hive Server。

2．HiveServer2服务

通过 Thrift 提供的服务（默认端口是 10000），客户端就可以在不启动 CLI 的情况下对 Hive 中的数据进行操作了，并且允许用不同语言如 Java、Python 语言编写的客户端进行访问。使用 Thrift、JDBC 和 ODBC 连接器的客户端需要运行 Hive 服务器来和 Hive 进行通信。

3．HWI（Hive Web Interface）服务

HWI 是通过浏览器访问 Hive 的方式，它是 Hive 的 Web 接口，默认端口是 9999。在没有安装任何客户端软件的情况下，这个简单的 Web 接口可以代替 CLI。另外，HWI 是一个功能更全面的 Hadoop Web 接口，其中包括运行 Hive 查询和浏览 Hive mestore 的应用程序。命令为 bin/hive --service hwi。

4．MetaStore服务

负责元数据服务 Metastore 和 Hive 服务运行在同一个进程中。使用这个服务，可以让 Metastore 作为一个单独的进程来运行。通过设置 METASTORE_PORT 环境变量可以指定服务器监听的端口号。

8.1.7 元数据存储 Metastore

Metastore 是 Hive 集中存放元数据的地方。Metastore 包括两部分：服务和后台数据的存储。Hive 有 3 种 Metastore 的配置方式，分别是内嵌模式、本地模式和远程模式。

内嵌模式使用的是内嵌的 Derby 数据库来存储数据，配置简单，但是一次只能与一个客户端连接，适用于做单元测试，不适用于生产环境。

本地模式和远程模式都采用外部数据库来存储数据，目前支持的数据库有 MySQL、Oracle、SQL Server 等，在这里我们使用 MySQL 数据库。本地元存储和远程元存储的区别是本地元数据不需要单独启动 Metastore 服务，因为本地元存储用的是和本地 Hive 在同一个进程里的 Metastore 服务。

在默认的情况下，Metastore 服务和 Hive 服务运行在同一个虚拟机中，它包含一个内嵌的以本地磁盘作为存储的 Derby 数据库实例，被称之为"内嵌 Metastore 配置"。

下面分别介绍内嵌模式、本地模式和远程模式，如图 8.2 所示。

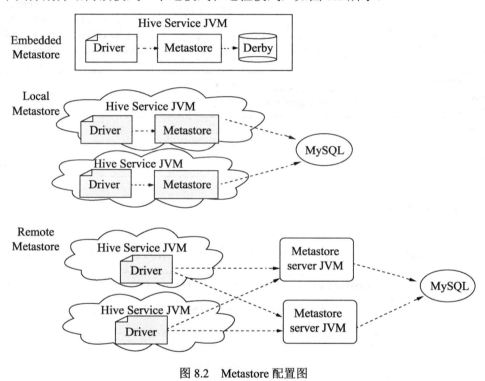

图 8.2　Metastore 配置图

8.1.8　Embedded 模式

Embedded 模式连接到一个 In-memory 的数据库 Derby，一般用于单元测试，如图 8.3 和表 8.3 所示。

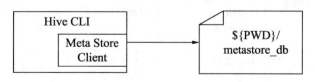

图 8.3　Embedded 模式

表 8.3 Embedded 模式的设置

Parameter	参数	Description	描述	Example	举例
Javax.jdo.option.ConnectionURL	JDBC连接字符串	JDBC connection URL along with database name containing metadata	JDBC连接URL及包含元数据的数据库名称	jdbc:derby:;databaseName=mestastore_db;create=true	申明derby数据库的连接URL路径，默认创建名称为mestastore_db的数据库
Javax.jdo.option.ConnectionDriverName	JDBC的driver	JDBC driver name. Embedded Derby fir Single user mode.	JDBC驱动程序的名称。嵌入式Derby fir单用户模式	org.apache.derby.jdbc.EmbeddedDriver	使用org.apache.derby.jdbc.EmbeddedDriver连接到Java DB的内容
Javax.jdo.option.ConnectionUserName	连接用户名	User name for Derby database	Derby数据库的用户名	App	应用
Javax.jdo.option.ConnectionPassword	连接密码	Password	密码	mine	mine

8.1.9 Local 模式

Local 模式通过网络连接到一个数据库中，是经常使用的模式，如图 8.4 所示。

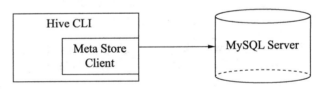

图 8.4 Local 模式

表 8.4 Local模式参数设置

Parameter	参数	Description	描述	Example	举例
Javax.jdo.option.ConnectionURL	JDBC连接字符串	JDBC connection URL along with database name containing metadata	JDBC连接URL及包含元数据的数据库名称	Jdbc:mysql://\<host name\>/\<database name\>?CreateDatabaseIfNotExist=true	指定JDBC连接的主机名和数据库名，如果数据库不存在，就创建一个数据库
Javax.jdo.option.ConnectionDriverName	JDBC的driver	Any JDD supported JDBC driver	任何JDD支持的JDBC驱动程序	com.mysql.jdbc.Driver	JDBC驱动

（续）

Parameter	参　　数	Description	描　　述	Example	举　　例
Javax.jdo.option.ConnectionUserName	连接用户名	User name	用户名	—	—
Javax.jdo.option.ConnectionPassword	连接密码	Password	密码	—	—

8.1.10　Remote 模式

Remote 模式用于非 Java 客户端访问元数据库，在服务器端会启动 MetaStoreServer，客户端通过 Thrift 协议及 MetaStoreServer 来访问元数据库，如图 8.5 和表 8.5 所示。

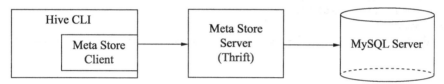

图 8.5　Remote 模式

表 8.5　Remote模式参数设置

Parameter	参　　数	Description	描　　述	Example
Hive.metastore.uris	远程连接元数据库	Location of the metastore server	元存储服务器的位置	thrift://<host_name>:9083
Hive.metastore.local	本地或者远程的元数据	— —	— —	false

8.2　Hive 的配置与安装

前面我们讲到有 3 种 Metastore 的配置方式，分别是内嵌模式、本地模式和远程模式。这里我们以本地模式为例，介绍 Hive 的配置与安装。首先，本地模式需要 MySQL 作为

Hive Metastore 的存储数据库,因此在安装 Hive 之前需要先安装 MySQL。

8.2.1 安装 MySQL

关于 MySQL 的安装比较容易,并且安装 MySQL 有多种方式,读者也可以下载不同的 MySQL 版本进行安装,本节的安装方式基于 YUM,安装和相关配置过程如下。

(1) 安装 MySQL 数据库,命令如下:

```
yum install -y mysql-server
```

(2) 对数据库字符集进行设置。

在/etc/my.cnf 文件中加入 default-character-set=utf8,代码如下:

```
[mysqld]
datadir = /var/lib/mysql.
socket=/var/lib/mysql/mysql.sock
user =mysql
# Disabling symbolic-links is recommended to prevent assorted security risks
symbolic-links=0

[mysqld_safe]
log-error=/var/log/mysqld.log
pid file=/var/run/mysqld/mysqld.pid
default-character-set=utf8
```

(3) 启动 MySQL 服务。代码如下:

```
[root@master~]# service mysqld start
Starting mysqld:                               [OK]
[root@master~]#
service mysqld status                          //查看 MySQL 是否启动
[root@master~]# service mysqld status
MySQLd (pid 2216) is running
[root@master~]#
chkconfig mysqld on                            //设置 MySQL 开机自动启动
[root@master~]# chkconfig mysqld on
[root@master~]#
```

(4) 设置 MySQL 的 root 密码为 123456,命令如下:

```
mysql -u root -p
```

接着会提示输入密码,默认密码为空,直接回车即可:

```
[root@master~]# mysql -u root -p
Enter password:
Welcome to the MySQL monitor. Commands end with ;or\g.
Your MySQL connection id is 5
Server version:5.1.73 Source distribution
Copyright(c)2000,2013,Oracle and/or its affiliates.All rights reserved.
Oracle is a registered trademark of Oracle Corporation and/or its
affiliates. Other names may be trademarks of their respective
```

```
owners.
Type'help;'or'\h' for help.Type '\c' to clear the current input statement.
mysql>
```

然后就可以完成密码的设置了，执行过程如下：

```
MySQL >set password for root@localhost=password('123456');
Query OK,0 rows affected (0. 00 sec )
mysql > "
```

退出 MySQL：

```
mysql >quit
Bye
[root@master~]#
```

如果远程无法访问，可以创建新的用户并且赋予其权限、修改密码、重启 MySQL，远端需要改成%，具体如下操作。

（5）接前面的操作，这里创建一个远程连接的用户 root1。代码如下：

```
CREATE USER 'root1'@'%' IDENTIFIED BY '123456';
mysql> CREATE USER 'root1'@'%' IDENTIFIED BY '123456';
Query OK,0 rows  affected (0.00 sec )
->
```

授权:GRANT ALL ON *.* TO 'root1'@'%'; 。代码如下：

```
mysql> GRANT ALL ON *.* TO'root1'@'%' ;
Query OK,0 rows affected(0.00 sec)
mysql>
```

修改密码：

```
SET PASSWORD FOR 'root1'@'%' = PASSWORD("123456");
mysql> SET PASWORD FOR ' root1'@'%' = PASSWORD("123456") ;
Query OK,0 rows  affected (0.00 sec)
mysql>
```

（6）安装 MyManager 连接 MySQL

在安装完 MySQL 之后，就可以通过客户端连接到 MySQL 了。比如这里我们通过 MySQL 的客户端工具 MyManager 连接到 MySQL，连接成功后提示如图 8.6 所示。在这里读者可以选择自己熟悉的客户端工具。

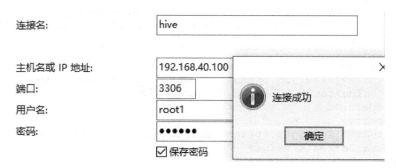

图 8.6　MyManager 连接到 MySQL

8.2.2 配置 Hive

8.2.1 节我们介绍了 MySQL 的安装步骤，主要目标是基于 MySQL 存储元数据，接下来安装 Hive。

（1）将 Hive 存放在服务器上，如图 8.7 所示。

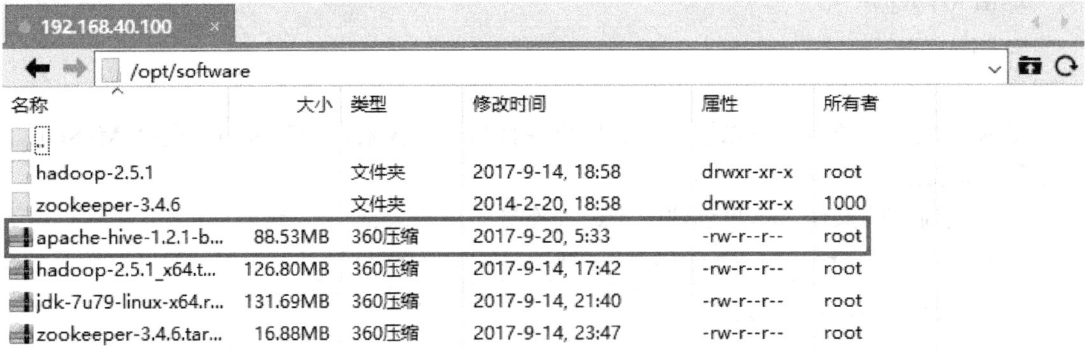

192.168.40.100

/opt/software

名称	大小	类型	修改时间	属性	所有者
..					
hadoop-2.5.1		文件夹	2017-9-14, 18:58	drwxr-xr-x	root
zookeeper-3.4.6		文件夹	2014-2-20, 18:58	drwxr-xr-x	1000
apache-hive-1.2.1-b...	88.53MB	360压缩	2017-9-20, 5:33	-rw-r--r--	root
hadoop-2.5.1_x64.t...	126.80MB	360压缩	2017-9-14, 17:42	-rw-r--r--	root
jdk-7u79-linux-x64.r...	131.69MB	360压缩	2017-9-14, 21:40	-rw-r--r--	root
zookeeper-3.4.6.tar...	16.88MB	360压缩	2017-9-14, 23:47	-rw-r--r--	root

图 8.7 把 Hive 存放到服务器的特定目录下

（2）解压缩后，在 Hive 的 conf 目录下创建一个 hive-site.xml 文件：

```
<?xml version="1.0"?>
<?xml-stylesheet type="text/xsl" href="configuration.xsl"?>
<configuration>
/*Hive 数据的存放路径*/
    <property>
        <name>hive.metastore.warehouse.dir</name>
        <value>/user/hive/warehouse</value>
    </property>
    <property>
        <name>hive.metastore.local</name>
        <value>true</value>
    </property>
    <property>
        <name>javax.jdo.option.ConnectionURL</name>
        <value>jdbc:MySQL://localhost/hive?createDatabaseIfNotExist=true
        </value>
    </property>
    <property>
        <name>javax.jdo.option.ConnectionDriverName</name>
        <value>com.MySQL.jdbc.Driver</value>
    </property>
    <property>
        <name>javax.jdo.option.ConnectionUserName</name>
        <value>root</value>
    </property>
    <property>
        <name>javax.jdo.option.ConnectionPassword</name>
```

```
            <value>123456</value>
        </property>
</configuration>
```

用 Hive 的 lib 目录下的 jline 包，替换掉 Hadoop 的 hadoop-2.5.2\share\hadoop\yarn\lib 下的低版本的 JAR 包，如图 8.8 所示。

```
[root@master lib]# ls -l jine*
-rw-rw-r--. 1 root  root 213854  Apr 30 2015 jline -2.12.jar
[root@master lib ]#
```

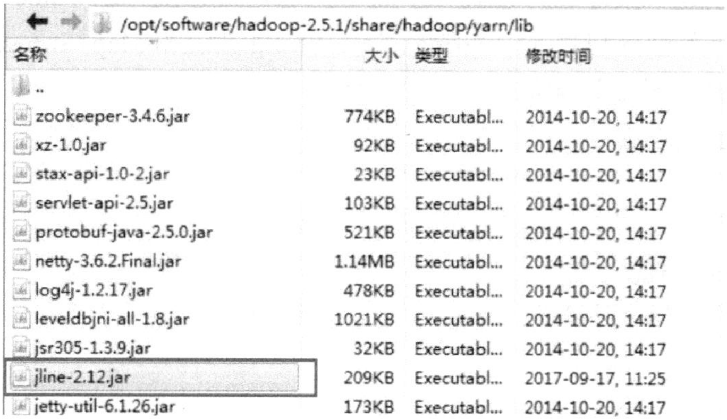

图 8.8　替换为 jline -2.12.jar 包

（3）接着可以把 Hive 路径配置到环境变量中，通过 hive shell 命令访问 Hive，命令如下：

```
[root@master~]# hive shell
17/09/18 12:02:55 WARN conf.HiveConf:HiveConf
Logging initialized using configuration in jarog4j.properties
hive> show databases ;
OK
default
userdb
Time taken:1.777 seconds,Fetched:2 row(s)
Hive>
```

8.3　Hive 表的操作

前面我们已经完成了 MySQL 的安装，用于存储 Hive 元数据，同时完成了 Hive 的配置，并通过 Shell 成功登录 Hive。接下来就可以创建 Hive 表并进行数据操作了。

Hive 是一个数据仓库，它可以将结构化的数据文件映射为一张数据库表，并具有 SQL 语言的查询功能，在这里需要再次强调的是对于数据仓库来说，往往存放的是历史数据，它的作用是完成查询分析，用于企业高层决策参考，往往不会完成单条记录的增加、修改和删除操作。

Hive 表的创建语法与传统的关系型数据库类似，但是 Hive 的类型可以更加复杂，比如可以是数组类型和 Map 类型。

8.3.1 创建 Hive 表

首先我们来创建一张普通的 Hive 表，表名为 person，包含 id、name、age、fav 和 addr 这 5 个字段，数据类型分别为 int、String、int、String 数组和 Map<String,String>，支持的数据文件格式为 txt。建表语句如下：

```
CREATE TABLE person(
id INT,
name STRING,
age  INT,
fav  ARRAY<STRING>,
addr MAP<STRING,STRING>
)
//表中的字段和数据类型
COMMENT 'This is the person table'
//表的简介，有无皆可
ROW FORMAT DELIMITED FIELDS TERMINATED BY '\t'
COLLECTION ITEMS TERMINATED BY '-'
MAP KEYS TERMINATED BY ':'
STORED AS TEXTFILE;
//定义这张表使用的数据文件格式，这里指定为 txt 类型
```

上述建表语句中，ROW FORMAT DELIMITED FIELDS TERMINATED BY '\t'是定义每一个字段的分隔符，这里的'\t'表示以 Tab 键作为分隔符分隔每行字段。

COLLECTION ITEMS TERMINATED BY '-' 是定义集合类型中每个对象的分隔符，fav 字段是 String 类型的数组，这里定义以'-'为分隔符，分隔每个字符串。

MAP KEYS TERMINATED BY ':' 是定义 Map 类型键值对的分隔符，这里以':'为分隔符，分隔键与值。

STORED AS TEXTFILE 是定义这张表使用的数据文件格式，这里指定为 txt 类型。

8.3.2 导入数据

使用上述建表语句建出的表只是一张空表，里面不包含任何数据，接下来要向表中导入数据。前面提到过 Hive 是将结构化的数据文件映射为一张表，所以我们先准备一个数据文件，在这里应该为一个 txt 格式的文件，并且这个 txt 文件的内容需要有一个固定的格式，这个格式就是我们建表时定义的那些字段和元素分隔符。

文件具体内容如下:

```
1    rod    18    study-game-driver    std_addr:beijing-work_addr:shanghai
2    tom    21    study-game-driver    std_addr:beijing-work_addr:beijing
3    jerry  33    study-game-driver    std_addr:beijing-work_addr:shenzhen
```

上面内容中每一行代表一条记录,每行字段中间隔一个'Tab'键,fav 字段中每个元素中间用'-'隔开,addr 字段中的键和值用':'隔开。

有了数据文件,就可以向表中导入数据了。下面指定导入数据文件的所在目录和要导入哪一张表,导入数据的语句如下:

```
LOAD DATA LOCAL INPATH person.txt OVERWRITE INTO TABLE person;
```

导入完成之后就可以使用如下语句进行查询,查看表中是否存在数据:

```
select * from person;
```

查询命令及查询结果如下:

```
 hive> select * from person;
OK
 1 rod  18 ["study","game","driver"] {"std_addr":"beijing","work_addr":"shanghai"}
 2 tom 21 ["study","game","driver"] {"std_addr":"beijing","work_addr":"beijing
jerry      33         ["study","game","driver"]    {"std_addr":"beijing","work_addr":"shenzhen"}
Time taken: 0.6 seconds,Fetched: 4 row(s)
```

8.4 表的分区与分桶

Hive 中存放的数据往往是以 PB 为单位的庞大的数据集,海量的数据需要耗费大量的时间去处理,若是每次查询都对全部数据进行检索,效率会极为低下。而且在许多场景下,我们并不需要对全部数据进行检索,因此引入分区和分桶的方法减少每一次扫描总数据量,这种做法可以显著地改善性能。

8.4.1 表的分区

把数据依照单个或多个列进行分区,通常按照时间、地域进行分区。比如一张统计了全国各地一年来不同时刻温度的表,可以按照地域来分区,也可以按照时间来分区,甚至可以将时间和地区都当做分区条件。为了达到性能表现的一致性,对不同列的划分应该让数据尽可能均匀分布。最好的情况下,分区的划分条件总是能够对应 where 语句的部分查询条件。

分区应当在建表时就设置好了,我们还是以前面创建的 person 表为例,加上分区的建表语句如下:

```
CREATE TABLE person(
  id INT,
  name STRING,
  age  INT,
  fav  ARRAY<STRING>,
  addr MAP<STRING,STRING>
)
COMMENT 'This is the person table'
ROW FORMAT DELIMITED FIELDS TERMINATED BY '\t'
PARTITONED BY (dt STRING)
COLLECTION ITEMS TERMINATED BY '-'
MAP KEYS TERMINATED BY ':'
STORED AS TEXTFILE;
```

以 String 类型的字段 dt 为依据进行分区。同时，导入数据的语句也有所变化，必须在导入数据时指定要导入哪一个分区中。

```
LOAD DATA LOCAL INPATH 'person.txt' OVERWRITE INTO TABLE person partition (dt='20180315');
```

除指定导入数据文件的所在目录和要导入哪一张表外，还要使用 partition 关键字指定导入哪一个分区，导入数据后可使用查询语句查看该分区内的数据。

```
select * from person where partition(dt='20180315');
```

具体操作步骤如下：

（1）启动 HDFS 并进入 Hive。

（2）查看数据库并选择一个数据库启用，如果不选择数据库，默认使 Deafult 数据库。

（3）输入建表语句，在已启用的数据库下建一张表。

```
Hives>CREATE TABLE person(
    >id INT,
    >name STRING,
    >age  INT,
    >fav  ARRAY <STRING >.
    >addr MAP<STRING ,STRING>
    >)
    >COMMENT "This is the person table'
    >PARTITIONED BY(dt STRING)
    >ROW FORMAT DELIMITED FIELDS TERMINATED BY "\t'
    > COLLECTION ITENS TERHINATED BY'-'
    > MAP KEYS TERMINATED BY':"
    > STORED AS TEXTFILE;
OK
  Time taken: 0 .097 seconds
```

（4）编写数据文件。

（5）向表中导入数据，注意不要输错数据文件的目录。

```
hive>LOAD DATA LOCAL INPATH '/root/person.txt' OVERWRITE INTO TABLE person partition(ldt='20170815');
Loading data to table hivel.person partition (dt=20180315)
Partition hivel.person{dt=20180315}stats:[numFiles=1,numRows=0.
totalSize=193,rawDataSize=0]
```

OK
Time taken: 0.537 seconds

（6）输入查询语句，查看分区内是否已插入数据。

```
hive> select addr['work_addr'] from person where dt=20180315
    >
OK
 shanghai
 beijing
 shenzhen
Time taken: 0.986 seconds,Fetched: 4 row(s)
```

8.4.2 表的分桶

分桶是相对分区进行更细粒度的划分。在分区数量过于庞大以至于可能导致文件系统崩溃时，我们就需要使用分桶来解决问题了。

分桶将整个数据内容按照某列属性值的 Hash 值进行区分。比如，如要按照 ID 属性分为 4 个桶，就是对 ID 属性值的 Hash 值对 4 取模，按照取模结果对数据分桶。例如，取模结果为 0 的数据记录存放到一个文件中，取模为 1 的数据存放到一个文件中，取模为 2 的数据存放到一个文件中。

分桶同样应当在建表时就建立，建表语句与之前建立分区表类似。我们还是创建表 person，其建表语句如下：

```
CREATE TABLE person(
id INT,
name STRING,
age  INT,
fav  ARRAY<STRING>,
addr MAP<STRING,STRING>
)
COMMENT 'This is the person table'
ROW FORMAT DELIMITED FIELDS TERMINATED BY '\t'
PARTITIONED BY(dt STRING)
CLUSTERED BY (id) into 3 buckets
COLLECTION ITEMS TERMINATED BY '-'
MAP KEYS TERMINATED BY ':'
STORED AS TEXTFILE;
```

以字段 id 为分桶依据，将每个分区分成 3 个桶，导入数据的语句与分区表时没有区别，还是向分区中导入数据。

```
LOAD DATA LOCAL INPATH 'person.txt' OVERWRITE INTO TABLE person partition (dt='20180315');
```

如果要查询 3 个桶中第 1 个桶中的全部数据，可以通过以下查询语句进行查询。

```
select * from person tablesample(bucket 1 out of 3 on id);
```

具体过程操作步骤如下（相同的步骤不再重复描述）：

（1）启动 HDFS，并进入 Hive。
（2）输入建表语句如下：

```
hive> CREATE TABLE person(
    >id INT,
    >name STRING,
    >age INT,
    >fav  ARRAY <STRING> ,
    >addr MAP<STRING , STRING>
    > )
    >COMHENT'This is the person table'
    >PARTITIONED BY(dt STRING)
    > clustered by (id) into 4 buckets
    >ROW FORMAT DELIMITED FIELDS TERMINATED BY'\t'
    > COLLECTION ITEMS TERMINATED BY': '
    >MAP KEYS TERMINATED BY':'
    > STORED AS TEXTFILE;
OK
Time taken: 0.073 seconds
```

（3）编写数据文件。
（4）向表中分区导入数据，注意不要输错数据文件的目录。
（5）输入查询语句，查看分桶内是否成功。

```
Hives>select * from person tablesample(bucket 1 out of 3 on id):
OK
3 jerry 33 ["study","game","driver] {"std_addr":"beijing","work_addr":
"shanghai} 20170315
Time taken:0.087 seconds,Fetched: I row(S)
```

从查询结果中可以看出，分桶中只有一条数据。

8.5 内部表与外部表

与传统的关系型数据库不同，Hive 创建的表分为内部表和外部表，对于内部表来说，在创建的时候会把数据移动到数据仓库所指向的位置；如果是外部表，则仅仅记录数据所在的位置。

同时，对于内部表来说，在删除的时候会将元数据和数据一起删除，而外部表仅仅是删除了元数据，真正的数据不会删除。所以，如果在共享源数据的情况下，可以选择使用外部表；如果仅仅是 Hive 内部使用，则可以使用内部表。接下来分别结合实例介绍内部表和外部表。

8.5.1 内部表

内部表顾名思义，数据文件的全部操作都由 Hive 来完成，也就是说除了 Hive 外不会再有其他的应用使用该数据文件。前面我们所创建的表都是内部表。内部表在创建后不仅默认会在"/user/hive/warehouse/数据库名"下生成表的目录，还会在目录下生成一份表的数据文件，并且在表删除后，该目录和数据文件也会删除。

在 Hive 下输入 dfs -ls /user/hive/warehouse/hive1.db; 查看表目录：

```
Hives >dfs -ls/user/hive/warehouse/hive1.db
Found 1 items
Drwxrwxrwx  - root   supergroup  0 2017-09-21 13:39 /user/hive/warehouse/
h1ve1.db/person
```

再输入 dfs -ls /user/hive/warehouse/hive1.db/person/分区名; 查看表的数据文件：

```
Hives > dfs -ls /user/hive/warehouse/hive1.db/person/dt=20180315
Found I items
-rwxrwxrwx
3rootsupergroup1922017-09-2113:39/user/hive/warehouse/hive.db/person/
dt-20180315/person.t
xt
```

输入 drop table person; 删除表 person，再输入 dfs -ls /user/hive/warehouse/hive1.db; 查看表和数据文件是否被删除掉：

```
Hive > drop table person;
OK
Time taken: 0.222 seconds
hives>dfs -ls/user/hive/warehouse/hive1.db;
```

查询结果显示没有任何内容，说明表目录和数据文件一同被删除了。

8.5.2 外部表

在实际应用过程中，数据文件往往并不是只由 Hive 来操作，其他应用或计算也会操作该文件，这种情况下往往不允许改变数据文件的格式或者位置。面对这种情况，我们一般选择创建外部表来实现 Hive 操作。外部表也会在"/user/hive/warehouse/数据库名"下生成表的目录，但是目录内不会生成数据文件，并且删除表不会影响到导入数据的源数据文件。

外部表的创建有两种方式，一种是创建一张空表，然后向表中导入数据，其创建语句如下：

```
CREATE external TABLE person(
id INT,
```

```
  name STRING,
  age  INT,
  fav  ARRAY<STRING>,
  addr MAP<STRING,STRING>
)
COMMENT 'This is the person table'
ROW FORMAT DELIMITED FIELDS TERMINATED BY '\t'
COLLECTION ITEMS TERMINATED BY '-'
MAP KEYS TERMINATED BY ':'
STORED AS TEXTFILE;
```

上述建表语句中，通过 external 关键字指明创建的表是外部表。创建外部表还有另外一种方式，在创建表的同时，指定数据文件的位置，其建表语句如下：

```
CREATE external TABLE person(
  id INT,
  name STRING,
  age  INT,
  fav  ARRAY<STRING>,
  addr MAP<STRING,STRING>
)
COMMENT 'This is the person table'
ROW FORMAT DELIMITED FIELDS TERMINATED BY '\t'
COLLECTION ITEMS TERMINATED BY '-'
MAP KEYS TERMINATED BY ':'
LOCATION 'hdfs://192.168.79.100:9000/root/ '
STORED AS TEXTFILE;
```

上述建表语句，添加 external 关键词创建外部表，使用 location 关键词指定数据文件的位置。

下面使用第一种方式创建一张外部表，首先在 Hive 下输入建表语句创建外部表 person，建表语句如下：

```
hive> CREATE external TABLE person(
    >id INT,
    >name  STRING,
    >age   INT,
    >fav ARRAY<STRING>,
    >addr MAP<STRING, STRING>
    > )
    >COMMENT'This is the person table'
    >ROW FORMAT DELIMITED FTELDS TERMINATED BY'\t'
    >COLLECTION ITEMS TERMINATED BY'-'
    >MAP KEYS TERMINATED BY':'
    > STORED AS TEXTFILE;
OK
Time taken; :0.146 seconds
```

表创建完后，就可以向表中导入数据了。

```
Hives>LOAD DATA LOCAL INPATH '/root/person.txt' OVERWRITE INTO TABLE person;
Loading data to table hivel. person
```

```
Table hivel.person stats:[numFiles=1,numRows=0,totalsize=192,ramDataSize=0]
OK
Time taken: 0.212 seconds
```

在 Hive 下输入 dfs -ls /user/hive/warehouse/hive1.db，查看表目录：

```
Hive>dfs -ls/user/hive/warehouse/hive.db
Found 1 items
drwxrwxrwx-root supergroup 0 2017-09-21 14:13 /user/hive/warehouse/hive1.db/person
```

在 Hive 下输入 dfs -ls /user/hive/warehouse/hive1.db/person，查看数据文件。然后删除表后再查看文件是否存在，这样就可以看出内部表和外部表的区别了。

8.6 内置函数与自定义函数

Hive 中包含很多内置的函数，如果这些函数不能满足我们的业务场景时，可以通过编写用户自定义函数（User-Defined Function，UDF）来实现，并在 Hive 中调用。

UDF 函数有 3 种类型，分别是 UDF、UDAF 和 UDTF。

- UDF 函数的特点是作用于单条数据，并且输出一个数据行。大多数的函数如字符串函数、日期函数等都属于这一类。
- UDAF 函数可以接受多个数据输入，同时输出一个数据行，比如 COUNT、MIN、MAX，这类函数都是聚集函数。
- UDTF 函数的特点是作用于单个数据行，同时产生多个输出数据行。

8.6.1 内置函数实例

在讨论实例之前，为了便于自定义函数的演示，我们先创建一个虚拟表，创建虚拟表的过程如下：

```
echo'X'> dual.txt
hive> create table dual (dummy String);
hive> load data local inpath 'dual.txt' overwrite into table dual;
```

上述代码中，echo 'X' >dual.txt 是代表把 X 写入 dual.txt 文件中，接着创建一个名为 dual 的表，表字段是 dummy，并把 dual.txt 中的 X 导入到表中，完成后，可以通过以下操作查看临时表的内容。

```
hive>select * from dual;
OK
X
Time taken:0.286 seconds,Fetched:1row(s)
```

临时表创建完后，我们就可以编写自定义函数了。下面分别举例来说明 UDF 函数、UDAF 函数和 UDTF 函数的用法。

1. UDF 函数实例

首先来看一个 UDF 函数实例，这个例子很简单，就是将'a'、'b'、'c'、'd'进行拼接起来，完成输出，需要用到 concat() 函数。代码如下：

```
hive> select concat('a','b','c','d') from dual;
OK
abcd
Time taken: 0.726 seconds,Fetched: 1row(s)
```

2. UDAF 函数实例

接着再看一个 UDAF 函数实例，就是统计 UID 的总和。代码及运行结果如下：

```
hive> select sum(uid) from user_dimension;
Query ID = root_20170921222228_391990bb-e12f-4a58-b1e8-3e1410f2a362
Total jobs=1
Launching Job 1 out of 1
Number of reduce tasks determined at compile time: 1
In order to change the average load for a reducer (in bytes);
  set hive.exec.reducers.bytes.per.reducer=<number>
In order to limit the maximum number of reducers;
  set hive.exec.reducers.max=<number>
In order to set a constant nuaber of reducers:
  set mapreduce.job.reduces=<number>
StartingJob=job_1505992163946_0001,TrackingURL=http://master1:8088,
/proxy/application_1505992163946_0001
Kill Cormand = /opt/software/hadoop-2.5.1/bin hadoop job -kill j0b_
1505992163946_0001
Hadoop job information for Stage-1:nunber of mappers: 1; nunber of reducers: 1
2017-09-21 22:22:51,222 Stage-1 map =0%, reduce=0%
2017-09-21 22:23:33,867 Stage-1 map =100%, reduce =0%,Cumulative CPU 7.11sec
2017-09-21 22:23:44,290 Stage-1 map =100% , reduce=100%,Cumulative CPU 8.88 sec
MapReduce Total cumulative CPU time: 8 seconds 880.msec
Ended Job=job_1505992163946_0001
MapReduce Jobs Launched:
stage-stage-1:Map: 1 Reduce: 1 Cunulative CPu:8.88sec HDFS Read: 44116 HDFS Write: 9 SUCCESS
Tatal MapReduce CPU Time Spent: 8 seconds 880 msec
OK
499500.0
Tine taken: 76.774 seconds,Fetched: 1 row(s)
hive>
```

3. UDTF 函数实例

接着再看一个 UDTF 函数实例，就是将'a, b, c, d, e'分解成'a'、'b'、'c'、'd'、'e'，并逐行输出。

```
hive> select explode(split('a,b,c,d,e', ',')) from dual;
OK
a
b
c
d
e
Time taken: 0.105 seconds,Fetched:5  row(s)
```

8.6.2 自定义 UDAF 函数实例

前面介绍了 Hive 自带的 UDF、UDAF 和 UDTF 函数，如果系统自带的函数无法满足业务需求时，可以自己定义一个函数，并在需要的时候进行调用。以下是一个计算一组整数中最大值的 UDAF 函数实例。

（1）编写 Java 代码如下：

```java
import org.apache.hadoop.hive.ql.exec.UDAF;
import org.apache.hadoop.hive.ql.exec.UDAFEvaluator;
import org.apache.hadoop.io.IntWritable;
public class Maximum extends UDAF {
public static class MaximumIntUDAFEvaluator implements UDAFEvaluator {
    private IntWritable result;
    public void init() {
        result = null;
    }
    public boolean iterate(IntWritable value) {
        if (value == null) {
            return true;
        }
        if (result == null) {
            result = new IntWritable(value.get());
        } else {
            result.set(Math.max(result.get(), value.get()));
        }
            return true;
    }
     public IntWritable terminatePartial() {
        return result;
    }
    public boolean merge(IntWritable other) {
        return iterate(other);
    }
    public IntWritable terminate() {
        return result;
    }
  }
}
```

（2）导出 JAR 包，并将 JAR 包添加到 Hive 中。

```
hive> add jar Maximum.jar;
```

(3) 用 JAR 包生成一个函数。

hive> create function maximumtest as 'com.firsthigh.udaf.Maximum';

(4) 运行函数并检查结果。

```
hive> select maximumtest(price) from record_dimension;
Query ID = root_20170921233816_45023bfd-b8a2-4d71-a188-d7cf5837360a
Total jobs=1
Launching Job 1 out of 1
Number of reduce tasks determined at compile time: 1
Starting
Job=job_1505992163946_0002,TrackingURL=http://master1:8088/pxoxy/
application_1505992163946_0002
kill Comnand = /opt/software/hadoop-2.5.1/bin/hadoop job -kill job_
1505992163946_0002/
Hadoop job information for Stage-1:number of mappers:1; number of reducers:1
2017-09-21 23:39:28,721 Stage-1 map= 0%,   reduce =0%
2017-09-21 23:40:01,925 Stage-1map=100%, reduce = 0%, Cumulative CPU 2.91 sec
2017-09-21 23:40:26,374 Stage-1map=100%, reduce=100%, Cumulative CPU 5.97 sec
MapReduce Total cumulative CPU time:5 seconds 970 msec.
Ended Job = job_1505992163946_0002
MapReduce Jobs Launched:
stage-stage-1: Map 1 Reduce: 1  Cumulative CPU: 5.97 sec  HDFS Read; 70184
HDFS Write; 5 SUCCESS
Total MapReduce CPU Time Spent: 5 seconds 970 msec
OK
1000
Tine  taken; 133.872 seconds,Fetched: 1row(s)
```

8.7 通过 Java 访问 Hive

前面已经实现了 Hive 的配置和安装，之后可以基于 Hive Shell 实现数据的基础分析了。本节来看一下如何通过 Java 访问 Hive 数据仓库。

通过 Java 访问 Hive 需要依赖于 Hive 提供的 JDBC 驱动，操作步骤与 Java 访问 MySQL 基本类似。下面是一个 Java 访问 Hive 的实例。代码如下：

```
package com.sendto.database;

import java.sql.Connection;
import java.sql.DriverManager;
import java.sql.ResultSet;
import java.sql.SQLException;
import java.sql.Statement;
public class HiveTest {
    final static String  DRIVER ="org.apache.hive.jdbc.HiveDriver";
    final static String  DATABASE_PATH
 ="jdbc:hive2://192.168.202.130:10000/userdb02";
    final static String  USER_NAME ="root";
    final static String  PASSWORD ="";
    public static void main(String[] args)throws SQLException {
```

```
        Connection conn =null;
        Statement stmt =null;
        ResultSet rs =null;
        String hql ="select count(*) from person";
        int count =0;
        try{
            //1.注册驱动
            Class.forName(DRIVER);
            //2.创建连接
            conn=DriverManager.getConnection(DATABASE_PATH,USER_NAME,
            PASSWORD);
            //3 创建 statement 对象
            stmt = conn.createStatement();
            //4 执行 HQL 语句
            rs = stmt.executeQuery(hql);
            //5.处理结果集
            if(rs.next()){
                count = rs.getInt(1);
            }
        }catch(Exception e){
            e.printStackTrace();
        }finally{
            //6.关闭连接
            if(rs !=null){  rs.close();  }
            if(stmt !=null){  stmt.close();  }
            if(conn !=null){  conn.close();  }
        }
        System.out.println(count);
    }
}
```

上述代码中，jdbc:hive2://192.168.202.130:10000 部分，代表 JDBC 协议，hive2 是子协议，192.168.202.130 是 IP 地址，10000 代表端口号。整体连接过程和 JDBC 类似，在此不再赘述。

前面我们介绍过基于 Hive Shell 访问 Hive，如果想通过 Java 访问 Hive 则需要通过 HiveServer2 的方式。

HiveServer2 可以使 Hive 以服务器形式运行，这样便于不同的客户端连接到 Hive，并进行相关操作，可以通过 hive.server2.thrift.port 配置属性来指明服务器所监听的端口号，默认为 10000。启动命令如下：

```
hive --service hiveserver2 &
```

以上命令是以 Hiveserver2 的方式启动 Hive，并在后台运行，启动成功后，就可以通过 Java 连接到 Hive 了。

8.8 Hive 优化

Hive 的底层是 MapReduce，当数据量太大时，往往可以通过并行来提高效率，比如通过 Partition 实现运行多个 Reduce，可是如果处理不当则容易引发数据倾斜，从而导致效率降低，这就涉及 Hive 的优化。Hive 的优化主要分为 3 个方面，即 MapReduce 优化、配置优化和程序优化，下面分别介绍这几种优化方式。

8.8.1 MapReduce 优化

前面说过 HQL 解析之后会成为 MapReduce 程序，所以可以从 MapReduce 运行的角度来考虑性能优化。

首先考虑的是要避免数据倾斜，比如，通过任务监控页面发现只有少量 Reduce 任务未完成，则可能出现了数据倾斜问题。在处理大批量数据时，可以通过 Partition 分区将数据分发到不同的 Reduce 中进行处理，可是如果分配不当，则可能造成某个 Reduce 中处理千万条数据，有的 Reduce 中只处理几十条数据，这就造成了数据倾斜。关于 Partition 的实例，在前面章节已介绍过，读者可以参考。

另外，尽量避免大量的 Job，因为 Job 数较多的作业往往运行效率比较低。比如有几张表，每个表中的数据有几百到几千条，如果通过多次关联，则可能会产生十几个甚至几十个 Job，而 Job 的初始化等操作都是比较耗资源的，在这种情况下就会造成性能低下。

在 Hive 中使用 SUM、COUNT、MAX、MIN 等 UDAF 函数时，不需要担心出现数据倾斜问题，因为 Hadoop 在 Map 端的汇总合并时已经优化过了，不会出现数据倾斜问题。

另外需要注意的是，在遇到 COUNT(DISTINCT)的情况下，如果数据量太大，则会效率低下，在遇到多个 COUNT(DISTINCT) 的情况下则效率会更低，主要是因为 COUNT(DISTINCT)是根据 GROUP BY 字段来分组的。

8.8.2 配置优化

除了使用 MapReduce 优化，还可以通过 Hive 相关配置来进行优化，主要分为以下几个方面。

1．列裁剪

列裁剪的意思是忽略不需要的列，Hive 在读取数据时，可以仅仅读取需要用到的列，而不需要把所有的列都读取出来。比如有一张表，包括 a、b、c、d、e 共 5 列，Hive 在查询时仅仅取出需要的 b、d、e，而不需要将所有的列都取出来。

2．分区裁剪

另外一个参数优化的方式是在查询过程中减少不必要的分区，分区参数为：hive.optimize.pruner=true（默认值为真）。

3. join 操作

在写 join 操作的代码时，最好将数目少的表或者子查询放在 join 操作符的左边，主要原因是在 Reduce 阶段，处在 join 左边的表内容会被加载进内存中。同时，对于同一个 key 来说，对应的 value 值小的放在前面，value 值大的放在后面，这也是"小表放前"原则。

4. GROUP BY 操作

在进行 GROUP BY 操作的时候，需要注意以下几方面：

（1）Map 端部分聚合。

由于并不是所有的聚合操作都必须要在 Reduce 端进行，很多聚合操作可以先在 Map 端进行聚合，接着在 Reduce 端得出最终结果。这时需要修改的参数是 hive.map.aggr=true，这个参数是用于设定是否需要在 Map 端进行聚合。另外一个参数是 hive.groupby.mapaggr.checkinterval，用于设定 Map 端进行聚合操作的条目数。

（2）在有数据倾斜时进行负载均衡。

与之相关的参数是 hive.groupby.skewindata，当设置为 true 时，生成的查询计划会有两个 MapReduce 任务。其中，在第一个 MapReduce 中，Map 的输出结果会被随机分布到 Reduce 中，接着每个 Reduce 做部分聚合操作，并输出结果。

8.9 小　　结

本章主要介绍了有关 Hive 的相关知识，包括 Hive 的理论基础、数据库异同、设计目的、应用、运行框架、执行原理，以及 Hive 环境搭建、数据操作、内部表、外部表，还讲解了如何通过 Java 访问 Hive 及 Hive 的优化。

第 9 章 大数据快速读写——HBase

HBase 属于列式非关系型数据库（NoSQL），最早起源于 Google 发布的 Bigtable，是由 Powerset 公司的 Chad walters 和 Jim Kelleman 在 2006 年末发起的。2007 年 7 月，由 Mike Cafarella 提供代码，形成了一个基本可用的系统。

HBase 的第一个发布版本是在 2007 年 10 月和 Hadoop 0.15.0 捆绑发布的。2010 年 5 月，HBase 从 Hadoop 子项目升级成 Apache 顶层项目。今天，HBase 已然成为一种广泛应用于各种行业生产中的成熟技术，它的用户包括 Adobe、StumbleUpon、Twitter 和雅虎等公司。

本章主要涉及如下知识点。
- 了解 NoSQL 数据库的分类与应用，以及关系型和非关系型数据库的区别。
- 掌握 HBase 数据模型及执行原理，了解 HBase 体系架构的组件。
- 掌握 HBase 的 Shell 操作，以及通过 Java API 访问 HBase 实例。

9.1 关于 NoSQL

有计算机编程经验的读者一定对关系型数据库比较熟悉。但是随着数据量的增大和对性能提升的要求，同时由于数据格式的多样性，关系型数据库已经无法满足需求，在这种情况下 NoSQL 应运而生。

本节主要介绍一下什么是 NoSQL 及其和关系型数据库的区别，从而为读者更好地理解 HBase 做好铺垫。

9.1.1 什么是 NoSQL

NoSQL = Not Only SQL，不仅仅是 SQL。NoSQL 是一个通用术语，即非关系型数据库，它不是以 SQL 作为其主要访问语言。现在有许多类型的 NoSQL 数据库，BerkeleyDB 就是本地 NoSQL 数据库的一个示例，而 HBase 是一个分布式数据库。从技术层面上来说，HBase 实际上是一个"数据存储"，而不是"数据库"，因为它缺乏关系型数据库的很多属性，如类型化列、辅助索引、触发器和高级查询语言等。

一个关系型数据库虽然可以很好地被扩展，但仅限于某个点，也就是一个数据库服务器的大小，并且为了实现最佳性能，需要专用的硬件和存储设备。随着互联网的发展，传统的关系型数据库在应付超大规模和高并发的系统上已经显得"力不从心"，非关系型数据库就是在这样的背景下产生的。

9.1.2 NoSQL 数据库的分类

NoSQL 数据库共分为 4 类，分别是键值（Key-Value）存储数据库、列存储数据库、文档型数据库和图形（Graph）数据库，具体介绍如下。

1．键值（Key-Value）存储数据库

键值存储数据库中的数据是以键值对格式进行存储的，类似于 Java 中的 Map，其数据库中的表有一个特定的 Key 键和键所指向的 Value 值。Key-Value 模型的优势在于简单且容易部署，可以将程序中的数据直接映射到数据库中，使程序中的数据和键值存储数据库中的数据存储方式很相近，比如 Redis。

2．列存储数据库

列存储数据库与传统的关系型数据库不同，关系型数据库按照行进行存储，而列数据库是每一列单独存放，仅仅查询所需要的列，查询速度大幅提高。在 HBase 中，这些列是由列家族来安排的。

3．文档型数据库

文档型数据库和键值存储类似。该类型的数据模型是将内容按照某些特定的格式进行存储，如 JSON 格式和 MongoDB 就属于文档型数据库。

4．图形（Graph）数据库

图形数据库与关系型数据库和列式数据库不同，它是基于灵活的图形模型，并且可以扩展到多个服务器上。另外，由于 NoSQL 数据库并没有标准的查询语言（SQL），所以在进行数据库查询时需要制定数据模型，如 Neo4j。

9.1.3 NoSQL 数据库的应用

NoSQL 数据库主要适用于以下场景：
- 数据量大、数据模型比较简单。
- 对数据库性能要求较高，需要节省开发成本和维护成本。
- 不需要高度的数据一致性。
- 对于给定 key，比较容易映射复杂值的环境，数据之间关系性不强。

9.1.4 关系型数据库与非关系型数据库的区别

关系型数据库与非关系型数据库的区别主要在成本、查询速度、存储数据格式和扩展性4个方面。

- 成本：NoSQL 数据库简单易部署，基本都是开源软件，不需要像使用 Oracle 那样花费大笔资金购买后使用，相比关系型数据库 NoSQL 价格便宜。
- 查询速度：NoSQL 数据库将数据存储于缓存中，关系型数据库将数据存储在硬盘中，自然查询速度远不及 NoSQL 数据库。
- 存储数据格式：NoSQL 的存储格式是 key/value 形式、文档形式、图片形式等，所以可以存储基础类型及对象或者集合等各种格式，而数据库则只支持基础类型。
- 扩展性：关系型数据库有类似于 JOIN 这样的多表查询机制的限制导致扩展很艰难。

9.2 HBase 基础

我们知道 HDFS 是大型数据集分析处理的文件系统，具有高延迟的特点，它更倾向于读取整个数据集而不是某条记录，因此当处理低延迟的用户请求时，HBase 是更好的选择，它能实现某条数据的快速定位，提供实时读写功能。下面我们详细介绍 HBase 的核心概念和应用。

9.2.1 HBase 简介

HBase 即 HadoopDataBase，是一个基于 HDFS 和 ZooKeeper 的列式数据库，是一个高可靠性、高性能、面向列、可伸缩、实时读写的分布式数据库。HBase 具有可伸缩性，它自底向上地进行构建，能够简单地通过增加节点来达到线性扩展。

HBase 利用 Hadoop HDFS 作为其文件存储系统，利用基于 YARN 的 MapReduce 来处理 HBase 中的海量数据，利用 ZooKeeper 作为其分布式协同服务，主要用来存储非结构化和半结构化的松散数据，也就是列存储的 NoSQL 数据库。HBase 在大数据生态体系中的位置如图 9.1 所示，读者可以结合前面讲过的章节好好理解每门技术在体系中的位置。

HBase 并不是关系型数据库，它并没有严格的结构，而且不支持传统的 SQL。它可以运用 key/value 方式进行数据的存储，同时在特定的情况下，能够做 RDBMS 不能做的事，基于 HBase 可以在廉价硬件构成的集群上管理超大规模的稀疏表。

当然，HBase 也并不是适用于所有情况，首先要确保有足够数据的情况下使用 HBase，如果数据达到数亿或者数十亿行，那么 HBase 是一个很好的选择。如果只有几百万或者几千万行的数据，则可以考虑使用传统的关系型数据库，主要原因是在数据量不是足够大时，所有的数据有可能在 HBase 的单个节点上,而其余的集群可能处于空闲状态。

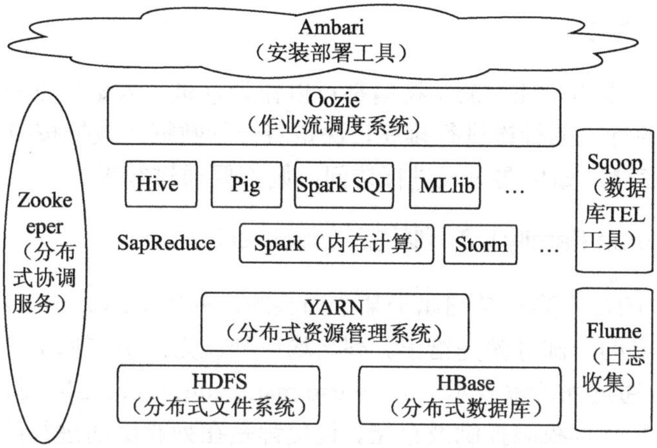

图 9.1 大数据生态系统图

另外,我们还要考虑的是 HBase 并没有 RDBMS 所提供的索引、事务、高级查询语言等功能,不可以将 RDBMS 构建的应用程序直接移植到 HBase 上。如果一定要将 RDBMS 移植到 HBase 上,则需要考虑重新设计。

HBase 还有一个显著的特点就是它有许多支持线性和模块化扩展的功能。HBase 集群可以通过添加商业性服务器的 RegionServers 来扩展。例如,如果集群从 10 个 RegionServers 扩展到 20 个 RegionServers,那么在存储和处理能力方面的性能也都会倍增。

9.2.2 HBase 数据模型

接着我们来看一下 HBase 数据模型,HBase 数据模型主要包括行键(Row Key)、列族、时间戳、单元格(Cell)和 HLog(WAL log),如图 9.2 所示。

Row Key	Time Stamp	CF1	CF2	CF3
"com.cnn.www"	t6		CF2:q1=va13	CF3:q4=va14
	t5			
	t3	CF1:q2=va12		

图 9.2 HBase 数据模型

下面对数据模型中的每个元素进行说明。

1. 行键（Row Key）

首先行键是字节数组，任何的字符串都可以作为行键，只是行键只能存储 64KB 的字节数据。图 9.2 中的行根据行键进行排序，决定着一行数据，数据按照 Row key 的字节序（byte order）排序存储。如果要对表进行访问，则都要通过行键。

2. 列族（Column Family）& 列标签（qualifier）

在 HBase 表中的每个列都是归属于某个列族的，列族 CF1、CF2 和 CF3 必须作为表模式（schema）定义的一部分预先给出。列名均以列族为前缀，每个"列族"均可以有多个列成员 column，通过列标签表示，如 course:math, course:english，新的列族成员以后可以动态的根据需要加入。权限控制及存储、调优都是在列族层面进行的；HBase 会将同一列族里面的数据存储在同一目录下，在几个文件中保存。

3. 时间戳（Timestamp）

在 HBase 中的每个 Cell 存储单元，会对同一份数据存储多个版本，HBase 引入了时间戳区分每个版本之间的差异，时间戳的类型是 64 位整型，不同版本的数据按照时间倒序排序，最新的数据版本排在最前面。

时间戳可以由 HBase 在数据写入的时候自动赋值，这个时候的时间戳是精确到毫秒的当前系统时间，时间戳也可以由客户显式地赋值。

4. 单元格（Cell）

单元格 Cell 内容是没有解析的字节数组，由行和列的坐标交叉决定，同时单元格是有版本的。单元格中的数据是没有类型的，都是由二进制字节码形式存储。

9.2.3　HBase 体系架构及组件

前面介绍了 HBase 的数据模型，接着来看一下 HBase 的体系架构及相关组件。HBase 体系架构如图 9.3 所示。

下面我们通过介绍 HBase 体系架构中的各个组件，来看一看每个组件的作用及组件之间是如何协作的。

1. 元数据存储ZooKeeper

首先来看一下 ZooKeeper 组件，该组件主要用于存储 HBase 的 schema 和 Table 元数据，它保障在任何时候，集群中只有一个 Master，同时，ZooKeeper 用来存贮所有 Region 的寻址入口，进行实时监控 RegionServer 的上线和下线信息，并且实时通知 HMaster。

第 9 章　大数据快速读写——HBase

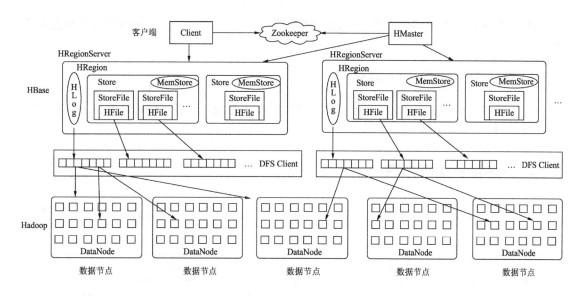

图 9.3　HBase 体系架构

2．资源分配HMaster

HMaster 组件类似 HDFS 中的 NameNode，它不存储数据，主要作用是为 RegionServer 分配 Region，同时负责 RegionServer 的负载均衡，如果发现失效的 RegionServer，它会重新分配上面的 Region，HMaster 组件还会管理用户对 Table 的增、删、改操作。

3．Region的处理HRegionServer

HMaster 组件类似 HDFS 中的 NameNode，而 HRegionServer 组件则类似 HDFS 中的 DataNode。HRegionServer 组件负责维护 Region，处理对 Region 的 I/O 请求，同时，HRegionServer 会负责切分在运行过程中变得过大的 Region。

4．保持访问性能Client

Client 组件主要包含访问 HBase 的接口，同时维护 Cache 来加快对 HBase 的访问性能。

5．分布式存储和负载均衡最小单元Region

在前面介绍组件时，已经说到 Region 了。Region 是 HBase 中分布式存储和负载均衡的最小单元，HBase 会自动把表水平划分成多个区域，也就是多个 Region。每个 Region 会保存一个表里面某段连续的数据，每个表一开始只有一个 Region，随着数据不断插入，Region 会不断增大，当增大到一个阈值的时候，Region 就会等分成两个新的 Region，这个等分的过程又称之为裂变。

当表中的行不断增多时,就会有越来越多的Region。这样一张完整的表被保存在多个RegionServer上。

6. MemStore与StoreFile组件

一个区域,也就是一个Region往往由多个Store组成,一个Store包括位于内存中的MemStore和位于磁盘的StoreFile。

9.2.4 HBase 执行原理

9.2.3节中介绍了HBase的体系架构,接着结合图9.4来看一下HBase的执行原理。

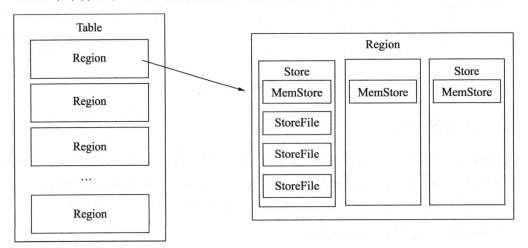

图 9.4 HBase 执行原理

从图9.4中可以看出,左边的Table包含多个Region,每个Region中又有多个Store,同时,每个Store又由一个MemStore和0至多个StoreFile组成。

为了更好地理解HBase执行原理,结合9.2.3节的HBase体系架构,可以看出StoreFile以HFile格式保存在HDFS上。

在执行写入操作时,会先将数据写入MemStore中,当MemStore中的数据达到某个阈值时,HRegionServer会启动Flash Cache进程写入StoreFile,每次写入会形成单独的一个StoreFile。当StoreFile文件的数量增长到一定阈值后,系统会进行合并,在合并过程中会进行版本合并和删除工作,形成更大的StoreFile。

当一个Region中所有的StoreFile大小之和超过一定阈值后,会把当前的Region分割为两个,并由HMaster分配到相应的RegionServer服务器上,实现负载均衡。

在检索数据时,会先从MemStore中进行查找,如果从MemStore中找不到数据,会再从StoreFile上查找。

9.3 HBase 安装

介绍了 HBase 的体系结构和执行原理,接下来在介绍如何应用 HBase 之前,先介绍 HBase 的安装。

安装 HBase 主要分为以下几步:

1. 下载HBase

首先从 HBase 官网 http://hbase.apache.org/ 下载 HBase 安装包,下载链接为 http://www-eu.apache.org/dist/,并解压 HBase 安装包。

2. 修改核心配置文件

HBase 的核心配置文件是 hbase-site.xml,可以搭建 HBase 单机版、完全分布式、高可用几种方式。根据不同的搭建方式,hbase-site.xml 的内容也不一样。

(1)单机版方式

```
<property>
    <name>hbase.cluster.distributed</name>
    <value>true</value>
</property>
<property>
    <name>hbase.rootdir</name>
    <value>file:///opt/software/hbase-0.98.12.1-hadoop2</value>
</property>
<property>
    <name>hbase.ZooKeeper.property.dataDir</name>
    <value>/opt/software/ZooKeeper-3.4.6</value>
</property>
```

(2)完全分布式集群版方式

```
<property>
    <name>hbase.cluster.distributed</name>
    <value>true</value>
</property>
<property>
    <name>hbase.rootdir</name>
    <value>hdfs://master:9000/hbase</value>
</property>
<property>
    <name> hbase.zookeeper.quorum </name>
    <value>master,slave</value>
</property>
```

(3)HDFS 高可用集群版方式

```xml
<property>
    <name>hbase.rootdir</name>
    <value>hdfs://firsthigh/hbase</value>
</property>
<property>
    <name>hbase.master</name>
    <value>8020</value>
</property>
<property>
    <name>hbase.zookeeper.quorum</name>
    <value>hadoopmaster,hadoop01,hadoop02</value>
</property>
<property>
    <name>hbase.zookeeper.property.clientPort</name>
    <value>2181</value>
</property>
<property>
    <name>hbase.zookeeper.property.dataDir</name>
    <value>/opt/software/zookeeper/conf</value>
</property>
<property>
<name>hbase.cluster.distributed</name>
    <value>true</value>
</property>
<property>
    <name>hbase.tmp.dir</name>
    <value>/var/hbase/tmp</value>
</property>
```

3. 配置regionservers文件

在里面输入主、从节点的主机名称，RegionServer 是 HBase 集群运行在每个工作节点上的服务，它是整个 HBase 系统的关键所在，提供了对于 Region 的管理和服务。修改如下：

```
master
slave
~
```

启动 ZooKeeper，HDFS 与 YARN 服务，命令如下：

```
zkServer.sh start
start-all.sh
```

在这里通过 JPS 命令确认服务已经正常启动后再启动 HBase

4. 启动HBase

启动 HBase 的命令如下：

```
start-hbase.sh
```

5. 查看相关服务

启动完 HBase 之后，可以通过 JPS 命令查看相关服务是否已经存在，如果启动成功，可以看到以下相关进程，其中 **HMaster** 和 **HRegionServer** 是不可少的。

```
10073 DataNode
9800  QuorumPeerMain
80025 HMaster
80152 HRegionServer
10421  NodeManager
10328 ResourceManager
80203 JPS
9983  NameNode
```

6. 启动HBase Shell

在启动之后，就可以通过 hbase shell 命令进入 Hbase，命令如下：

```
[root@master1~]# hbase shell
SLF4J: Class path contains multiple SLF4J bindings.
SLF4J: Found binding in [jar:file:/opt/software/hbase-]
SLF4J:Found binding in [jar:file:/opt/software/hadoop-]
SLF4J:See http://www.slf4j.org/codes.html#multiple _bin
SLF4]: Actual binding is of type [org.slf4j.impl.Log4jL
HBase SHell ; enter 'help<RETURN>' for list of supported
Type"exit<RETURN>" to leave the HBase Shell
version 1.1.3,r72bc50f5fafeb105b213942bbee3d61ca72498
```

7. 测试

通过 list 命令显示相关表，只要不出现异常，则代表 HBase 安装成功。

```
hbase(main) :001:0> list
TABLE
test
```

8. 通过浏览器查看HBase

在浏览器中通过 16010 访问 HBase 主页面，如图 9.5 所示。

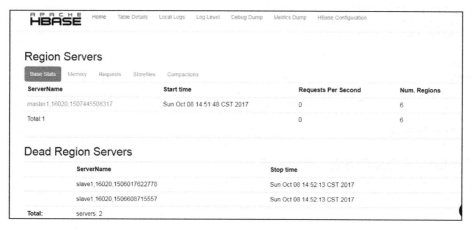

图 9.5　通过浏览器查看 HBase

9.4 HBase 的 Shell 操作

HBase Shell 是 HBase 的命令行工具，可以通过它对 HBase 进行操作。HBase 的常用 Shell 命令如下：

1. 创建表

create '表名称', '列族名称1','列族名称2','列族名称N'

例如，输入 create 'test3','cf1','cf2'，显示如下：

```
hbase(main):002:0> create 'test3','cf1','cf2'
1 row(s) in 3.0610 seconds
=>Hbase::Table-test3
hbase(main):003:0> list
TABLE
Like
test
test1
test2
test3
5 row(s)in 0 .0240 seconds
```

2. 删除表

删除表之前，需要先将表进行 disable 操作，然后再进行 drop 操作。

```
disable '表名称'
drop '表名称'
```

示例代码如下：

```
hbase(main):007:0> disable'test3'
1 row(s) in 0.0350 seconds
hbase(main):008:0> drop'test3'
1 row(s) in 1.4180 seconds
hbase(main):009:0> list
TABLE
Like
Test
test1
test24row(s) in 0.0310 seconds
```

3. 显示所有表

通过 list 命令显示所有表。代码如下：

```
hbase(main):003:0>list
TABLE
Like
test
test1
```

```
test2
test3
5 row(s) in 0.0240 seconds
```

4. 查询数据

查询数据的命令如下:

```
scan '表名称'
scan 'table1'
```

get '表名称','ROW 值'

```
get 'table1','111'
```

5. 增加数据

增加数据的命令如下:

```
put '表名称','ROW 值','列族名称:列名','列值'
put 'test','123','cf1:name','rod'
```

6. 删除数据

删除数据的命令如下:

```
delete '表名称','ROW 值','列族名称:列名'
delete 'test','110','cf1:age'
```

7. 修改数据

修改数据的命令如下:

```
put '表名称','ROW 值','列族名称:列名','新列值'
put 'test','123','cf1:name','Zoumaru'
```

9.5　Java API 访问 HBase 实例

HBase Shell 可以基于后台访问 HBase，本节将介绍如何基于 Java API 实现对 HBase 的远程操作，包括创建表、增加数据和查询数据。

9.5.1　创建表

我们可以通过 Java 远程创建一张表，进行 ZooKeeper 的设置并建立连接，定义表名和列族，最后创建表、添加列族。具体代码如下:

```java
public static void createTable(){                        //创建表
    Configuration conf = HBaseConfiguration.create();
    conf.set("hbase.zookeeper.quorum",
```

```java
"masternode001,slavenode001");
        try {
    //建立连接
            Connection conn = ConnectionFactory.createConnection(conf);
    //获取 Admin
            Admin admin = conn.getAdmin();
            //添加表的描述
            HTableDescriptor tableDes = new HTableDescriptor(TableName.
            valueOf("test11"));
            //添加列族
            HColumnDescriptor colDesc = new HColumnDescriptor("cf1");
            tableDes.addFamily(colDesc);
            admin.createTable(tableDes);
        } catch (IOException e) {
            e.printStackTrace();
        }
    }
```

调用 createTable()方法，执行完毕后，可在 HBase 上通过 list 指令查看生成的表，结果如下：

```
hbase(main):001:0> list
TABLE
test
test11
2 row(s) in 0.4780 seconds
=>["test" "test11" ]
```

结果显示新创建的表 test11 已生成。

9.5.2 插入数据

在创建表成功后，我们可以通过 put()方法向表中添加数据。添加数据时首先需要选择相应的表，再添加相应的列和数据，具体代码如下：

```java
public static void addData(){ //添加数据
Configuration conf = HBaseConfiguration.create();
    conf.set("hbase.zookeeper.quorum", "masternode001,slavenode001");
    try {
        Connection conn = ConnectionFactory.createConnection(conf);
    //确定连接的表
        Table table = conn.getTable(TableName.valueOf("test11"));
    //设置 Row Key
        Put put = new Put("110".getBytes());
    //写入列名及数据
        put.addColumn("cf1".getBytes(),"name".getBytes(),"lisa".getBytes());
        table.put(put);
    } catch (IOException e) {
        e.printStackTrace();
    }

}
```

调用 addData()方法，执行完毕后，可在 HBase 上通过 scan 指令查看表中的数据，结果如下：

```
hbase(main) :005:0> scan 'test11
ROW                       COL UMN+CE LL
110                       column=cf1:name,timestamp=1506007927066 ,value-lisa
I row(s) in 0.0630 seconds
```

由结果可知，通过 add 方法添加的列和数据都已添加成功。

9.5.3 查询数据

增加数据之后，可以通过 Java API 直接查看 HBase 中的数据。查询数据分为两种方式，一种是通过 get 方法获取指定 cell 的数据值，另一种是通过 Scan()方法遍历查询表中的全部数据，下面分别介绍两种方法。

（1）查询指定 cell 的值，在这里指定查询的表名、rowkey、列族和属性。

```java
public static void getData(){
    Configuration conf = HBaseConfiguration.create();
    conf.set("hbase.zookeeper.quorum", "masternode001,slavenode001");
    try {
        Connection conn = ConnectionFactory.createConnection(conf);
        Table table = conn.getTable(TableName.valueOf("test11"));
        //通过 Row Key 指定行
        Get get = new Get("110".getBytes());
        Result rs = table.get(get);
        //指定列名，确定 cell
        Cell cell = rs.getColumnLatestCell("cf1".getBytes(), "name".getBytes());
        //输出 cell 的 Row Key 和值
        System.out.println(new String(CellUtil.cloneRow(cell),"utf-8"));
        System.out.println(new String(CellUtil.cloneValue(cell),"utf-8"));
    } catch (IOException e) {
        e.printStackTrace();
    }
}
```

（2）调用 getData()方法，执行完毕后，控制台会直接输出如下查询结果：

```
110
lisa
```

（3）遍历查询指定表的全部数据，在这里通过 Scan 方法遍历查询表中的全部数据，得到结果集后输出。

```java
public static void getDataByScan(){
    Configuration conf = HBaseConfiguration.create();
    conf.set("hbase.ZooKeeper.quorum", "masternode001,slavenode001");
    try {
        Connection conn = ConnectionFactory.createConnection(conf);
        Table table = conn.getTable(TableName.valueOf("test11"));
```

```
        //调用Scan()方法
        Scan scan = new Scan();
        //取得遍历查询的结果
        ResultScanner resultscanner = table.getScanner(scan);
        //对结果集进行迭代
        Iterator its = resultscanner.iterator();
        //循环输出查询结果
        while(its.hasNext()){
            Result rs = (Result) its.next();
            Cell cell = rs.getColumnLatestCell("cf1".getBytes(), "name".
            getBytes());
            System.out.println(new String(CellUtil.cloneRow(cell),
            "utf-8"));
            System.out.println(new String(CellUtil.cloneValue(cell),
            "utf-8"));
        }
    } catch (IOException e) {
        e.printStackTrace();
    }
}
```

（3）调用 getDataByScan()方法，执行完毕后，控制台会直接如下输出查询结果：

```
110
lisa
```

9.6 小　　结

本章主要介绍了什么是 HBase、HBase 的架构及各个组件，并通过 HBase Shell 和基于 Java API 访问 HBase，最终实现数据的增加、删除、修改和查询操作。HBase 属于列式数据库，当数据量达到亿级别时可以考虑使用 HBase，当数据量上百万或者千万级别时，使用关系型数据库是一个不错的选择。

第 10 章　海量日志采集工具——Flume

在大数据技术架构中，主要包括数据采集、数据存储、数据计算、数据分析、数据可视化等核心步骤。其中数据采集至关重要，只有将数据源的数据采集过来，才可以进行计算和分析等工作。但是由于数据源很分散，导致数据的收集变得越发复杂。

目前的数据采集主要分为结构化数据采集和非结构化数据采集，采集的方式也略有区别。本章主要介绍非结构化的数据收集工具 Flume。

本章主要涉及的知识点如下：

- Flume 的特点、架构及主要组件。
- 如何安装和配置 Flume。
- Flume 的典型应用实例。
- 以 Exec 方式实现数据收集。

10.1　什么是 Flume

Flume 是一种用于高效收集、聚合和移动大量日志数据的分布式的可靠可用服务。它具有基于流式数据的简单灵活架构，基于它的可靠性机制和许多故障切换及恢复机制来说，它具有健壮性和容错性，使用了一个允许在线分析应用程序的简单可扩展数据模型。

Apache Flume 是一种分布式、可靠和可用的系统，用于高效收集、聚合，以及将大量日志数据从许多不同的来源移动到集中式数据存储上。

使用 Apache Flume 不仅限于日志数据的聚合。由于数据源是可定制的，因此可以使用 Flume 来传输大量的事件数据，包括但不限于网络流量数据、社交媒体生成的数据、电子邮件消息和其他数据源。

10.2　Flume 的特点

1. 事务

Flume 使用两个独立的事务负责从 Source 到 Channel 及从 Channel 到 Sink 的事件传递。

在 Source 到 Channel 的过程中，一旦所有事件全部传递到 Channel 并且提交成功，那么 Source 就将该文件标记为完成；从 Channel 到 Sink 的过程同样以事务的方式传递。由于某种原因使得事件无法记录时，事务将回滚，所有事件仍保留在 Channel 中重新等待传递。

2．可靠性

Channel 中的 File Channel 具有持久性，事件写入 File Channel 后，即使 Agent 重新启动，事件也不会丢失。Flume 中还提供了一种 Memory Channel 的方式，但它不具有持久存储的能力，数据完整性不能得到保证；与 File Channel 相比，Memory Channel 的优点是具有较高的吞吐量。

3．多层代理

使用分层结构的 Flume 代理，实现了 Flume 事件的汇聚，也就是第一层代理采集原始 Source 的事件，并将它们发送到第二层，第二层代理数量比第一层少，汇总了第一层事件后再把这些事件写入 HDFS。

将一组节点的事件汇聚到一个文件中，这样可以减少文件数量，增加文件大小，减轻施加在 HDFS 上的压力。另外，因为向文件输入数据的节点变多，所以文件可以更快地推陈出新，从而使得这些文件可用于分析的时间更接近于事件的创建时间。

10.3　Flume 架构

Flume 由一组以分布式拓扑结构相互连接的代理构成。Flume 中有多个 Agent，即 Flume 代理。Agent 是由 Source、Sink 和 Channel 共同构成的 Java 进程，Flume 的 Source 产生事件后将其传给 Channel，Channel 存储这些事件直至转发给 Sink。Flume 主体架构如图 10.1 所示。

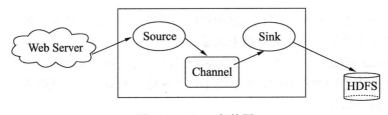

图 10.1　Flume 架构图

10.4　Flume 的主要组件

Flume 的主要组件有 Event、Client、Agent、Source、Channel 和 Sink 等。接下来我们

一起认识一下各组件的作用。

10.4.1 Event、Client 与 Agent——数据传输

事件 Event 是 Flume 数据传输的基本单元，Flume 以事件的形式将数据从源头传送到最终目的。

假设我们需要进行日志传输，Client 把原始需要收集的日志信息包装成 Events 并且发送到一个或多个 Agent 上。这样做的主要目的是从数据源系统中将 Flume 解耦。

代理 Agent 是 Flume 流的基础部分，一个 Agent 包含 Source、Channel、Sink 和其他组件，它基于这些组件把 Event 从一个节点传输到另一个节点或最终目的地上，由 Flume 为这些组件提供配置、生命周期管理和监控支持。

10.4.2 Source——Event 接收

Source 的主要职责是接收 Event，并将 Event 批量地放到一个或者多个 Channel 中。接下来我们分别介绍一下常用的两类 Source：Spooling Directory Source 和 Exec Source。

1．Spooling Directory Source获取数据

Spooling Directory Source 是通过读取硬盘上需要被收集数据的文件到 spooling 目录来获取数据，然后再将数据发送到 Channel。该 Source 会监控指定的目录来发现新文件并解析新文件。在给定的文件已被读完之后，它被重命名为指示完成（或可选地删除），属性列表如表 10.1 所示。

表 10.1 属性列表

属 性 名	默 认 值	含 义
type	--	类型，spooldir
spoolDir		监听目录
fileSuffix	COMPLETE	当数据读取完后添加的文件后缀
deletePolicy	never	当文件读取完后，文件是否删除，可选值有:never, Immediate
fileHeader	false	是否把路径加入Header

与 Exec 源不同，即使 Flume 重新启动或死机，这种 Source 也是可靠的，不会丢失数据。同时需要注意的是，产生的文件不能进行任意修改，否则会停止处理；在实际应用中可以将文件写到一个临时目录下之后再统一移动到监听目录下。

2. Exec Source收集数据

Exec源在启动时运行给定的UNIX命令,并期望该进程在标准输出上连续生成数据。如果进程由于任何原因退出,则源也将退出并且不会继续产生数据。这意味着诸如 cat [named pipe]或tail -F [file]的配置将产生期望的结果。Exec Source 属性列表如表 10.2 所示。

表 10.2 Exec Source属性

属 性 名	默认值	含 义
type	—	Exec类型
command	—	执行的命令

10.4.3 Channel—Event 传输

Channel 位于 Source 和 Sink 之间,用于缓存 Event,当 Sink 成功将 Event 发送到下一个 Agent 或最终目的处之后,会将 Event 从 Channel 上移除。不同的 Channel 提供的持久化水平也是不一样的,并且 Channel 可以和任何数量的 Source 和 Sink 工作。

1. Memory Channel内存中存储

Memory Channel 是指 Events 被存储在已配置最大容量的内存队列中,因此它不具有持久存储能力,Memory Channel 的配置属性如表 10.3 所示。

表 10.3 Memory Channel配置属性

属 性 名	默认值	含 义
type	—	类型名称,memory
Capacity	10000	存放的Event最大数目
transactionCapacity	10000	每次事务中,从Source服务的数据,或写入Sink的数据(条数)
byteCapacityBufferPercentage	20	Header中数据的比例
byteCapacity	—	存储的最大数据量(byte)

在使用 Memory Channel 时,如果出现问题导致虚拟机宕机或操作系统重新启动,事件就会丢失,在这种情况下,数据完整性不能保证,这种情况是否可以接受,主要取决于具体应用。与 File Channel 相比,Memory Channel 的优势在于具有较高的吞吐量,在要求高吞吐量并且允许 Agent Event 失败所导致数据丢失的情况下,Memory Channel 是理想的选择。

2. File Channel持久化存储

File Channel 具有持久性，只要事件被写入 Channel，即使代理重新启动，事件也不会丢失，能保障数据的完整性，File Channel 属性如表 10.4 所示。

表 10.4 File Channel属性

属 性 名	默 认 值	含 义
Type	--	类型名称，file
checkpointDir	~/.flume/file--channel/ checkpoint	Checkpoint文件存放位置
dataDirs	~/.flume/file--channel/data	数据目录，分隔符分割

10.4.4 Sink—Event 发送

Sink 的主要职责是将 Event 传输到下一个 Agent 或最终目的处，成功传输完成后将 Event 从 Channel 中移除。Sink 主要分为两大类：File Roll Sink 和 Hdfs Sink。

1. File Roll Sink写入本地

File Roll Sink 是指将事件写入本地文件系统中，首先我们要在本地文件系统中创建一个缓冲目录，新增文件是由手工添加的，如表 10.5 所示。

表 10.5 File Roll Sink属性

Property Name	属性名	Default	默认	Description	描述
channel	通道	--	--		
Type	类型	--	--	The component type name,needs to be file_roll	组件类型名称，应为 file_roll
sink.directory	sink.directory	--	--	The directory where files will be stored	存储文件的目录
sink.path Manager	sink.path Manager	DEFAULT	DEFAULT	The PathManager implementation to use.	应用的路径管理器的实现
sink.path Manager. extension	sink.path Manager.extension	--	--	The file extension if the default PathManager is used.	使用默认路径管理器时的文件扩展名
sink.path Manager. prefix	sink.path Manager. prefix	--	--	A character string to add to the begging of the file name if the default PathManager is used	如果使用默认路径管理器，则添加到请求文件名的字符串中
sink.roll Interval	sink.roll Interval	30	30	Roll the file every 30 seconds.Specifying 0 will disable rolling and cause all events to be written to a single file.	每30秒滚动一次。指定0将禁用滚动，并导致将所有事件写入单个文件中

(续)

Property Name	属性名	Default	默认	Description	描述
sink.serializer	sink.serializer	TEXT	TEXT	Other possible options include avro_event or the FQCN of an implementation of EventSerializer.Builder interface.	其他可能的选项包括 avro_event 或实现 EventSerializer.Builder 接口的FQCN
batchSize	batchSize	100	100	---	---

2. HDFS Sink写入HDFS

HDFS Sink 是指将事件写入 Hadoop 分布式文件系统（HDFS）。它可以根据经过的时间、数据大小或事件数量定期滚动文件，也就是关闭当前文件并创建新文件。对于正在进行写操作处理的文件，其文件名会添加一个后缀".tmp"，以表明文件处理尚未完成，具体属性如表 10.6 所示。

表 10.6 HDFS Sink属性

属 性 名	默 认 值	含 义
Type	---	类型名称，hdfs
Hdfs.path	---	HDFS 目录
Hdfs.flePrefix	FlumeData	Flume写入HDFS的文件前缀
hds.rollinterval	30	文件滚动时间间隔(单位：秒)
Hdf.rollSize	1024	文件滚动大小(单位：byte)
Hdf.rolCount	10	文件滚动事件数目

10.4.5 其他组件

Interceptor 组件主要作用于 Source，可以按照特定的顺序对 Events 进行装饰或过滤。Sink Group 允许用户将多个 Sink 组合在一起，Sink Processor 则能够通过组中的 Sink 切换来实现负载均衡，也可以在一个 Sink 出现故障时切换转到另一个 Sink。

10.5 Flume 安装

通过 10.4 节的学习，我们已经知道了 Flume 的特点和主要组件。接下来安装 Flume，具体步骤如下：

（1）下载 Flume。

从 Flume 下载页面（http://flume.apache.org/download.html）下载一个稳定版本的 Flume

二进制发行包,比如 apache-flume-1.6.0-bin.tar.gz 版本。

(2)上传包并安装。将其放入适当的位置,这里的路径为/opt/software,然后将其解压,命令如下:

```
tar -zxvf apache-flume-1.6.0-bin.tar.gz
```

(3)把 Flume 配置到环境变量中,这样可以在任意目录下启动 Flume,配置文件如下:

```
JAVA_HOME=/usr/java/jdk1.7.0_79
HADOOP_HOME=/opt/software/hadoop-2.5.1
HIVE_HOME=/opt/software/apache-hive-1.2.1-bin
FLUME_HOME=/opt/software/apache-flume-1.6.0-bin
ZOOKEEPER_HOME=/opt/software/zookeeper-3.4.6
PATH=$PATH:$HOME/bin:$HADOOP_HOME/bin:$HADOOP_HOME/sbin:$JAVA_HOME/bin:
$HIVE_HOME/bin:$ZOOKEEPER_HOME/bin:$FLUME_HOME/bin
export PATH
```

(4)使环境变量生效:

```
source .bash_profile
```

10.6　Flume 应用典型实例

前面已经完成了 Flume 的安装,接下来通过典型实例来演示如何应用 Flume,读者可以根据自身项目的需求进行选择。

在配置使用 Flume 时,最核心的是编写配置文件,在配置文件中配置好 Source、Sink 和 Channel 属性,在启动 Flume 时会读取这个配置文件,并根据配置完成数据的采集工作。

10.6.1　本地数据读取(conf1)

首先,我们看一个从本地目录下收集数据的实例。首先在目录(如/opt)下分别创建 flume 和 sink 文件夹,flume 文件夹作为数据源,sink 文件夹作为输出目录,输入 Flume 代理的启动命令,此时我们在/opt 目录下上传并编辑 conf1 文件,之后上传到 flume 的文件会被收集到 sink 文件夹中。具体步骤如下。

1. 编辑配置文件

在/opt 文件夹下编辑 conf1 配置文件,内容如下:

```
a1.sources = r1
a1.sinks = k1
a1.channels = c1

# 配置 Source
a1.sources.r1.type =spooldir
a1.sources.r1.spoolDir=/opt/flume
```

```
#配置Sink
a1.sinks.k1.type =file_roll
a1.sinks.k1.sink.directory=/opt/sink

# 设置Channel 类型为Memory
a1.channels.c1.type = memory
a1.channels.c1.capacity = 1000
a1.channels.c1.transactionCapacity = 100

# 把Source 和 Sink 绑到 Channel 上
a1.sources.r1.channels = c1
a1.sinks.k1.channel = c1
```

2．创建相应的文件夹

在/opt 文件夹下分别建立 flume 和 sink 文件夹，分别作为 Source 和 Sink 的目录。

```
mkdir flume
mkdir sink
```

3．启动Flume代理

使用 flume-ng 命令启动 Flume 代理，同时在 console 控制台打印日志，命令如下：

```
flume-ng agent --conf-file ./opt/conf1 --name a1-Dflume.root.logger=INFO,
console
```

4．测试数据收集

向/opt/flume 文件夹下导入一个 test.txt 文本文件，Flume 客户端会有如下显示：

```
17/08/1721:07:44INFOinstrumentation.MonitoredCounterGroup:Componenttype:
SOURCE,name:r1 started
17/08/17 21:09: 38 INFO avro.ReliableSpoolingFileEventReader:
 Last read took us just up to a file boundary.Rolling to the next file,if
there is one.
17/08/1721:09:38INFOavro.ReliableSpoolingFileEventReader:Preparingtomov
efile/opt/fluno/test.txt to/opt/flume/test.txt.
COMPLETED
```

一旦 Flume 成功启动并完成日志收集，再查看 flume 文件夹中的 test.txt 文件时，则变成了 text.txt.COMPLETED，sink 文件夹中会收集到文件。这个实例中演示了 Flume 的工作流程，实现 test.txt 的收集工作。

10.6.2 收集至 HDFS

本节我们看一个从本地目录下收集数据到 HDFS 的实例。首先在目录（如/opt）下创建一个 flume 文件夹，flume 作为数据源，输入 Flume 代理的启动命令，在/opt 目录下创建并编辑 conf2 文件，最终实现上传的 flume 的文件会被收集到 HDFS 中。conf2 文件内容如下：

```
a1.sources = r1
a1.sinks = k1
a1.channels = c1

# 配置 Source
a1.sources.r1.type =spooldir
a1.sources.r1.spoolDir=/opt/flume/

# 配置 Sink
a1.sinks.k1.type =hdfs
#配置收集后的文件，放在 HDFS 的哪个位置
a1.sinks.k1.hdfs.path=hdfs://master:9000/flume/data/
a1.sinks.k1.hdfs.rollInterval=0
a1.sinks.k1.hdfs.rollSize=10240000
a1.sinks.k1.hdfs.rollCount=0
a1.sinks.k1.hdfs.idleTimeout=3
a1.sinks.k1.hdfs.fileType=DataStream
a1.sinks.k1.hdfs.round=true
a1.sinks.k1.hdfs.roundValue=10
a1.sinks.k1.hdfs.roundUnit=minute
a1.sinks.k1.hdfs.useLocalTimeStamp=true
#a1.sinks.k1.type =hdfs

# 配置 channel 为 Memory
a1.channels.c1.type = memory
a1.channels.c1.capacity = 1000
a1.channels.c1.transactionCapacity = 100

# 绑定 Source 和 Sink 到 Channel 上
a1.sources.r1.channels = c1
a1.sinks.k1.channel = c1
```

完成配置之后，就可以启动 Flume 代理了，启动命令如下：

```
flume-ng agent --conf-file conf2 --name a1 -Dflume.root.logger=INFO,console
```

接着，我们上传文件到/opt/flume 目录下，如图 10.2 所示。文件处理完毕，如图 10.3 所示。

图 10.2　上传文件到/opt/flume 目录下

图 10.3　文件处理完毕

接着打开 NameNode 节点，可以看到在 HDFS 上生成了多个文件，这些文件就是 Flume 收集过来的，如图 10.4 所示。

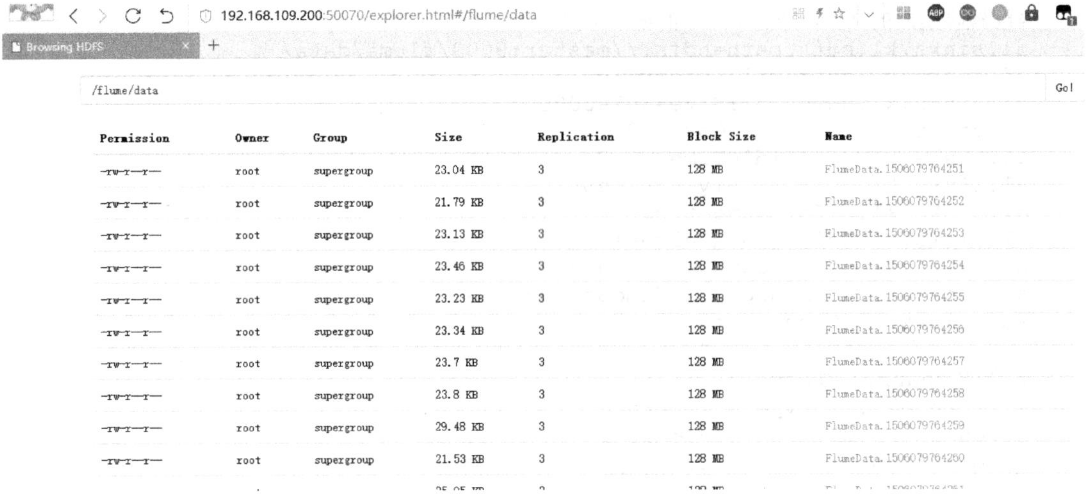

图 10.4　文件被收集到 HDFS 上

10.6.3　基于日期分区的数据收集

有些时候，由于收集的文件很大，或者因业务需求，需要将收集到的数据按照日期分开存储。本节实例就是基于 HDFS 收集，再根据年、月、日、时分分别生成文件，具体步骤如下。

（1）在目录（如/opt）下创建一个 flume 文件夹，flume 作为数据源。

（2）在/opt 目录下创建并编辑 conf 3 文件，并在 conf 3 文件中设置 HDFS 的存储路径，并设置以年月日的格式作为分区条件。

（3）将在本地 flume 文件夹下生成的文本日志收集到 HDFS 里，并设置以日期 %Y-%m-%d/%H%M 格式进行分类储存。

conf3 的主要内容如下：

```
a1.sources = r1
a1.sinks = k1
a1.channels = c1
```

```
# 配置 Source
a1.sources.r1.type =spooldir
a1.sources.r1.spoolDir=/opt/flume/

# 配置 Sink
a1.sinks.k1.type =hdfs
#按照%Y-%m-%d/%H%M 格式分开存储文件
a1.sinks.k1.hdfs.path=hdfs://master:9000/flume/data/%Y-%m-%d/%H%M
a1.sinks.k1.hdfs.rollInterval=0
a1.sinks.k1.hdfs.rollSize=10240000
a1.sinks.k1.hdfs.rollCount=0
a1.sinks.k1.hdfs.idleTimeout=3
a1.sinks.k1.hdfs.fileType=DataStream
a1.sinks.k1.hdfs.round=true
a1.sinks.k1.hdfs.roundValue=10
a1.sinks.k1.hdfs.roundUnit=minute
a1.sinks.k1.hdfs.useLocalTimeStamp=true
#a1.sinks.k1.type =hdfs

# Use a channel which buffers events in memory
a1.channels.c1.type = memory
a1.channels.c1.capacity = 1000
a1.channels.c1.transactionCapacity = 100

# 绑定 Source 和 Sink 到 Channel 上
a1.sources.r1.channels = c1
a1.sinks.k1.channel = c1
```

启动 Flume 代理：

```
flume-ng agent --conf-file conf3 --name a1 -Dflume.root.logger=INFO,console
```

随后，可以查看上传的文件是否已经被收集到 HDFS。

10.7 通过 exec 命令实现数据收集

还有一种收集文件的方法是使用 exec 命令。比如，如果想要读取文件内容持续增加的文件，可以使用 tail -f 命令来实现。本实例演示 Flume 如何结合 tail –f 命令完成数据收集，通过 tail –f 命令查看新增内容，通过 Flume 持续将新增内容收集到 HDFS 中。这个实例中，我们使用了一个 py 脚本来持续生成模拟数据。

10.7.1 安装工具

在真实项目中可以直接读取真实数据，或者配合其他组件读取。这里为了便于演示，我们以 Python 脚本的方式生成模拟数据，读者可根据自己的业务情况，自行选择数据来

源。如果想要运行脚本生成模拟数据,首先需要先安装所依赖的模块,安装过程如下:

(1) 安装 YUM。

如果通过 wget 工具从互联网上下载文件,需要先安装 YUM,安装过程如下:

```
[root@master opt]# yum -y install wget
Loaded plugins : fastestmirror
Loading mirror speeds  from  cached hostfile
*base: mirrors.aliyun.com
*extras: mirrors.aliyun.com
*updates: mirrors.sohu.com
base
extras
updates
Setting up Install Process
Resolving Dependencies
-->Running transaction check
--->Package wget.x86_64 0: 1.12-10.el6 will be installed
--> Finished Dependency Resolution
Dependencies Resolved
```

(2) 安装插件 setuptools

接着安装依赖插件 setuptools:

```
wget http://pypi.python.org/packages/source/s/setuptools/setuptools-2.0.tar.gz --no-check-certificate
tar -zxvf setuptools-2.0.tar.gz
cd setuptools-2.0
python setup.py build
python setup.py install
```

(3) 安装 pip 工具

后面需要用 pip 工具安装其他组件,这里先要安装 pip:

```
wget https://pypi.python.org/packages/source/p/pip/pip-1.3.1.tar.gz --no-check-certificate
tar -zxvf pip-1.3.1.tar.gz
cd pip-1.3.1
python setup.py install
```

(4) 安装生成模拟数据的相关模块

```
pip install Faker
pip install importlib
pip install ipaddress
```

(5) 生成并上传 py 脚本

建立 /opt/resource/hadoop-master/datasource 文件夹,将 py 脚本上传到此文件夹中。

建立 opt/data/loganalysis/ 文件夹,用于存放生成的数据,命令如下:

```
cd /opt/resource/hadoop-master/datasource
sh command.sh
```

结果如图 10.5 所示。

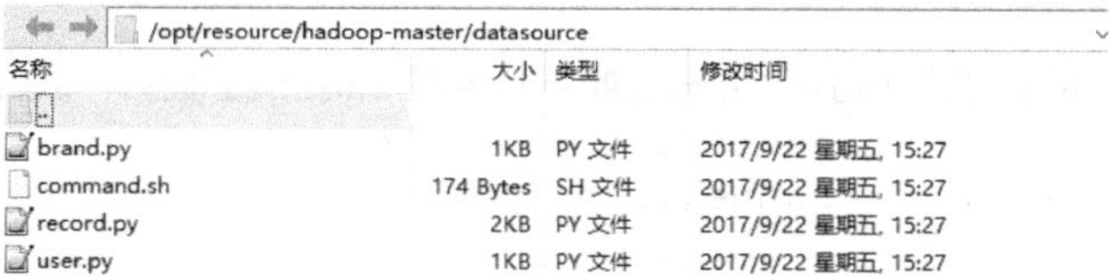

图 10.5 生成的数据文件

```
[root@master datasource]# sh command.sh
start to generate transaction data...
```

运行生成数据的脚本，这时会形成 record.list 文件，并会持续追加数据到文件中，执行 ll 命令，发现文件已经生成。

```
[root@master loganalysis]# ll
total 64
- rw-r--r--.1 root root 36947 Sep 22 16:47 record.list
```

10.7.2　编辑配置文件 conf4

配置文件 conf4 如下：

```
logAgent.sources = logSource
logAgent.channels = fileChannel
logAgent.sinks = hdfsSink
#指定 Source 的类型是 exec
logAgent.sources.logSource.type = exec
#指定命令是 tial -F,持续监测/opt/data/loganalysis/record.list 中的数据
logAgent.sources.logSource.command = tail -F /opt/data/loganalysis/record.list

# 将 Channel 设置为 fileChannel
logAgent.sources.logSource.channels = fileChannel

# 设置 Sink 为 HDFS
logAgent.sinks.hdfsSink.type = hdfs
#文件生成的时间
logAgent.sinks.hdfsSink.hdfs.path = hdfs://master:9000/flume/record/%Y-%m-%d/%H%M
logAgent.sinks.hdfsSink.hdfs.filePrefix= transaction_log
logAgent.sinks.hdfsSink.hdfs.rollInterval= 600
logAgent.sinks.hdfsSink.hdfs.rollCount= 10000
logAgent.sinks.hdfsSink.hdfs.rollSize= 0
logAgent.sinks.hdfsSink.hdfs.round = true
logAgent.sinks.hdfsSink.hdfs.roundValue = 10
logAgent.sinks.hdfsSink.hdfs.roundUnit = minute
logAgent.sinks.hdfsSink.hdfs.fileType = DataStream
logAgent.sinks.hdfsSink.hdfs.useLocalTimeStamp = true
#Specify the channel the sink should use
logAgent.sinks.hdfsSink.channel = fileChannel
```

```
# 设置 Channel 的类型为 file, 并设置断点目录和 channel 数据存放目录
logAgent.channels.fileChannel.type = file
logAgent.channels.fileChannel.checkpointDir=
/opt/software/dataCheckpointDir
logAgent.channels.fileChannel.dataDirs= /opt/software/dataDir
```

10.7.3 运行 Flume

开始运行 Flume,执行命令如下:

```
flume-ng agent --conf-file conf4 --name logAgent -Dflume.root.logger=
INFO,console
```

10.7.4 查看生成的文件

查看生成的文件,这里查看 record.list 文件:

```
[root@master loganalysis]# tail-f record.list
000000028,00000648,00000363,1506078682,281,JiLin,XiZang,JUHUASUAN,
630407715931,EMS, 240.175.115. 210, eo
000000029,00000167,00000210,1506078682,604,XiZang,NeiMengGu,TAOBAO,
869945894191333,ZHONGTONG, 63.43.250.156,oc000000030,00000312,00000797,
1506078682,432,TianJin,JiangSu,JUHUASUAN, 639090853859,YUANTONG,233.19.
130.189,pl
000000031,00000545,00000856,1506078684,304,QingHai,JiangXi,TIANMAOCHAOSHI,
3112127795095553,EMS, 85.228.163.127,br000000032,00000563,00000654,
1506078684,694,GuangXi,XiangGang,JUHUASUAN,3112088517312769,SHUNFENG 116.
21.136.2,zh
000000033,00000864,00000137,1506078686,107,NeiMengGu,ShangHai,JUHUASUAN,
5590557721235400,EMS,97.0.74.148,th000000034,00000925,00000800,1506078686,
892,HeNan,JiangSu,TAOBA0,3528046691385801,YUANTONG,147.30.250.35.gl
000000035,00000896,00000387,1506078686,124,ShanXi3,LiaoNing,TIANMA0,
3096688681772746,SHENTONG,180.118.166.214,tt
000000036,00000154.00000137,1506078686,402,HeBei,JiLin,TAOBAO,601112389
9895233,SHENTONG,236.157.180.130, ia
000000037,00000381,00000131,1506078686,651,GuangXi,GuangDong,TIANMAOCHA
OSHI,4561742855445,YUNDA,244.158.206.146,bhb000000038,00000819,00000423
,1506078686,661,GuiZhou,GuangDong, TIANMAOCHAOSHI,4286097066931,YUNDA,
157.8.171.204,om
000000039,00000248,00000903,1506078686,90,JiangXi,XinJiang.TIANMA,
6011388767453668,YUNDA,66.191.134.104,my
000000040,00000024,00000321,1506078686,717,HeBei,HuBei,TIANMAOCHAOSHI,
869982871579289,ZHONGTONG,151.191.47.30,1g
000000041,00000528,00000090,1506078686,969,NingXia,ShanXi1,TIANMAOCHAOSHI,
3158174217237137,ZHONGTONG,116.159.183.68,ur
000000042,00000881,00000591,1506078686,186,XiangGang,HeiLongJiang,
JUHUASUAN,869934148863327.YUANTONG.68.23.103.130,uk
```

```
000000043,00000360,00000288,1506078686,815,GuangXi,HuNan,TIANMAOCHAOSHI,
374930886095625,SHENTONG,108.76.153.118,pl
000000044,00000386,00000041,1506078686,379,TaiWan,JiangXi,TAOBAO,
4015181734504,SHUNFENG,250.214.23.215,sq
```

10.7.5 查看 HDFS 中的数据

查看 HDFS 中生成的数据，HDFS 生成的文件如下，比如 2017-09-22/1910 代表是 2017 年 9 月 22 日 19 点 10 分生成的数据。

```
[root@master opt]# hdfs dfs-ls /flume
Found 2 items
drwxr-xr-x -root supergroup  0 2017-09-22 14:16 /flume/data
drwxr-xr-x -root supergroup  0 2017-09-22 19:14 /flume/record
[root@master opt]# hdfs dfs-ls /flume/record
Found litems
drwxr-xr-x  root supergroup  0 2017-09-22 19:14 /flume/record/2017-09-22
[root@master opt]# hdfs dfs-ls /flume/record/2017-09-22
Found 1 items
drwxr-xr-x  root supergroup  0 2017-09-22 19:17 /flume/record/2017-09-22/1910
[root@master opt]# hdfs dfs-ls /flume/record/2017-09-22/1910
Found 26 items
-rw-r--r--3rootsupergroup63172017-09-2219:14/flume/record/2017-09-22/1910/transaction_log.1506078870217
-rw-r--r--3rootsupergroup17832017-09-2219:14/flume/record/2017-09-22/1910/transaction_log.1506078870218
-rw-r--r--3rootsupergroup21982017-09-2219:14/flume/record/2017-09-22/1910/transaction_log.1506078870219
-rw-r--r--3rootsupergroup18892017-09-2219:14/flume/record/2017-09-22/1910/transaction_log.1506078870220
-rw-r--r--3rootsupergroup23132017-09-2219:14/flume/record/2017-09-22/1910/transaction_log.1506078870221
-rw-r--r--3rootsupergroup18502017-09-2219:14/flume/record/2017-09-22/1910/transaction_log1506078870222
-rw-r--r--3rootsupergroup22202017-09-2219:14/flume/record/2017-09-22/1910/transaction_log.1506078870223
-rw-r--r--3rootsupergroup18682017-09-2219:15/flume/record/2017-09-22/1910/transaction_log.1506078870224
-rw-r--r--3rootsupergroup23072017-09-2219:15/flume/record/2017-09-22/1910/transaction_log.1506078870225
-rw-r--r--3rootsupergroup17572017-09-2219:15/flume/record/2017-09-22/1910/transaction_log1506078870226
-rw-r--r--3rootsupergroup22232017-09-2219:15/flume/record/2017-09-22/1910/transaction_log.1506078870227
-rw-r--r--3rootsupergroup18772017-09-2219:15/flume/record/2017-09-22/1910/transaction_log.1506078870228
-rw-r--r--3rootsupergroup22112017-09-2219:15/flume/record/2017-09-22/1910/transaction_log.1506078870229
-rw-r--r--3rootsupergroup41102017-09-2219:15/flume/record/2017-09-22/
```

```
1910/transaction_log.1506078870230
-rw-r--r--3rootsupergroup18862017-09-2219:15/flume/record/2017-09-22/
1910/transaction_log.1506078870231
-rw-r--r--3rootsupergroup23172017-09-2219:15/flume/record/2017-09-22/
1910/transaction_log.1506078870232
-rw-r--r--3rootsupergroup17892017-09-2219:15/flume/record/2017-09-22/
1910/transaction_log.1506078870233
```

10.8 小　　结

本章主要通过实例学习了 Flume 及其特点、架构，以及主要组件 Event、Client、Agent、Source、Channel、Sink 的作用。同时又通过实例介绍了基于 Flume 实现本地数据读取、日期分区的数据收集和自动收集数据等方法。

第 11 章　Hadoop 和关系型数据库间的数据传输工具——Sqoop

数据采集是大数据项目中的一个重要步骤，关于数据清洗和数据分析等操作都是基于数据采集。目前的数据采集主要分为结构化和非结构化数据，采集的方式也略有区别，第 10 章中我们学习了 Flume，本章主要讨论收集结构化数据的 Sqoop。

当我们需要将数据从外部结构化数据存储导入 Hadoop 分布式文件系统 HDFS 或 Hive 时，可以使用 Sqoop 来完成。

本章主要涉及的知识点如下：
- Sqoop 架构和安装配置。
- 通过实例介绍 Sqoop 数据导入和导出操作。

11.1　什么是 Sqoop

Sqoop 是一种用于在 Hadoop 和结构化数据存储（如关系数据库）之间高效传输大批量数据的工具。它允许用户将数据从结构化存储器抽取到 Hadoop 中，用于进一步处理。

例如，可以利用 Sqoop 从关系型数据库管理系统（如 MySQL 或者 Oracle 或者主机）向 Hadoop 分布式文件系统（HDFS）中导入数据，可以通过 MapReduce 处理这些数据，同时这些数据也可以被 Hive 使用，甚至可以使用 Sqoop 将数据从数据库转移到 HBase 上。同时 Sqoop 可以将数据从 Hadoop 中（如 Hive 中）导入到关系型数据库中，Sqoop 会基于配置文件，完成数据的导入和导出工作。

11.2　Sqoop 工作机制

Sqoop 的底层原理本质上是 MapReduce 任务。Sqoop 是通过一个 MapReduce 作业从数据库中导入一个表，这个作业从表中逐行抽取数据，接着将一行行的数据写入 HDFS。

那么底层是如何实现的呢？下面我们具体介绍 Sqoop 如何将数据从数据库中导入到 HDFS 中，以及从 HDFS 中导入到数据库中的过程，以此来更好地了解 Sqoop 的工作机制。如图 11.1 所示为 Sqoop 将数据从关系型数据库导入到 HDFS 中的原理图。

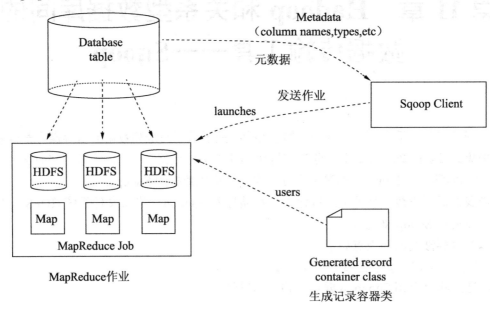

图 11.1　Sqoop 数据导入机制

Sqoop 的底层是 Java，Java 提供了 JDBC API，通过 JDBC API，应用程序可以访问存储在关系型数据库中的数据。Sqoop 导入、导出数据时都需要用到 JDBC。在导入之前，Sqoop 会通过 JDBC 查询出表中的列和列的类型，同时这些类型会与 Java 的数据类型相匹配，而底层运行的 MapReduce 会根据这些 Java 类型来保存字段对应的值。

在导入过程中，有两个 DBWritable 接口的序列化方法，通过这两个方法可以实现对 JDBC 的交互，其中 readFields()方法将 ResultSet 中的数据填充到对象的字段中。

方法一：
```
Public void readFields(resultSet _dbResults)throws SQLException;
```
方法二：
```
Public void write(PreparedStatement _dbstmt)throws SQLException
```

在导入过程中，Sqoop 会启动相应的 MapReduce 作业，而 MapReduce 作业会通过 InputFormat 以 JDBC 的方式从数据库中读取数据，Hadoop 中的 DataDrivenDBInputFormat 能够为不同的 Map 任务将查询结果进行划分。

在生成反序列化代码和配置了 InputFormat 之后，Sqoop 就会将作业发到 MapReduce 集群，Map 任务会把查询到的 ResultSet 数据填充到类的实例中。在导入数据时，如果不想取出全部数据，可以通过类似于 where 的语句进行限制，如图 11.2 所示。

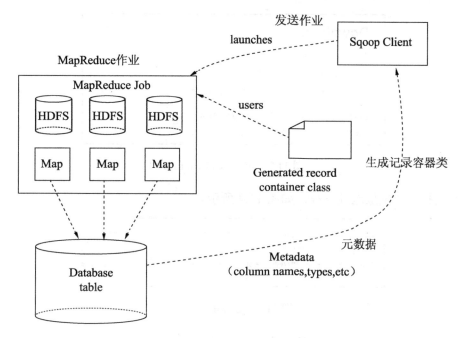

图 11.2　Sqoop 数据导出机制

　　Sqoop 的导出通常是将 HDFS 的数据导入到关系型数据库中，关系型数据库中的表必须提前创建好。在底层方面，导出之前，Sqoop 会选择导出方法，通常是 JDBC，接着 Sqoop 会生成一个 Java 类，这个类可以解析文本中的数据，并将相应的值插入表中。在启动 MapReduce 之后，会从 HDFS 中读取并解析数据，同时执行选定的导出方法。

　　基于 JDBC 的导出方法会生成多条 insert 语句，每条 insert 语句均会向表中插入多条数据，同时为了确保不同 I/O 的操作可以并行执行，在从 HDFS 读取数据并与数据库通信时，是多个单独线程同时运行的。

11.3　Sqoop 的安装与配置

　　在介绍了 Sqoop 及 Sqoop 的底层运行原理之后，我们开始进行 Sqoop 的安装与配置。Sqoop 的安装与配置比较简单，主要是在配置文件中定义数据源和导出目标，下面是安装与配置步骤。

11.3.1　下载 Sqoop

　　（1）访问 Sqoop 官网 http://sqoop.apache.org，然后单击 nearby mirror 下载镜像，如图 11.3 所示。

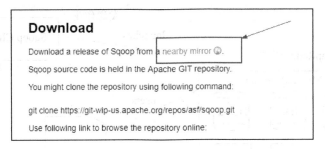

图 11.3 单击 nearby mirror

（2）选择一个镜像地址下载，如图 11.4 所示。

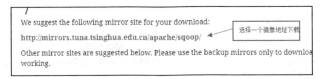

图 11.4 选择镜像地址

（3）选择版本及下载链接，如图 11.5 和图 11.6 所示。

图 11.5 选择版本

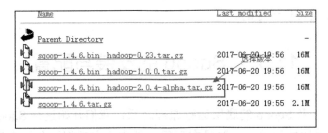

图 11.6 选择要下载的文件

11.3.2 Sqoop 配置

在下载完后，就可以进行配置了，Sqoop 的核心配置文件是 sqoop-env.sh 文件，配置内容如下：

```
export HADOOP_COMMON_HOME=/opt/software/hadoop-2.5.1
export HADOOP_MAPRED_HOME=/opt/software/hadoop-2.5.1

# Set Hadoop-specific environment variables here.

#Set path  to where bin/hadoop is available
export HADOOP_COMMON_HOME=/opt/software/hadoop-2 5.1

#Set path to where hadoop-*-core.jar is available
export HADOOP_MAPRED_HOME=/opt/software/hadoop-2.5.1

#set the path to where bin/hbase is available
#export HBASE_HOME=
```

同时由于 Sqoop 要用到 JDBC 链接库，因此需要把 MySQL 的驱动复制到 Sqoop 的 lib 目录下面，否则会报错。

11.4　Sqoop 数据导入实例

配置完以后，就可以进行数据的导入和导出了。Sqoop 可以在 HDFS/Hive 和关系型数据库之间进行数据的导入导出，其中主要使用了 import 和 export 这两个命令。

import 命令用来将关系型数据库中的表导入到 HDFS 或者 Hive 中，表中的每一行在 HDFS 中被表示为分开的记录，记录可以被存储为 txt 文件，或者二进制形式的 Avro 和 SequenceFiles。

接下来将通过实例分别实现使用 import 命令将数据导入 HDFS 和 Hive 中。由于数据的导入和导出操作涉及关系型数据库 MySQL，所以首先要先打开 MySQL 服务。代码如下：

```
[root@master~]# service mysqld start
Starting MySQLd :                              [OK]
[root@master~]# service mysqld status
MySQLd (pid 1172) is  running...
[root@master~]# 。
```

然后登录 MySQL。

```
[root@master~]# mysql -u root -p
Enter password:
Reading table information for completion of table and column names
You can  turn off this feature to get a quicker startup with-A

Welcome to the MySQL monitor. Commands end with ;or \g.
Your MySQL connection id is  45
Server version: 5.1.73 Source distribution

Copyright (c) 2000,2013,Oracle and/or its affiliates.All rights reserved.
Oracle is a registered trademark of Oracle Corporation and/or its
affiliates.Other names may be trademarks of their respective
owners.
```

Type'help; 'or'\h' for help.Type'\c' to clear the current input statement.
MySQL>

查看所有的数据库：

```
mysql>show databases
+-------------------+
| Database          |
+-------------------+
| information schemahive |
| mysql             |
| test              |
| rows in set (0.01sec) |
+-------------------+
```

使用 test 数据库：

mysql > use test;
Reading tabte Info matlon for completion of table and column names
You can turn off this feature to get a quicker startup with-A

Database chanqed

在数据库中创建表，并插入 4 条数据。

mysql> create table t_user(id int,name varchar(20) ,age int) ;
Query OK, 0 rows affected (0.01sec)

mysql> insert into t_user values(1, 'rod',20);
Query OK,1 row affected (0.00 sec)

mysql> insert into t_user values(2,'tom' ,21);
Query OK,1row affected (0.00 sec)

mysql> insert into t_user values(3,'lucy' ,22);
Query OK,1 row affected (0.00 sec)

mysql> insert into t_user values(4,'jet',23);
Query OK,1 row affected (0.00 sec)
mysql>

查看表结构和表中的数据：

```
mysql> desc t_user;
+--------+------------+------+-----+---------+-------+
|Field   |Type        |NullI |Key  |Default  |Extra  |
+--------+------------+------+-----+---------+-------+
| id     |int(11)     | YES  |NULL |         |       |
| name   |varchar(20) | YES  |NULL |         |       |
| age    |int(11)     | YES  |NULL |         |       |
+--------+------------+------+-----+---------+-------+
3rows in set (0.00 sec)
mysql> select * from t_user;
+---+----+---+
```

```
|id      | name  | age   |
+- - -+— - -+- - - +
|  1     | rod   | 20    |
|  2     | tom   | 21    |
|  3     | lucy  | 23    |
|  4     | good  | 24    |
|  5     | leo   | 28    |
+- - -+— - -+- - - +
5 rows in set (0.00 sec)
```

在关系型数据库的表和数据准备好后，就可以编写相应实例了。

11.4.1 向 HDFS 中导入数据

首先，在/opt/sqoopconf 目录下创建一个 conf1 文件，在这个文件中配置数据导入的相关信息，其内容如下：

```
import                                  //使用 import 工具
--connect                               //指定链接的目标数据库
jdbc:mysql://master:3306/test
--username                              //数据库的登录用户名
root
--password                              //数据库的登录密码
123456
--table                                 //数据源的表名
t_user
--columns                               //指定要导入的字段
id,name,age
--where                                 //判断条件，这里只导入 ID 大于 0 的记录
id>0
--target-dir                            //指定数据导入到 HDFS 上的位置
hdfs://master:9000/sqoop
--delete-target-dir                     //若目标目录已存在，删除它
-m
1
--as-textfile                           //导入数据以文本格式存放在 HDFS 上
--null-string
```

基于配置文件执行 Sqoop ，命令格式是:sqoop --options-file 文件名，执行命令和执行结果如下：

```
[root@master sqoopconf]# sqoop --options-file conf1
Warning: /opt/software/sqoop-1.4.6.bin__hadoop-2.0.4-alpha/bin/../../
hbase does
not exist! HB
ase imports will fail.
17/09/22 23:17:42 INFO sqoop.Sqoop: Running Sqoop version: 1.4.6
17/09/22 23:17:42 WARN tool.BaseSqoopTool: Setting your password on the
command-line is insecure.Consider using-P instead.
17/09/22 23:17:43 INFO manager.MySQLManager: Preparing to use a MySQL
streaming resultset.
```

```
17/09/22 23:17:43 INFO tool.CodeGenTool: Beginning code generation
17/09/22 23:17:43 INFO manager.SqlManager: Executing SQL statement: SELECT
t.* FROM、t._user'AS t LIMIT 1
17/09/22 23:17:43 INFO manager.SqlManager: Executing SQL statement: SELECT
t.* FROM、t._userAS t LIMIT 1
17/09/22 23:17:43 INFO orm.CompilationManager: HADOOP_MAPRED_HOME is
/opt/software/hadoop-2.5.1
Note: /tmp/sqoop- root/compile/8ffb4c4d3918756c64f573322c920ef6/t_user.
java uses or overridesa deprecated API.
Note: Recompile with-Xlint:deprecation for details.
17/09/22 23:17:46 INFO orm.CompilationManager: Writing jar file: /tmp/
sqoop- root/compile/8ffb4c4d3918756c64f573322c920ef6/t_user.jar
17/09/22 23:17:48 INFO tool.ImportTool: Destination directory
hdfs://master:9000/sqoop is not present,hence not deleting.


        FILE: Number of write operations=0
        HDFS: Number of bytes read=87
        HDFS: Number of bytes written=47
        HDFS: Number of read operations=4
        HDFS: Number of large read operations=0
        HDFS: Number of write operations=2
Job Counters
        Launched map tasks=1
        Other local map tasks=1
        Total time spent by all maps in occupied slots (ms )=5366
        Total time spent by all reduces in occupied slots (ms)=0
        Total time spent by all map tasks (ms)=5366
        Total vcore-seconds taken by all map tasks=5366
        Total megabyte-seconds taken by all map tasks=5494784
    Map-Reduce Framework
        Map input records=5
        Map output records=5
        Input split bytes=87
        Spilled Records=0
        Failed Shuffles=0
        Merged Map outputs=0
        GC time elapsed (ms )=132
        CPU time spent (ms )=1270
        Physical memory (bytes) snapshot=113975296
        Virtual memory (bytes) snapshot=866390016
        Total committed heap usage (bytes)=15597568
    File Input Format Counters
        Bytes Read=0
    File Output Fomat Counters
        Bytes Written=47
17/09/22 23:18:14 INFO mapreduce.ImportJobBase: Transferred 47 bytes in
26.2846 seconds (1.7881bytes/sec )17/09/22 23:18:14 INFO mapreduce.
ImportJobBase: Retrieved 5 records.
```

根据上述运行结果读者会发现，在基于 **Sqoop** 执行时，底层是运行的 **MapReduce**，运行完后，就可以查看导入结果了。比如，基于浏览器，登录 HDFS 查看 Sqoop 目录，结果如图 11.7 所示，说明数据已经导入成功。

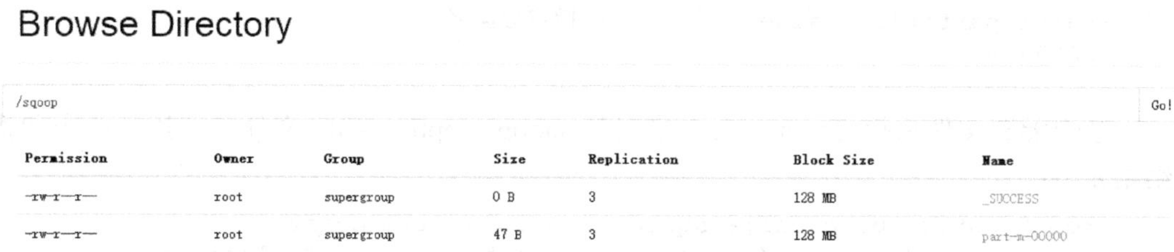

Browse Directory

Permission	Owner	Group	Size	Replication	Block Size	Name
-rw-r--r--	root	supergroup	0 B	3	128 MB	_SUCCESS
-rw-r--r--	root	supergroup	47 B	3	128 MB	part-m-00000

图 11.7　数据导入成功

11.4.2　将数据导入 Hive

前面介绍了导入数据到 HDFS 中的过程，导入数据到 Hive 的过程也与之类似，首先在目录/opt/sqoopconf 下创建 conf2 文件，在这个文件中保存将数据导入 Hive 的配置信息，内容如下：

```
import                                //使用 import 工具
--connect                             //指定链接的目标数据库
jdbc:mysql://master:3306/test
--username                            //数据库的登录用户名
root
--password                            //数据库的登录密码
123456
--table                               //数据源的表名
t_user
--columns                             //指定要导入的字段
id,name,age
--where                               //判断条件，这里只导入 ID 大于 0 的记录
id>0
--target-dir                          //指定数据导入到 HDFS 上的位置
hdfs://master:9000/sqoop2
--delete-target-dir                   //若目标目录已存在，删除它
-m
1
--as-textfile                         //导入的数据以文本格式存放在 HDFS 上
--null-string

--hive-import                         //向 Hive 中导入数据
--hive-overwrite                      //如数据已存在则覆盖它
--create-hive-table                   //创建 Hive 表
--hive-table                          //指定表名
t_user
--hive-partition-key                  //指定分区字段
dt
```

```
--hive-partition-value              //指定分区名
'2017-05-19'
```

基于配置文件执行 Sqoop，命令格式是:sqoop --options-file 文件名，执行命令和执行结果如下：

```
[root@master sqoopconf]# sqoop --options-file conf2
warning:/opt/software/sqoop-1.4.b.b1n_hadoop-2.0.4-alpha/bin/../../hbasedoesnotexist!Hbaseimports will fall.
17/09/22 23:59:06 INFO sqoop.Sqoop: Running Sqoop version: 1.4.6
17/09/2223:59:06WARNtool.BaseSqoopTool:Settingyourpasswordonthecommand-lineisinsecure.Consider using-P instead.
17/09/22 23:59:06INFOtool.BaseSqoopTool :UsingHive-specificdelimiters foroutput.Youcanoverride
17/09/22 23:59:06 INFO tool.BaseSqoopTool :delimiters with--fields-terminated-by,etc.
17/09/22 23:59:07 INFO manager.MySQLManager: Preparing to use a MySQL streaming resultset.
17/09/22 23:59:07 INFO tool.CodeGenTool :Beginning code generation
17/09/2223:59:07INFOmanager.SqlManager:ExecutingSQLstatement:SELECTt.*FROM't.user'AStLIMIT 1
17/09/2223:59:07INFOmanager.SqlManager:ExecutingSQLstatement:SELECTt.*FROM't_user'AStLIMIT 1
17/09/2223:59:07INFOorm.CompilationManager:HADOOP_MAPRED_HOMEis/opt/software/hadoop-2.5.1
Note:/tmp/sqoop-root/compile/3b3c6746c3clee9ae8225d9628855b5/t_user.javausesoroverridesadeprecated API.
Note: Recompile with-Xlint :deprecation for details.
17/09/2223:59:09INFOorm.CompilationManager:Writingjarfile:/tmp/sqoop-root/compile/3b3c6746c3clee9ae8225d9628855ab5/t_user.
17/09/2223:59:11INFOtool.ImportTool:Destinationdirectoryhdfs ://master:900/sqoop2isnotpresent,hence not deleting.
17/09/22 23:59:11WARN manager.MySQLManager: It looks like you are importing from mysql.

17/09/22 23:59:40 INFO mapreduce.Job: Job job_1505633089581_0002 completed successfully
17/09/22 23:59:40 INFO mapreduce.Job: Counters: 30
        File System Counters
        FILE: Number of bytes read=0
        FILE :Number of bytes written=115911
        FILE :Number of read operations=0
        FILE: Number of large read operations=0
        FILE: Number of write operations=0
        HDFS: Number of bytes read=87
        HDFS: Number of bytes written=47
        HDFS: Number of read operations=4
        HDFS: Number of large read operations=0
        HDFS: Number of write operations=2
Job Counters
        Launched map tasks=1
        Other local map tasks=1
        Total time spent by all maps in occupied slots (ms)=10420
```

```
                Total time spent by all reduces in occupied slots (ms)=0
                Total time spent by all map tasks (ms)=10420
                Total vcore-seconds taken by all map tasks=10420
                Total megabyte-seconds taken by all map tasks=10670080
        Map-Reduce Framework
                Map input records=5
                Map output records=5
                Input split bytes=87
                Spilled Records=0
                Failed Shuffles=0
                Merged Map outputs=0
                GC time elapsed (ms)=72
                CPU time spent (ms)=1220
                Physical memory (bytes) snapshot=115281920
                Virtual memory (bytes) snapshot=863027200
                Total committed heap usage (bytes)=15597568
            File Output Fomat Counters
                    Bytes Written=47
17/09/22 23:59:40 INFO mapreduce.ImportJobBase:Transferred 47 bytes in 29.4639
seconds (1.5952 bytes/sec) 17/09/22 23:59:41 INFO mapreduce.ImportJobBase:
Retrieved 5 records.
17/09/22 23:59:41 INFO manager.SqlManager:Executing SQL statement:SELECT t.
*FROM 't_user' AS t LIMIT 1 17/09/22 23:59:41 INFO hive.HiveImport: Loading
uploaded data
into Hive
17/09/22 23:59:44 INFO hive.HiveImport: 17/09/22 23:59;44 WARN conf.HiveConf:
HiveConf of name hive.metastore.local does not exist
17/09/22 23:59:44 INFO hive.HiveImport:
17/09/22 23:59:44 INFO hive.HiveImport:Logging initialized using configuratio
n in jar:file:/opt/software/apache-hive-common-1.2.1jar!/hive-log4j.pro
17/09/22 23:59:57 INFO hive.HiveImport: OK
17/09/22 23:59:57 INFO hive.HiveImport: Time taken: 2.608 seconds
17/09/22 23:59:57 INFO hive.HiveImport: Loading data to table default.t_
userpartition(dt=2017-05-19)
17/09/22 23:59:57 INFO hive.HiveImport:Partition default.t_user{dt=2017-05-
19}stats:[numFiles=1,numRows=0,totalSize=47,rawDataSize=0]
17/09/22 23:59:58 INFO hive.HiveImport: OK
17/09/22 23:59:58 INFO hive.HiveImport: Time taken: 1.066 seconds
17/09/22 23:59:58 INFO hive.HiveImport: Hive import complete.
17/09/22 23:59:58 INFO hive.HiveImport:Export directory is contains the_
SUCCESS file only,removing the directory.
```

导入完成之后，即可进入 Hive Shell 检查表结构和表内数据：

```
[root@master sqoopconf]# hive shell
17/09/23 00:03:58 WARN conf.HiveConf: HiveConf of name hive.metastore.local
does not exist

Logging initialized using configuration in jar:file:/opt/softwareapache-hive
1.2.1-bin/lib/hive-common-2.1.1.jar!hive-log4j.properties
hive> show tables;
OK
t_user
Time taken: 1.594 seconds,Fetched: 1row(s)
hive> select * from t_user
1       rod    20   2017- 05-19
```

```
2         tom     21   2017-05-19
3         lucy    23   2017-05-19
4         good    24   2017-05-19
5         leo     28   2017-05-19
Time taken :0.815 seconds,Fetched: 5 row(s)
hives>desc t_user;
OK
id                      int
name                    string
age                     int
dt                      string

# Partition Infomation
# col._name  data_ type    comment

dt              string
Time taken: 0.169 seconds,Fetched: 9 row(s)
```

以上结果说明数据已经导入 Hive 中。

11.4.3 向 HDFS 中导入查询结果

除了前面介绍的导入数据的方式之外，Sqoop 也支持导入 SQL 查询的结果集，可以用 --query 参数指定 SQL 语句的方式导入数据。在导入查询的时候，需要用--target-dir 参数指定目的文件。

如果想并行地导入查询结果，每个 Map 需要执行一个查询副本，查询必须要有一个 $CONDITIONS 符号，表示每个 Sqoop 进程被唯一的条件语句替换，同时必须要用--split-by 参数选择一个列来进行划分，例如：

```
$ sqoop import \
  --query 'SELECT a.*, b.* FROM a JOIN b on(a.id == b.id) WHERE $CONDITIONS' \
  --split-by a.id --target-dir /user/zoo/joinresults
```

或者，通过-m 1 参数指定单个 Map 来确保查询仅执行一次且被连续导入：

```
$ sqoop import \
  --query 'SELECT a.*, b.* FROM a JOIN b on (a.id == b.id) WHERE $CONDITIONS' \
  -m 1 --target-dir /user/zoo/joinresults
```

接下来，我们看一个基于 SQL 查询结果集导入 HDFS 的例子。

（1）在/opt/sqoopconf 目录下创建 conf 3 文件，内容如下：

```
import
--connect
jdbc:mysql://master:3306/test
--username
root
--password
```

```
123456
--target-dir
hdfs://master:9000/sqoop1
--delete-target-dir
-m
1
--as-textfile
--null-string
''
--query
"select p.id,p.name,p.age,c.id as cid,date_format(c.create_date,'%Y-%m-
%d') from t_user p join t_id_card c on p.id=c.p_id where p.name is not null
and $CONDITIONS"
```

(2) 在 MySQL 中创建相应的表和数据：

```
mysql -u root -p                              //登陆 MySQL,
Enter password:                               //登录密码
MySQL> show databases;
MySQL> use test;                              //把表建在 test 中
MySQL> show tables;
```

创建 t_user 表，并插入数据：

```
MySQL> create table t_user(id int, name varchar(20),age int);
                                              //创建名为 t_user 的表
MySQL> insert into t_user values(1,'rod',20);  //在 t_user 中插入数据
MySQL> insert into t_user values(2,'tom',21);
MySQL> desc t_user;                           //查看表结构
MySQL> select * from t_user;                  //查看表内容
/********************************************************/
```

创建 t_id_card 表：

```
MySQL> create table t_id_card(id int,p_id int,create_date date);
                                              //创建名为 t_id_card 的表
MySQL> insert into t_id_card values(001,1,'2012-03-11');
                                              //在 t_id_card 中插入数据
MySQL> insert into t_id_card values(002,2,'2013-05-13');
MySQL> insert into t_id_card values(003,3,'2014-03-15');
MySQL> insert into t_id_card values(004,4,'2015-06-27');
MySQL> insert into t_id_card values(005,5,'2016-01-23');
MySQL> select * from t_user;                  //查看表内容
MySQL> exit                                   //退出
```

(3) 在/opt/sqoopconf 下执行 sqoop --options-file optfile4。

(4) 查看结果：

```
[root@master sqoopconf]# hdfs dfs -ls /sqoop1/
Found 2 items
- rw-r--r--   3 root supergroup          0 2017-09-24 10:40 /sqoop1/_SUCCESS
- rw-r--r--   3 root supergroup        112 2017-09-24 10:40 /sqoop1/part-m-00000
```

```
[root@master sqoopconf]# hdfs  dfs -ls /sqoopl/part-m-00000
- rw-r--r--  3 root supergroup  112 2017-09-24 10:40 /sqoopl/part-m-00000
[root@master sqoopconf]#hdf dfs -cat /sqoopl/part-m-00000
1,rod,20,1,2012-03-11
2,tom,21,2,2013-05-13
3,lucy,23,3,2014-03-15
4,good,24,4,2015-06-27
5,leo,28,5,2016-01-23
```

根据查询结果，说明操作成功。

11.5　Sqoop 数据导出实例

前面介绍了如何将数据从关系型数据库中导入 HDFS 或者 Hive 中，接下来介绍如何将数据从 HDFS 中导出到关系型数据库中。

数据导出操作可以用 export 命令，在执行数据导出之前，数据库中必须已经存在要导入的目标表。在导出过程中，HDFS 或者 Hive 上的文件会根据用户指定的分隔符被读取解析并写入到 MySQL 相应的表中。

接下来我们看一个具体的实例，这个实例中，将 HDFS 上的数据根据配置文件导出到 MySQL 数据库的 t_user 表中。

（1）在/opt/sqoopconf 下编辑 conf4 文件，内容如下：

```
export
--connect
jdbc:mysql://master:3306/test
--username
root
--password
123456
--table
t_user
--columns
id,name,age
--export-dir
hdfs://master:9000/sqoop
-m
1
```

（2）在/opt/sqoopconf 下执行 sqoop --options-file conf4 命令。

执行完之后，通过 select * from t_user 查看结果，结果如下。

```
+------+------+------+
|id  |name| age |
+------+------+------+
|1   |rod |20   |
|2   |tom |21   |
|3   |lucy|23   |
|4   |good|24   |
|5   |leo |28   |
```

```
+------+------+------+
5 rows in set (0.00 sec)
```

根据以上结果,说明数据导出成功。

11.6 小　　结

本章主要介绍了数据采集工具 Sqoop,包括 Sqoop 的安装和配置,以及如何将关系型数据库中的数据导入 HDFS 或者 Hive 中。归纳起来说,Sqoop 的主要作用就是从关系型数据库中采集数据。

第 12 章 分布式消息队列——Kafka

前面我们讲过基于 MapReduce 可以处理大批量的离线数据（往往是历史数据）。如果项目需要在线实时处理，比如在线商城，实时处理生成的订单，在这种情况下就可以使用 Kafka。Kafka 的一个典型应用是结合 Storm 等流式处理系统进行数据的在线处理。

本章主要涉及的知识点如下：
- Kafka 的基本概念及架构。
- Kafka 中的主要组件和安装配置。

12.1 什么是 Kafka

Kafka 是 Apache 开发的一个开源流处理平台，它是一个可持久化的分布式的消息队列，最初由 LinkedIn 公司开发，是一个分布式、分区的、多副本的、多订阅者，基于 ZooKeeper 协调的分布式日志系统（也可以当做 MQ 系统），常见可以用于 Web/Nginx 日志、访问日志和消息服务等。

Kafka 和 Flume 都能处理日志消息，但它们主要有以下区别：
- Kafka 是一个通用的系统，它可以有多个生产者和多个消费者共享多个 Topic。
- Flume 被设计为向 HDFS 和 HBase 发送数据，它对 HDFS 有特殊的优化。
- 如果数据被设计给 Hadoop 使用，那么应使用 Flume；如果是在流式处理系统，如 Storm 中，则可以使用 Kafka。

12.2 Kafka 的架构和主要组件

下面我们来介绍 Kafka 的各个组件和架构。先来看 Kafka 的架构，如图 12.1 所示。

Kafka 的组件主要有：Producer（消息生产者）、Consumer（消息消费者）、Broker（Kafka 集群的 Server，负责处理消息读、写请求，存储消息）、Topic（消息队列/分类）、Partition、Offset 和 Segment。下面对 Kafka 的各个组件进行详细说明。

第 12 章 分布式消息队列——Kafka

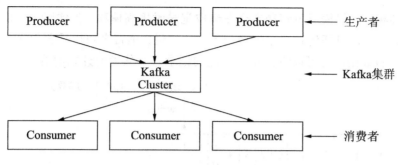

图 12.1 Kafka 的架构

12.2.1 消息记录的类别名——Topic

Topic 是发布记录的类别或名称。Kafka 的 Topic 是多用户的,即一个 Topic 可以有 0 个、1 个或多个消费者订阅写入 Topic 的数据。

对于每个 Topic,在物理上会被分成 1 个或多个 Parition,即分区,如图 12.2 所示。

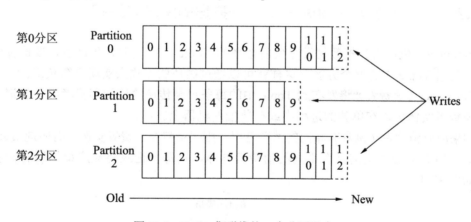

图 12.2 Kafka 集群维护一个分区日志

每个分区是一个有序的、不可变的记录序列,不断附加到一个结构化的提交日志中。每个分区中的记录都被分配一个顺序的 ID 号,称为唯一标识分区中每个记录的偏移量,如图 12.3 所示。

不论写入的数据是否已被消费,Kafka 集群使用一个设置好的保存期限保留所有已发布的记录。例如,如果保留策略设置为三天,则在发布记录后的三天内,该策略可用于消费,之后将以丢弃的方式来释放空间。Kafka 的性能在数据大小方面是不敏感的,因此长时间存储数据并不成问题。

实际上,在每个消费者保留的唯一的元数据是消费者在日志中的偏移位置。这个偏移量由消费者控制：一般情况下消费者会在读取记录时提高偏移，但实际上，由于位置由消费者控制的，它可以以任何顺序来消费记录。比如说，消费者可以重置为之前的偏移量来重新处理过去的数据，也可以直接跳到最近的记录，并从当前开始消费。

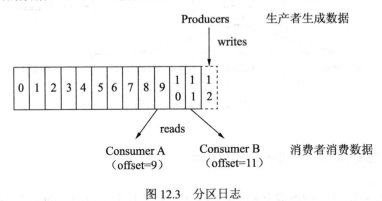

图 12.3　分区日志

所有的这些特征都意味着 Kafka 消费者非常"廉价"，因为他们的来去对集群没有大的影响，对于其他消费者也是这样。

12.2.2　Producer 与 Consumer——数据的生产和消费

Producer（生产者）把数据发布到所选择的 Topic 上，同时，生产者需要选择将哪个记录分配给 Topic 中的哪个分区，这样就可以通过循环的方式来实现平衡负载。

Consumer（消费者）将发布到 Topic 中的每条记录传递给订阅消费者组中的消费者实例。消费者实例可以在单独的进程中或在单独的机器上。

如果所有消费者实例具有相同的消费者组，则记录将在消费者实例上有效地负载平衡。

如果所有消费者实例具有不同的消费者群体，则每个记录将被广播给所有消费者进程，如图 12.4 所示。

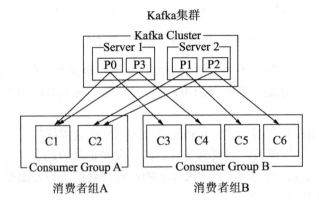

图 12.4　每个记录将被广播给所有消费者进程

两个服务器 Kafka 集群托管 4 个分区（P0-P3）与 2 个消费群体。消费者组 A 有 2 个消费者实例，消费者组 B 有 4 个消费者实例。

12.2.3 其他组件——Broker、Partition、Offset、Segment

- Broker：Kafka 节点，一个 Broker 就是一个 Kafka 节点，多个 Broker 组合在一起可以构成一个 Kafka 集群。
- Partition：每个 Partition 是一个有序的队列，它是 Topic 物理上的分组，一个 Topic 可以分成多个 Partition。
- Offset：每个消息在 Partition 中都有一个连续的序列号，这个序列号被称之为 Offset，可以用于 Partition 标识一条消息。
- Segment：一个 Partition 由多个 Segment 组成。

12.3 Kafka 的下载与集群安装

Kafka 下载与集群安装主要有以下步骤：
（1）上传包并安装。
（2）配置 config/server.properties。
（3）将 Kafka 发送到子节点（slave）上，并修改配置。
（4）启动 Kafka。
（5）创建一个 Topic。
（6）创建 Producer 和 Consumer，并进行测试。
在搭建 Kafka 集群之前，请保证 ZooKeeper 已安装并启动。
启动 ZooKeeper 指令：

```
zkServer.sh start
zkServer.sh status
```

其中，ZooKeeper 集群的部署如下：

```
clientPort=2181
server.1=master:2888:3888
server.2=slave1:2888:3888
server.3=slave2:2888:3888
```

12.3.1 安装包的下载与解压

首先从官网下载 Kafka 安装包，下载链接如下：

http://mirrors.tuna.tsinghua.edu.cn/apache/kafka/0.10.1.1/kafka_2.11-0.10.1.1.tgz

上传至/opt/software 目录下（根据自己的软件存放目录），输入解压命令如下：

```
tar -zxvf kafka_2.11-0.11.0.0.tgz
```

上传安装包操作如图 12.5 所示。

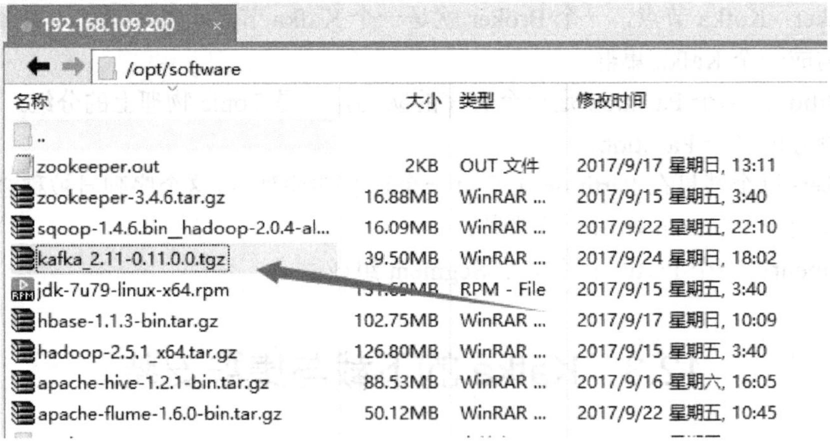

图 12.5　上传至/opt/software 目录下解压

解压 Kafka 安装包的命令和结果如下代码所示：

```
-rw-r--r-- . 1 root root 1762 Sep 17 13:11 ZooKeeper.out
[rootmaster software]# tar -zxvf kafka_2.11-0.11.0.0.tgz
kafka_2.11-0.11.0.0/
kafka_2.11-0.11.0.0/LICENSE
kafka_2.11-0.11.0.0/NOTICE
kafka_2.11-0.11.0.0/bin/
kafka_2.11-0.11.0.0/bin/connect-distributed.sh
kafka_2.11-0.11.0.0/bin/connect-standalone.sh
katka_2.11-0.11.0.6/b1n/connect-standalone.sh
kafka_2.11-0.11.0.0/bin/kafka-acls.sh
kafka_2.11-0.11.0.0/bin/kafka-broker-api-versions.sh
kafka_2.11-0.11.0.0/bin/kafka-configs.sh
kafka_2.11-0.11.0.0/bin/kafka-console-consumer.sh
kafka_2.11-0.11.0.0/bin/kafka-console-producer.sh
kafka_2.11-0.11.0.0/bin/kafka-consumer-groups.sh
kafka_2.11-0.11.0.0/bin/kafka-consumer-offset-checker.sh
kafka_2.11-0.11.0.0/bin/kafka-consumer-nerf-test sh
```

12.3.2　Kafka 的安装配置

下面要修改 Kafka 配置文件，即位于 kafka/config/下的 server.properties 文件，在配置文件中，需要修改以下内容。

1. broker.id设置Kafka节点

这里注意修改每个 Kafka 集群中的 broker.id，从 0 开始设置，表示为一个 Kafka 节点，后面其他节点配置为 1、2、3 等，直至设置好一个 Kafka 集群。

broker.id=0，如下：

```
# The id of the broker.This must be set to a unique integer for each
broker.id=0
```

2. 打开监听端口listeners

打开监听端口 Listeners，指令如下：

```
listeners=PLAINTEXT://master:9092
```

在配置文件中，修改内容如下：

```
# listeners = PLAINTEXT://your.host.name:9092
listeners=PLAINTEXT://master:9092
```

3. 修改log的目录，在指定的位置创建logs文件夹

```
log.dirs=/usr/local/kafka/logs
```

在配置文件中，修改内容如下：

```
# A comma separated lis of directories under which to store
log files
log.dirs=/usr/local/kafka/logs
```

4. 配置ZooKeeper集群的地址

配置 ZooKeeper 集群的地址，指令如下：

```
ZooKeeper.connect=master:2181,slave:2181,slave2:2181
```

在配置文件中，修改内容如下：

```
# root directory for all kafka znodes.
ZooKeeper.connect=master:2181,slave1:2181,slave2:2181
```

5. 将Kafka发送到子节点（如slave）上，并修改配置

master 节点中：

```
scp -r kafka_2.11-0.11.0.0 root@slave1:/opt/soft/
scp -r kafka_2.11-0.11.0.0 root@slave2:/opt/soft/
```

slave1，slave2 节点修改 server.properties。

6. 在主节点和从节点上启动Kafka

在/opt/software/kafka_2.11-0.11.0.0 路径下输入以下命令：

```
./ bin/kafka-server-start.sh config/server.properties
```

后台启动:

```
./bin/kafka-server-start.sh config/server.properties 1>/dev/null 2>&1 &
```

7. 测试

在主节点和从节点中创建一个名为 test 的 Topic 测试。

在主节点/opt/software/kafka_2.11-0.11.0.0 路径下输入以下命令:

```
./bin/kafka-topics.sh -zookeeper master:2181,slave1:2181,slave2:2181 -topic test -replication-factor 2 -partitions 5 --create
```

查看当前的 Topic:

```
./bin/kafka-topics.sh -zookeeper master:2181,slave1:2181,slave2:2181 -list
```

8. 创建Producer和Consumer,并进行测试

在主节点上创建生产窗口 Producer,在 master/opt/software/kafka_2.11-0.11.0.0 路径下输入以下命令:

```
./bin/kafka-console-producer.sh --broker-list master:9092 --topic test
```

在从节点上创建一个 Consumer,在 slave/opt/software/kafka_2.11-0.11.0.0 路径下输入以下命令:

```
./bin/kafka-console-consumer.sh --zookeeper slave1:2181,slave2:2181 --topic test --from-beginning
```

进行测试:现在可以在 master 节点上输入一些字,可以在 slave 节点上立刻收到:

```
encoder_class>                  implementation to use for
                                serializing values.(default: kafka.
                                serializer.DefaultEncoder)
[root@master kafka 2.11-0.11.0.0]#./bin/kafka-console-producer.sh --broker-list master:9092, slave:9092--topic topic100
>top
>second
>1
>2
>3
>4
>
[root@slave kafka_2.11-0.11.0.0]#./bin/kafka-console-consumer.sh -zookeeper master:2181,slave:2181 --from-beginning -topic topic100
UsingtheConsoleConsumerwitholdconsumerisdeprecatedandwillberemovedinafuturemajorrelease.Consider using the new consumer by passing [bootstrap-server] instead of [ZooKeeper].
top
second
```

1
2
3
4

12.4　Kafka 应用实例

在这里我们将介绍一个通过 Kafka 实时收集数据的实例，实例中包括 Constants、KafkaProperties、MyProducer 和 MyConsumer 类。首先要创建一个 Topic 并指定所要连接的 ZooKeeper 和 Broker 的属性。代码如下：

```
package com.sendto.collectdata;

public class Constants {
    //连接 ZooKeeper, broker(每个 server 都是一个 broker)
    public static final String ZOOKEEPER_LIST="192.168.158.55:2181,192.168.158.56:2181";
    public static final String BROKER_LIST="192.168.158.55:9092,192.168.158.56:9092";
    //设置 ZooKeeper
    public static final String ZOOKEEPERS="192.168.158.55,192.168.158.56";
}
```

这里主要定义了 BROKER_LIST 和 ZOOKEEPER_LIST 和 ZOOKEEPERS。

通过 KafkaProperties 来指向设置的属性。代码如下：

```
package com.sendto.collectdata;

public interface KafkaProperties {
    //把 ZooKeeper、Broker 的属性指向 Constant 内设置的属性
        final static String zkConnect = Constants.ZOOKEEPER_LIST;
        final static String broker_list = Constants.BROKER_LIST;
        final static String hbase_zkList = Constants.ZOOKEEPERS;
        //指定在已经建好的 Topic 上进行 Producer 和 Consumer
        final static String groupId = "group1";
        final static String topic = "topic100";
}
```

12.4.1　Producer 实例

在这里首先要创建一个 Producer，将 ZooKeeper 上的 Topic 作为参数传入并进行连接。

```
package com.sendto.collectdata;

import java.util.Properties;
import java.util.Random;
import kafka.producer.KeyedMessage;
```

```java
import kafka.producer.ProducerConfig;
import kafka.utils.Utils;

public class MyProducer extends Thread {

    //Kafka 提供的 JavaAPI 的 Producer
    private final kafka.javaapi.producer.Producer<Integer, String> producer;
    private final String topic;
    private final Properties props = new Properties();
    public MyProducer(String topic) {
        //字符串消息
        props.put("serializer.class", "kafka.serializer.StringEncoder");
        //告诉 broker_list 路径
        props.put("metadata.broker.list", KafkaProperties.broker_list);
        //构造 Producer 来发布消息
        producer = new kafka.javaapi.producer.Producer<Integer, String>
        (new ProducerConfig(props));
        this.topic = topic;
    }
    //随机产生一些数据并放入 Producer 中
    public void run() {
    // order_id,order_amt,create_time,province_id
    Random random = new Random();
    String[] order_amt = { "12.10", "35.10", "62.2", "71.0", "93.1" };
    String[] province_id = { "1", "2", "3", "4", "5", "6", "7", "8" };
    int i = 0;
    while (true) {

        String messageStr=i+"\t"+order_amt[random.nextInt(5)] + "\t"+
        province_id[random.nextInt(8)];
        System.out.println("hello , I am producer I am producing data:"+
        messageStr);
        producer.send(new KeyedMessage<Integer, String>(topic, messageStr));
            Utils.sleep(1000);
        }
    }
    public static void main(String[] args) {
        MyProducer producerThread = new MyProducer("topic100");
        producerThread.start();
    }
}
```

12.4.2 Consumer 实例

首先创建一个 Consumer，通过 createConsumerConfig()方法设置所连接的 ZooKeeper、Group 和会话超时时间等。代码如下：

```
package com.sendto.collectdata;
```

```java
import java.util.HashMap;
import java.util.List;
import java.util.Map;
import java.util.Properties;
import kafka.consumer.ConsumerConfig;
import kafka.consumer.ConsumerIterator;
import kafka.consumer.KafkaStream;
import kafka.javaapi.consumer.ConsumerConnector;

public class MyConsumer extends Thread {
    private final ConsumerConnector consumer;
    private final String topic;
    public MyConsumer(String topic) {
    //基于 Kafka 提供的 Consumer 来构建一个 JavaConsumerConnector
consumer = kafka.consumer.Consumer.createJavaConsumerConnector(create
ConsumerConfig());
        this.topic = topic;
}
//构造属性
private static ConsumerConfig createConsumerConfig() {
    Properties props = new Properties();
    props.put("ZooKeeper.connect", KafkaProperties.zkConnect);
    //Kafka 在 zkConnect 当中
    props.put("group.id", KafkaProperties.groupId);       //Kafka 组的数字 ID
    props.put("ZooKeeper.session.timeout.ms", "6000");
//ZooKeeper 会话超时时间
    props.put("ZooKeeper.sync.time.ms", "200");
//ZK follower 可以落后 ZK leader 的最大时间
    props.put("auto.commit.interval.ms", "1000");
//Consumer 向 ZooKeeper 提交 Offset 的频率,单位是秒
    return new ConsumerConfig(props);
}

//从 Topic 中取得数据进行遍历

// push 消费方式,服务端推送过来
public void run() {
    Map<String, Integer> topicCountMap = new HashMap<String, Integer>();
    topicCountMap.put(topic, new Integer(1));
    //流处理
Map<String, List<KafkaStream<byte[], byte[]>>> consumerMap = consumer
            .createMessageStreams(topicCountMap);
KafkaStream<byte[], byte[]> stream = consumerMap.get(topic).get(0);
ConsumerIterator<byte[], byte[]> it = stream.iterator();
while (it.hasNext()) {
        //逻辑处理
        System.out.println("I am consumer, I get data from producer:"
                + new String(it.next().message()));
    }
```

```
    }
    public static void main(String[] args) {
        MyConsumer consumerThread = new MyConsumer(KafkaProperties.topic);
        consumerThread.start();
    }
}
```

12.5 小　　结

本章主要介绍了在大数据背景下的分布式消息队列 Kafka，包括 Kafka 的基本概念和核心组件，以及 Kafka 集群的安装和应用案例。Kafka 的一个典型应用是与流式框架组合使用，完成实时数据的处理。

第 13 章　开源的内存数据库——Redis

随着互联网的快速发展，人们对高性能读写的需求也越来越多，比如微博，我们想在微博上第一时间得到最新资讯，常常会刷新页面。在刷新页面的过程中，其实就是要加载数据。在这种情况下，如果在关系型数据库中取值，页面刷新则会变得非常慢。为了提升速度，可以从内存中取数据，通过这种方式来提升查询效率。Redis 就是在这样的背景下应运而生的。

本章主要涉及的知识点如下：
- 了解什么是 Redis，知道 Redis 的特点，并掌握其安装与配置。
- 掌握 Redis 常见的数据类型，如 String、List、Hash、Set。
- 熟悉 Redis 的常用函数及应用。

13.1　Redis 简介

Redis 是内存数据库，它作为热门的 NoSQL 数据库之一被很多互联网公司所使用。它是一个 key-value 存储系统，类似于 Java 中的 Map；同时 Redis 还可以通过实现数据的持久化来防止数据丢失；Redis 可以周期性地将更新的数据写入磁盘中。

13.1.1　什么是 Redis

Redis 是一个具有支持 Lua 脚本、LRU 内存回收、内置复制机制、事务和不同级别磁盘持久性特点的开源内存数据结构存储库。它支持有散列（Hash）、字符串（String）、集合（Set）、列表（List）、范围查询的有序集、位图和具有半径查询的地理空间索引等数据结构，通常用作数据库、缓存和消息代理。另外，它还可以通过 Redis Sentinel 提供高可用性，并通 Redis Cluster 进行自动分区。

Redis 和 Memcached 类似，但与 Memcached 相比，Redis 支持更多的存储类型，比如 String、List、Set、Sorted Set 和 Hash。同时，Redis 是一个高性能的、开源的、基于 key-value 键值对的缓存与存储系统。Redis 使用起来更方便，它可以提供多种键值数据来适应不同场景下的缓存和存储要求。

13.1.2 Redis 的特点

了解了什么是 Redis 之后，我们接着介绍 Redis 的特点。首先，Redis 是内存数据库，从内存中读取数据的速度远远快于从硬盘中读取。Redis 具有以下特点。

- 数据读取性能高，速度快，主要适合存储一些读取频繁、变化较小的数据。
- 支持丰富的数据类型。Redis 支持二进制的 String、List、Hash 和 Set 等数据类型操作，并且对不同的数据类型提供了非常方便的操作方式。例如，Redis 的 String 存储类型可以包含任何数据，可以是 JPG 图片，也可以是序列化的对象，同时 Hash 类型可以通过 key 对应多个 value 的存储形式，这样就可以避免 key 值过多而消耗大量内存的情况。
- 原子性：Redis 的操作都是原子性的，支持事务。所谓原子性就是指一系列对数据的操作要么都成功，要么都失败。原子性的特点使我们不用去考虑并发的问题。
- 其他特性：Redis 还有一个特点就是可以设置 key 的过期时间，一旦到了过期时间，Redis 会自动删除缓存信息。另外，Redis 也可以将内存中的数据以异步的方式写入硬盘中，从而可以避免内存数据丢失的问题。

13.2 Redis 安装与配置

前面我们介绍了什么是 Redis 及 Redis 的特点，接下来介绍 Redis 的安装与配置。

1. 下载Redis压缩包

（1）第一种方法：可以通过在 Redis 官网 https://redis.io/ 下载 Redis，这里我们下载 redis-4.0.2.tar.gz，并将其上传到/opt/software/目录下。

（2）第二种方法：可以在 Linux 系统中下载 redis-4.0.1.tar.gz，命令如下：
wget http://download.Redis.io/releases/redis-4.0.1.tar.gz

2. 将上传的文件进行解压

将文件 tar -zxvf redis-4.0.1.tar.gz 进行解压，解压后的文件如下：
rwxrwxr-x. 6 root root   4096   Sep 26 22:49 redis-4. 0. 1

3. 进入redis-4.0.1 目录

进入 redis-4.0.1 目录：
[root@master software]# cd redis-4.0.1

通过 ll 命令查看文件夹中的文件，命令和执行结果如下：

```
[root@master redis -4.0.1]# ll
total  280
-rw-rw-r--. 1 root   root   127778  Jul 24 21:58  00- RELEASENOTES
-rw-rw-r-- . 1 root   root       53  Jul 24 21:58  BUGS
-rw-rw-r-- . 1 root   root     1815  Jul 24 21:58  CONTRIBUTING
```

4．make编译

编译 Redis，编译命令和编译结果如下：

```
[root@master redis -4.0.1]# make
cd src && make all
make[1]: Entering directory.'/opt/software/redis-4.0.1/src "
  CC Makefile. dep
make[1]: Leaving directory'/opt/software/redis -4 .0.1/src '
make[ 1]: Entering directory'/opt/software/redis-4.0.1/src'
Hint : It's a good idea to run  'make test' ;)
make[1]: Leaving directory.'/opt/software/redis-4.0.1/src'
```

5．启动Redis

首先进入 src 目录，执行./redis-server 来启动 Redis。

> 注意：这种方式启动 Redis 使用的是默认配置。也可以根据需要更改这个默认的 redis.conf 配置文件。

6．Redis测试

测试客户端程序和 Redis 服务的交互，进入 src 下，执行：./redis-cli。

```
[root@master src]#./redis-cli
127.0.0.1:6379>
```

127.0.0.1 是本机的 IP，Redis 服务器默认会使用 6379 端口，通过—port 参数可以自定义端口号。以上表示 Redis 安装成功，同时通过 set key1 value1 和 get key1 来测试 Redis 的运行是否正常。

13.3　客户端登录

前面已经介绍了 Redis 的安装和测试，本节接着介绍 Redis 的客户端登录，登录有设置密码登录和未设置密码登录两种方式。

13.3.1　密码为空登录

由于在 redis-4.0.1/redis.conf 文件中默认是没有 requirepass 参数，在这种情况下可以实现无密码登录，无密码登录过程如下：

开启 Redis 服务，在 redis-4.0.1 目录下输入：

src/redis-cli

即可登录客户端：

root@master01 redis-4.0.1]# src/redis-cli
127.0.0.1:6379>

13.3.2 设置密码登录

Redis 默认是没有密码保护的，但为了降低安全风险，我们需要通过密码进行登录，首先需要设置 Redis 密码。设置密码步骤如下：

Redis 的核心配置文件是 redis.conf，通常在/opt/software/redis-4.0.1 路径下面，redis.conf 配置文件中有个 requirepass 参数，它是配置 Redis 访问密码的参数；在文件中找到 requirepass 参数并将其修改为 requirepass 123456，"123456"是密码，注意此时需重启 Redis 才能生效。设置完密码后，就可以登录了，登录方式如下：

[root@master01 redis-4.0.1]#src/redis-cli -h 127.0.0.1 -p 6379 -a '123456'
127.0.0.1:6379>

13.4 Redis 的数据类型

在 Redis 中主要支持 5 种类型：String、List、Hash、Set 及 Sorted Set，其中 Sorted Set 是有序集合，如图 13.1 所示。

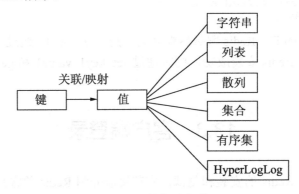

图 13.1 键值对类型

13.4.1 String 类型

字符串 String 是一种基本的类型，一个字符串类型的值最多能存储 512MB 数据量，它是二进制安全的,也就是说 String 可以包含任意类型的数据。下面我们看一个 String 类型的实例,通过 set 和 get 完成赋值和取值操作,具体如下:

```
127.0.0.1:6379> set key "123456"
OK
127.0.0.1:6379> get key
"123456"
```

关于操作 String 类型数据的 Redis 命令,详细用法如下。

1. 设置单个键的字符串值

SET 命令中,EX 设置过期时间(单位 ex 是秒,px 是毫秒),NX 表示键不存在才能赋值,XX 表示键存在时才能设置。

```
set key value [EX seconds] [PX milliseconds] [NX|XX]
```

例如:存储一个过期时间为 15 秒的键。

```
SET key1 "value1" EX 15
```

例如:存储一个键 key2,当这个键不存在时才赋值。

```
SET key2 "value2" NX
```

例如:存储一个键 key3,当这个键存在时才能设置新值。

```
SET key3 "value3" XX
```

```
127.0.0.1:6379> SET  key1 "value1" EX 100
OK
127.0.0.1:6379>  TTL  key1
(integer) 93         #剩余过期时间
127.0.0.1:6379> SET  key2 "value2" NX
OK                   #创建不存在时设置成功
127.0.0.1:6379> GET key2
"value2"
127.0.0.1:6379> SET key2 "new-value2" NX
(nil)                #键 key2 已存在,赋值失败
127.0.0.1:6379> EXISTS key3
(integer) 0
127.0.0.1:6379> SET key3 "value3" XX
(nil)                #因为键 key3 不存在,设置新值失败
```

2. 设置多个键的字符串值

前面我们已经学习了单个键的字符串存储,但单个值的存储在 key 值很多时工作量较大,因此对于多个 key 的存储,我们可以通过 mset 的方式批量进行赋值、存储,提高存储效率。设置多个字符串值,语法如下:

```
MSET key value [key value ...]
```

例如:在这里分别对 s1、s2 和 s3 赋值:

```
    MSET s1 "key1" s2 "key2" s3 "key3"
```

```
127.0.0.1;6379> MSET s1 "key1" s2 "key2" s3 "key3"
OK
127.0.0.1:6379> MGET s1 s2 s3
1)"key1"
2)"key2"
3)"key3"
```

当所有给定键都不存在时，为所有给定键设置字符串值，语法如下：

```
MSETNX key value [key value ...]
```

例：在这里分别对不存在的 s4、s5 和 s6 赋值。

```
MSETNX s4 "key4" s5 "key5" s6 "key6"
```

MSETNX 是原子性操作，即所有给定键要么就全部被赋值，要么全部不赋值，没有第三种状态。我们之前已对 s1、s2 和 s3 进行赋值操作，键已经存在，因此这里无法对其进行 MSETNX 操作。结果如下：

```
127.0.0.1:6379> MSETNX s1 "key11" s2 "key22" s3 "key33"
(integer) 0              #s1,s2 和 s3 键已经存在，操作失败
127.0.0.1:6379> MSETNX s1 "key11" s4 "key4" s5 "key5"
(integer) 0              #s1 键已经存在，操作失败
127.0.0.1:6379> MGET s1 s2 s3
1) "key1"                #MSETNX 命令未成功执行，s1 键没有被修改
2) "key2"
3) "key3"
127.0.0.1: 6379> EXISTS s4
(integer) 0              #MSETNX 命令未成功执行，s4 键没有被设置
```

13.4.2 List 类型

Redis 列表是简单的字符串列表，可以向列表的头部或者尾部插入数据。List 的应用场景比较多，是 Redis 重要的数据结构之一。比如，如果想从社交网站的几千万条新鲜事件中迅速提取 100 条最新数据，可以使用 List，包括 twitter 的关注列表。List 的内部运用是双向链表，支持反向查找和遍历，更方便操作。List 列表类型常用命令如下。

（1）在项目列表两端增加元素，命令如下：

```
LPUSH key value [value ...]
RPUSH key value [value ...]
```

LPUSH 命令用于向列表的左边增加元素，LPUSH 的返回值表示增加元素后列表的长度。

```
Redis> LPUSH numbers 1
(integer) 1
```

LPUSH 命令还支持同时增加多个元素，示例如下：

```
Redis> LPUSH numbers 2 3
(integer) 3
```

如果想向列表的右边增加元素，就可以使用 RPUSH 命令，RPUSH 的用法和 LPUSH 命令类似。

```
Redis> RPUSH numbers 0 -1
(integer) 5
```

（2）从列表两端弹出元素，命令如下：

```
LPOP key
RPOP key
```

如果想要从两端弹出元素，可以使用 LPOP 和 RPOP 命令，LPOP 命令执行包括两步操作：第 1 步是把列表左边的元素从列表中移除，第 2 步是返回被移除的元素值。

```
Redis > RPUSH list1 "zoo"
(integer) 1
Redis > RPUSH list1 "par"
(integer) 2
Redis > LPOP list1
"zoo"
```

（3）获取列表中元素的个数，命令如下：

```
LLEN key
```

可以通过 LLEN 命令获得某个 key 所对应的元素个数，当键不存在时 LLEN 命令会返回 0。

（4）获得列表片段，命令如下：

```
LRANGE key start stop
```

LRANGE 命令是列表类型的常用命令，通过 LRANGE 命令可以获得列表中的某一片段。LRANGE 命令会返回索引从 start 到 stop 之间的所有元素，Redis 列表的起始索引为 0。

```
Redis> LRANGE numbers 0 2
1)    "2"
2)    "1"
3)    "0"
```

LRANGE 命令同时支持负索引，会从右边开始计算，例如"-1"表示最右边的元素，"-2"则表示从右向左的第二个元素，比如：

```
Redis> LRANGE numbers -2 -1
```

- 在 start 的索引位置比 stop 的索引靠后的情况下，则会返回空列表。
- 在 stop 的索引大于实际索引范围的情况下，则会返回列表最右边的元素。

13.4.3 Hash 类型

散列类型 Hash，是存储了 key 和 value 映射的一种字典结构，这里的 value 只能是字符串,不能是其他的数据类型,同时 Hash 类型不能嵌套其他的数据类型。

散列类型适合存储对象,可以用对象类别和 ID 构成键名,使用字段表示对象属性,字段值就存储属性值。下面我们来看一下 Hash 类型的应用实例。

Hash 的主要命令如下。

(1) 设置单个字段。语法及代码如下:

```
HSET key field value
127.0.0.1:6379> HSET student name "Tom"
(integer) 1                        #为哈希表中字段赋值

HSETNX key field value
```

此命令仅当 key 的 field 不存在时执行,key 不存在时直接创建

```
127.0.0.1:6379> HSETNX student name "Jack"
(integer) 0                        #给定域 name 已经存在于哈希表中,赋值成功
```

(2) 设置多个字段。代码如下:

```
HMSET key field value [field value ...]

127.0.0.1:6379> HMSET student gender male score 98
OK
```

(3) 返回字段个数。代码如下:

```
HLEN key
127.0.0.1:6379> HLEN Student
(integer) 3
127.0.0.1:6379>
```

(4) 判断字段是否存在。代码如下:

```
HEXISTS key field

127.0.0.1:6379> HEXISTS Student name
(integer) 1
127.0.0.1:6379>
```

key 或者 field 不存在时返回 0

(5) 返回字段值。代码如下:

```
HGET key field
127.0.0.1:6379> HGET Student gender
"male"
127.0.0.1:6379>
```

(6) 返回多个字段值。代码如下:

```
HMGET key field [field ...]

127.0.0.1:6379> HMGET Student name gender hobby
1)"Tom"
2)"male"
3)(nil)                            #不存在的域返回 nil 值
```

（7）返回所有的键值对。代码如下：

```
HGETALL key

127.0.0.1:6379> HGETALL Student
1)"name"                                    #域
2)"Tom"                                     #值
3)"gender"
4)"male"
5)"Score"
6)"98"
```

（8）返回所有字段名。代码如下：

```
HKEYS key
127.0.0.1:6379>HKEYS student
1)"name"
2)"gender"
3)"score"
127.0.0.1:6379>
```

（9）返回所有值。代码如下：

```
HVALS key

127.0.0.1:6379>HVALS Student
1)"Tom"
2)"male"
3)"98"
127.0.0.1:6379>
```

（10）在字段对应的值上进行整数的增量计算。代码如下：

```
HINCRBY key field increment
127.0.0.1:6379> HSET student age 15
OK
127.0.0.1:6379> HGET student age
"15"
127.0.0.1:6379> HINCRBY student age 5
(integer)20
127.0.0.1:6379> HGET student age
"20"
```

（11）在字段对应的值上进行浮点数的增量计算。代码如下：

```
HINCRBYFLOAT key field increment
127.0.0.1:6379> HGET student score
"98"
127.0.0.1:6379> HINCRBYFLOAT student score 0.1
"98.1"
127.0.0.1:6379>
```

（12）删除指定的字段。代码如下：

```
HDEL key field [field1]
127.0.0.1:6379> HDEL key field [field1]
(integer) 1
127.0.0.1:6379>
```

13.4.4 Set 类型

集合 Set 是 String 类型的无序集合，集合元素是唯一的，不可以出现重复的元素。Set 底层是通过 Hash 表实现。Set 的常见命令如下。

1. SADD（元素添加）

SADD 命令示例如下：

```
SADD key member [member ...]
```

SADD 命令将一个或者多个 member 元素加入到集合中，如果 member 元素已经存在，则忽略该元素。如果 key 不存在，则会创建一个仅包含 member 元素作为成员的集合，添加单个元素的实例如下：

```
127.0.0.1:6379> SADD myset "zoo"
(integer) 1
```

添加重复元素。代码如下：

```
127.0.0.1:6379> SADD myset " zoo "
(integer) 0
```

添加多个元素。代码如下：

```
127.0.0.1:6379> SADD myset "koo" "zoo1"
(integer) 2
```

查看元素。代码如下：

```
127.0.0.1:6379> SMEMBERS myset
1) "zoo"
2) "koo"
3) "zoo1"
```

2. SCARD（元素数获取）

SCARD 命令用于获取存储在集合中的元素数量，当 key 不存在时，则返回 0。代码如下：

```
127.0.0.1:6379> scard myset
(integer) 3
```

3. SMEMBERS（返回所有key成员）

SMEMBERS 命令返回 key 中的所有成员，如果 key 不存在，则返回空集合。代码如下：

```
127.0.0.1:6379> SMEMBERS myset
1) "zoo"
2) "koo"
3) "zoo1"
```

4. SINTER（返回集合交集）

SINTER key 命令返回给定集合的交集，对于不存在的集合 key，会被视为空集。代码如下：

```
redis 127.0.0.1:6379> SADD myset "zoo"
(integer) 1
redis 127.0.0.1:6379> SADD myset "foo"
(integer) 1
redis 127.0.0.1:6379> SADD myset "bar"
(integer) 1
redis 127.0.0.1:6379> SADD myset2 "zoo"
(integer) 1
redis 127.0.0.1:6379> SADD myset2 "world"
(integer) 1
redis 127.0.0.1:6379> SINTER myset myset2
1) "zoo"
```

5. SDIFF（返回集合差集）

SDIFF key 命令返回所有给定集合之间的差集，对于不存在的 key，会被视为空集。代码如下：

```
redis> SADD key1 "a"
(integer) 1
redis> SADD key1 "b"
(integer) 1
redis> SADD key1 "c"
(integer) 1
redis> SADD key2 "c"
(integer) 1
redis> SADD key2 "d"
(integer) 1
redis> SADD key2 "e"
(integer) 1
redis> SDIFF key1 key2
1) "a"
2) "b"
redis>
```

6. SMOVE（元素移动）

SMOVE source destination member 命令的作用是将 member 元素从 source 集合移动到 destination 集合中。当 destination 集合已经包含 member 元素的时候，SMOVE 命令则只是简单地将 Source 集合中的 member 元素删除。代码如下：

```
redis 127.0.0.1:6379> SADD myset1 "hello"
(integer) 1
redis 127.0.0.1:6379> SADD myset1 "world"
(integer) 1
redis 127.0.0.1:6379> SADD myset1 "bar"
(integer) 1
```

```
redis 127.0.0.1:6379> SADD myset2 "foo"
(integer) 1
redis 127.0.0.1:6379> SMOVE myset1 myset2 "bar"
(integer) 1
redis 127.0.0.1:6379> SMEMBERS myset1
1) "World"
2) "Hello"
redis 127.0.0.1:6379> SMEMBERS myset2
1) "foo"
2) "bar"
```

7. SPOP（元素移除）

SPOP key 命令移除并返回集合中的一个随机元素。代码如下：

```
127.0.0.1:6379> SDIFF myset
1) "zoo"
2) "foo"
3) "koo"
127.0.0.1:6379> SPOP myset
"foo"
127.0.0.1:6379> SDIFF myset
1) "zoo"
2) "koo"
```

8. SRANDMEMBER（返回元素或数组）

SRANDMEMBER key 命令执行时，如果只是提供了 key 参数，则返回集合中的一个随机元素。如果提供了 count 参数，则会返回一个数组。下面是示例代码：

```
redis 127.0.0.1:6379> SADD myset1 "hello"
(integer) 1
redis 127.0.0.1:6379> SADD myset1 "world"
(integer) 1
redis 127.0.0.1:6379> SADD myset1 "bar"
(integer) 1
redis 127.0.0.1:6379> SRANDMEMBER myset1
"bar"
redis 127.0.0.1:6379> SRANDMEMBER myset1 2
1) "Hello"
2) "world"
```

9. SREM（删除集合元素）

Redis 的 SREM 命令用于删除集合中的一个或多个元素。代码如下：

```
redis 127.0.0.1:6379> SADD myset1 "hello"
(integer) 1
redis 127.0.0.1:6379> SADD myset1 "world"
(integer) 1
redis 127.0.0.1:6379> SADD myset1 "zoo"
(integer) 1
redis 127.0.0.1:6379> SREM myset1 "hello"
(integer) 1
```

```
redis 127.0.0.1:6379> SREM myset1 "foo"
(integer) 0
redis 127.0.0.1:6379> SMEMBERS myset1
1) "zoo"
2) "world"
```

10．SUNION（返回集合所有成员）

SUNION key 命令返回一个集合中的所有成员，也就是说这个集合是所有给定集合的并集。代码如下：

```
redis> SADD key1 "a"
(integer) 1
redis> SADD key1 "b"
(integer) 1
redis> SADD key1 "c"
(integer) 1
redis> SADD key2 "c"
(integer) 1
redis> SADD key2 "d"
(integer) 1
redis> SADD key2 "e"
(integer) 1
redis> SUNION key1 key2
1) "a"
2) "c"
3) "b"
4) "e"
5) "d"
```

13.5 小 结

本章主要学习了 Redis，了解了它的特点及安装与配置，并通过实例介绍了如何基于客户端登录 Redis，以及 Redis 的数据类型，包括 String、List、Hash 和 Set 等内容。

第 14 章　Ambari 和 CDH

我们知道，在开发大数据项目时需要有大数据环境。在企业中安装与配置 Hadoop 时，有的是基于原生 Hadoop 完成搭建，有的是基于工具来完成。目前使用的工具主要有两个，一个是 Ambari，另一个是 CDH。本章我们介绍基于 Ambari 和 CDH 工具完成大数据环境的搭建。

本章主要涉及的知识点如下：

- 认识 Ambari 与 CDH。
- Ambari 的安装与集群管理。
- CDH 的安装与集群管理。

14.1　Ambari 的安装与集群管理

Ambari 是 Apache 的顶级项目，它是一个基于 Web 页面的工具，可以用于安装、配置、管理和监视 Hadoop 集群，同时支持 HDFS、MapReduce、Hive、HBase、ZooKeeper 和 Oozie 等。Ambari 还提供了仪表盘，通过仪表盘可以查看集群的运行状况，而且提供了友好的用户界面，便于对运行性能进行诊断。

14.1.1　认识 HDP 与 Ambari

HDP（Hortonworks Data Platform）是 Hortonworks 公司基于 Apache Hadoop 项目进行维护的 Hadoop 版本，是一款可用于企业的开源 Apache Hadoop 分布式系统。

Ambari 是一种基于 Web 的工具，帮助管理自己的 HDP 集群，通过 Ambari 快速地自动部署 HDP 和其相关组件，包括搭建、管理和监视 Hadoop 整个生态圈（例如，Hadoop、Hive、Hbase、Sqoop 和 ZooKeeper 等）。

Ambari 提供了一个监视 Hadoop 集群健康和状态的仪表板 UI 页面进行系统警报，并在需要注意的时候通知用户（如节点掉电、剩余磁盘空间不足等），对集群中的 Hadoop、Hive 和 Spark 等服务的安装与配置管理做了极大简化。Ambari 主要具有以下优势：

- Hortonworks Ambari 的产品都是百分百开源的，可以免费使用。
- 用户界面友好，可以高效地查看信息并进行集群控制。

- 用户可以根据自身情况自定义监控视图。
- 支持二次开发。
- 社区活跃。

14.1.2　Ambari 的搭建

了解了 Ambari 的特点之后，接下来介绍如何搭建 Ambari。搭建过程以 3 台机器为例，分别是 master、slave1 和 slave2，其中，Ambari 的版本是 2.5.0.3。相关配置如表 14.1 所示。

表 14.1　Ambari的构建

Ambari集群规划	master　192.168.4.10
	slave1　192.168.4.11
	slave2　192.168.4.12
所用系统	CentOS 6.5
Ambari版本	2.5.0.3
HDP版本	2.6.0.3
HDP-UTILS版本	1.1.0.21
JDK版本	JDK 1.7

Ambari 的整个搭建步骤比较多，读者可以根据自身情况进行适当调整，如果有些环境已经配置好，即可跳过。下面具体介绍。

14.1.3　配置网卡与修改本机名

（1）配置网卡，配置文件和配置内容如下：

```
# vi /etc/sysconfig/network-scripts/ifcfg-eth0
写入：
DEVICE=eth0
TYPE =Ethernet
ONBOOT=yes
NM_CONTROLLED=yes
BOOTPROTO=static
IPADDR=192.168.4.10
GATEWAY=192.168.4.1
NATMASK=255.255.255.0
```

（2）修改主机名，文件和修改内容如下：

```
# vi /etc/sysconfig/network
把 HOSTNAME 改成集群节点的名字
NETWORKING=yes
HOSTNAME=master
```

14.1.4 定义 DNS 服务器与修改 hosts 主机映射关系

（1）在所有节点定义 DNS 服务器。

vi /etc/resolv.conf
写入：nameserver 114.114.114.114

重启网络指令：# service network restart。

```
[root@master ~]# service network restart
Shutting down interface eth0:                [OK]
Shutting down loopback intertace:            [OK]
Bringing up loopback interface:              [OK]
Bringing up interface eth0: Determining if ip address 192.168 .72,
10isalreadyinusefordeviceeth0...
                                             [OK]
```

接着通过 ping 测试是否可以连接到外网。

（2）在所有节点修改 hosts 主机映射关系。

hosts 文件中包含了 IP 地址与主机名之间的映射。

vi /etc/hosts

写入：

```
192.168.4.10 master
192.168.4.11 slave1
192.168.4.12 slave2
```

14.1.5 关闭防火墙并安装 JDK

1. 关闭所有节点的防火墙

临时关闭指令：# service iptables stop。

```
[root@master-]# service iptables stop
iptables:Setting chains to policy ACCEPT: filter    [OK]
iptables:Flushing firewall rules                    [OK]
iptables:Unloading modules:                         [OK]
```

重启后生效指令：# chkconfig iptables off。

[root@master~]#chkconfig iptables off

查看防火墙状态指令：# service iptables status。

```
[root@master~]# service iptables status
iptables: Firewall is not running.
```

2. 在所有节点安装JDK

安装 JDK 指令：# rpm -ivh jdk-7u79-linux-x64.rpm。

```
[root@ambari software]# rpm -ivh jdk-7u79-linux-x64.rpm
Preparing...               ###################################[100%]
  1:jdk                    ###################################[100%]
Unpacking JAR files...
        rt.jar...
        jsse.jar...
        charsets.jar...
        tools.jar...
        localedata.jar...
        jfxrt.jar...
```

修改.bash_profile 配置环境变量，内容如下：

```
JAVA_HOME=/usr/java/jdk1.7.0_79
JER_HOME=/usr/java/jdk1.7.0_79/jre
PATH=$PATH:$HOME/bin:$JAVA_HOME/bin
export CLASSPATH=$JAVA_HOME/lib/dt.jar:$JAVA_HOME/lib/tools.jar:JRE_HOME/bin
export PATH
```

使配置生效。

```
# source .bash_profile
```

14.1.6　升级 OpenSSL 安全套接层协议版本

对于所有节点来说，很多加密算法用到了 OpenSSL 这个组件，低版本的 OpenSSL 存在很多漏洞和风险需要升级到新版本。查看 OpenSSL，如果版本过低，则进行升级。

```
# rpm -qa | grep openssl
```

结果是：

```
# openssl-1.0.1e-15.el6.x86_64
```

这个版本比较旧，需要进行升级，升级命令如下：

```
# yum upgrade openssl
```

14.1.7　关闭 SELinux 的强制访问控制

由于所有节点的访问控制体系限制，进程只能访问某些在其任务中所需要的文件，如果不关闭 SELinux 可能因为权限原因冲突。

直接关闭 SELinux 命令如下：

```
# setenforce 0
```

不是立即关闭，而是重启机器后生效的关闭 SELINUX 命令如下：

```
# vi /etc/sysconfig/selinux
```

将 SELINUX=enforcing 改为 SELINUX=disabled

14.1.8　SSH 免密码登录

1. SSH免密码登录

在登录服务器时，免密码登录能大大提高工作效率，配置 SSH 免密码登录可以按照以下步骤操作。

在主节点和从节点都生成密钥：

```
# ssh-keygen -t rsa -P ''
```

主节点发送密钥到从节点：

```
# scp .ssh/id_rsa.pub root@slave1:~
```

在从节点操作：

```
# cat id_rsa.pub >> .ssh/authorized_keys
```

主节点：

```
# cat .ssh/id_rsa.pub >> .ssh/authorized_keys
```

如果提示没有 wget，则运行下面代码进行安装：

```
# yum -y install wget
```

2. 更改YUM

如果从国外站点安装更新速度太慢，可以更改 YUM 仓库更新源，并修改成阿里云镜像站点。

```
# mv /etc/yum.repos.d/CentOS-Base.repo/etc/yum.repos.d/CentOS-Base.repo.bk
```

下载 Repo：

```
# wget -O /etc/yum.repos.d/CentOS-Base.repo http://mirrors.aliyun.com/repo/Centos-6.repo
```

14.1.9　同步 NTP

在集群中，如果时间不一致则往往导致数据混乱，可以通过在所有节点上安装 NTP（Network Time Protocol，网络时间协议），用于同步系统时间，也可以通过自己搭一台 NTP 服务器来保持整个集群的时间一致性，过程如下：

```
# yum install -y ntp
# vi /etc/ntp.conf
restrict 127.0.0.1
restrict-6 ::1
restrict 192.168.127.0 mask 255.255.255.0 nomodify notrap
server 210.72.145.44     #中国授时中心
server 202.112.10.36     #1.cn.pool.ntp.org
```

```
server  59.124.196.83        #@0.asia.pool.ntp.org
server  192.168.72.10        #局域网 NTP 本机 IP
server  127.127.1.0          #本地时间
fudge server 127.127.1.0 stratum 10
```

首先,"restrict 192.168.127.0 mask 255.255.255.0 nomodify notrap"的意思是从 IP 地址 192.168.217.1-192.168.217.254,默认网关 255.255.255.0 的服务器都可以使用 NTP 服务器进行时间同步。

其次,指定互联网和局域网中作为 NTP 服务器的 IP 地址,在上述配置文件中相关内容如下:

```
server  210.72.145.44
server  202.112.10.36
server  59.124.196.83
server  192.168.72.10
```

另外,当服务器与公用的时间服务器失去联系时以本地时间为客户端提供时间服务,在上述配置文件中相关内容如下:

```
server  127.127.1.0  #本地时间
fudge server 127.127.1.0 stratum 10
```

配置文件修改完成后,保存退出,启动服务。

```
[root@master yum.repos.d]# chkconfig --list ntpd
ntpd            0:off 1:off 2:off 3:off 4:off 5:off 6:off

# chkconfig ntpd on
# service ntpd start
```

启动后,一般需要 5~10 分钟左右的时间才能与外部时间服务器开始同步时间。可以通过命令查询 NTPD 服务情况,查看服务连接和监听如下:

```
[root@master-]# netstat -tlunp | grep ntp
udp     0     0 192.168.72.100:123       0.0.0.0:*       1416/ntpd
udp     0     0 127.0.0.1:123            0.0.0.0:*       1416/ntpd
udp     0     0 0.0.0.0:123              0.0.0.0:*       1416/ntpd
udp     0     0 fe80::20C:29ff:fe28 8fc1:123  :::*       1416/ntpd
udp     0     0 :1:123                   :::*            1416/ntpd
udp     0     0 :::123                   :::*            1416/ntpd
```

重新启动服务:

```
# service ntpd restart
```

将其他节点的时间与 Master 同步,在其他所有节点运行命令,就可以将节点间的时间同步了。

```
# ntpdate master
```

14.1.10 关闭 Linux 的 THP 服务

Ambari 和 CM 管理平台都建议关闭 THP，否则 CPU 使用率会很高，HDFS 性能也会严重受影响。因此在所有节点关闭 THP 时，可以通过以下内核参数优化关闭系统 THP 特性。

直接关闭：

```
# echo never > /sys/kernel/mm/redhat_transparent_hugepage/enabled
```

永久关闭：

```
# vi /etc/grub.conf
```
kernel 项增加 transparent_hugepage=never

```
# grub.conf generated by anaconda
#
# Note that you do not have to rerun grub after making changes to this file
# NOTICE : You have a /boot partition.  This means that
#          all kernel and initrd paths are relative to /boot/,eg.
#          root (hd0,0)
#          kernel /vmlinuz- version ro root=/dev/mapper/vg_ master- lv_ root
#          initrd /initrd- [generic- ]version.img
#boot=/dev/sda
default=0
timeout = 5
splashimage=(hd0,0)/grub/splash.xpm.gz
hiddenmenu
title CentOS (2.6.32-431.el6.x86_ 64)
        root (hd0,0)
        kernel /vmlinuz-2.6.32-431.el6.x86_64roroot=/dev/mapper/vg_
        master-lv_rootrd.NO_LUKSKEYBOARDTYPE=pcKEYTABLE=usLANG=en_US.
        UTF-rd_NO_MDrd_LLV=vg_master/lv_swapSYSFONT=latarcyrheb-sun16
        crashkernel=auto rd_LVM_LV=vg_master/lv_root rd_NO_DM rhgb quiet
        transparent_hugepag
  e=never
   initrd /initramfs-2.6.32-431.el6.x86_64.img
```

重启，检查 THP 是否被禁用。

```
# cat /sys/kernel/mm/redhat_transparent_hugepage/enabled
```

如有[never]则表示 THP 已被禁用。

```
[root@master~]# cat /sys/kernel/mm/redhat_transparent_hugepage/eabled
always madvise [never]
```

14.1.11 配置 UMASK 与 HTTP 服务

（1）配置 UMASK，设定用户所创建目录的初始权限。

```
# umask 0022
```

(2) 在主节点配置 HTTP 服务。

HTTP（Hyper Text Transfer Protocol，超文本传输协议），在 Linux 下实现 Web 服务，配置 HTTP 服务到系统层，使其随系统自动启动。

```
# yum install httpd
# chkconfig httpd on
# service httpd start
```

14.1.12　安装本地源制作相关工具与 Createrepo

（1）在主节点安装本地源制作相关工具，命令如下：

```
# yum install yum-utils createrepo yum-plugin-priorities -y
```

（2）安装 Createrepo 主节点

Createrepo 用以创建 YUM 源（软件仓库），即为存放于本地特定位置的众多 RPM 包建立索引，描述各包所需的依赖信息，并形成元数据，命令如下：

```
# yum install -y createrepo
```

14.1.13　禁止离线更新与制作本地源

禁止 PackageKit。

1. 在主节点禁止离线更新

PackageKit 是一款以方便 Linux 软件安装与升级为目的的系统，这是因为在使用 YUM 安装期间有可能导致 YUM 被锁，使安装过程报错。Agent 会用到 YUM 安装 Hadoop 及上层组件，而 Server 也要用 YUM 安装 Ambari。

```
# vi /etc/yum/pluginconf.d/priorities.conf
```
把 enabled 改成 0
添加 gpgcheck 改成 0

```
[main]
enabled=0
Gpgcheck=0
```

enabled=1 说明启用这个更新库，为 0 表示不启用。
Gpgcheck=1 表示使用 gpg 文件来检查软件包的签名。
Gpgkey 参数表示 gpg 文件所存放的位置或 HTTP 形式的网址。

2. 在主节点制作本地源

配置 Ambari 源，制作本地源是因为在线安装 Ambari 太慢，制作本地源只需在主节点上进行。

```
# mkdir -p /var/www/html/ambari
```

Ambari 安装包下载网址如下：

http://public-repo-1.hortonworks.com/ambari/centos6/2.x/updates/2.5.0.3/ambari-2.5.0.3-centos6.tar.gz

将 ambari 放在/var/www/html/ambari/目录下并通过 tar -zxvf ambari-2.5.0.3-centos6.tar.gz 命令解压。代码如下：

```
tar -zxvf ambari-2.5.0.3-centos6.tar.gz
```

创建 Repodata。代码如下：

```
[root@master ambari]# createrepo /var/www/html/ambari/
Spawning worker 0 with 12 pkgs
Workers Finished
Gathering worker results
Saving Primary metadata
Saving file lists metadata
Saving other metadata
Generating sqlite DBs
Sqlite DBs complete
```

配置 HDP 源：

```
# mkdir -p /var/www/html/HDP
```

HDP 安装包下载网址如下：

http://public-repo-1.hortonworks.com/HDP/centos6/2.x/updates/2.6.0.3/HDP-2.6.0.3-centos6-rpm.tar.gz

在/var/www/html/HDP 目录下放入 HDP：

通过 tar -zxvf HDP-2.6.0.3-centos6-rpm.tar.gz 命令进行解压。代码如下：

```
tar -zxvf HDP-2.6.0.3-centos6-rpm.tar.gz
```

创建 Repodata。代码如下：

```
[root@master ambari]# createrepo /var/www/html/HDP
Spawning worker 0 with 232 pkgs
Workers Finished
Gathering worker  results

Saving Primary metadata
Saving filelists metadata
Saving other metadata
Generating sqlite DBs
Sqlite DBs complete
```

配置 HDP-UTILS 源：

```
# mkdir -p /var/www/html/HDP-UTILS
```

HDP-UTILS-1.1.0.21 安装包下载网址如下：

http://public-repo-1.hortonworks.com/HDP-UTILS-1.1.0.21/repos/centos6/HDP-UTILS-1.1.0.21-centos6.tar.gz

在 /var/www/html/HDP-UTILS 下放入 UTILS：

解压：`# tar -zxvf HDP-UTILS-1.1.0.21-centos6.tar.gz`
创建 repodata：`# createrepo /var/www/html/HDP-UTILS`

修改 ambari.repo，配置为本地源：

`# vi /var/www/html/ambari/ambari/centos6/ambari.repo`

更改后的内容如下：

```
#VERSION_NUMBER=2.5.0.3-7
[ambari-2.5.0.3]
name=ambari Version - ambari-2.5.0.3
baseurl=http://192.168.4.10/ambari/ambari/centos6/
gpgcheck=0
enabled=1
priority=1
```

修改 HDP.repo，配置为本地源：

`# vi /var/www/html/HDP/HDP/centos6/hdp.repo`

更改后的内容如下：

```
#VERSION_NUMBER=2.6.0.3-8
[HDP-2.6.0.3]
name=HDP Version - HDP-2.6.0.3
baseurl=http://192.168.4.10/HDP/HDP/centos6/
gpgcheck=0
enabled=1
priority=1
```

修改 HDP-UTILS.repo，配置为本地源：

`# vi /var/www/html/HDP-UTILS/hdp-util.repo`

更改后的内容如下：

```
[HDP-UTILS-1.1.0.21]
name=Hortonworks Data Platform Version - HDP-UTILS-1.1.0.21
baseurl=http://192.168.4.10/HDP-UTILS
gpgcheck=0
enabled=1
priority=1
```

将 repo 复制到 yum.repos.d 中：

```
# cp /var/www/html/ambari/ambari/centos6/ambari.repo /etc/yum.repos.d/
# cp /var/www/html/HDP/HDP/centos6/hdp.repo /etc/yum.repos.d/
# cp /var/www/html/HDP-UTILS/hdp-util.repo /etc/yum.repos.d/
```

发送到各个节点：

```
# scp /etc/yum.repos.d/ambari.repo root@slave:/etc/yum.repos.d/
# scp /etc/yum.repos.d/hdp.repo root@slave:/etc/yum.repos.d/
# scp /etc/yum.repos.d/hdp-util.repo root@slave:/etc/yum.repos.d/
```

运行 YUM Repolist：

```
[root@master yum.repos.d]# yum repolist
Loaded plugins: fastestmirror
Loading mirror speeds from cached hostfile
*base: mirrors.allyun.com
*epel: mirrors.tuna.tsinghua.edu.cn
*extras: mirrors.aliyun.com
*updates: mirrors.aliyun.com
ambari-2.5.0.3                                            |2.9 KB   00:00
ambari-2.5.0.3/primary_db                                 |8.5 KB   00:00
repo id             repo name                                      status
HOP.-2. 6. 0. 3     HOP version.HOP-2.6.0.3                        232
HOP-UTTLS-1.1.0.21  Hortonworks Data Platfom Version.HOP-UTILS-1.1.0.21 56
ambari-2.5.0.3      ambari Version.anbari-2.5.0.3                  12
base                CentO5-6.Base-mirrors.aliyun.com               6706
*epe1               Extra Packages for Enterprise Linux 6-X86_64   12.402
extras              CentOS-6.Extras.mirrors.aliyun.com             46
updates             CentOS-6- Updates- mirrors.aliyun.com          721
repolist: 20,175
```

```
# yum clean all
```

YUM 会把下载的软件包和 Header 存储在缓存中，并不会自动删除。如果开发者觉得占用磁盘空间，可以使用 Yum clean 命令进行清除。

```
# yum makecache
```

把服务器的包信息下载到本地计算机上缓存起来，结合 yum -C search subversion 使用，不用上网检索就能查找软件信息。

```
# yum list | grep ambari
```

如果可以看到 Ambari 的对应版本的安装包列表，说明公共库已配置成功。然后就可以安装 Ambari 的 Package 了。Ambari 私有库配置完成后，可以用网址加路径进行查看。

```
http://IP/ambari/ambari/centos6/
```

14.1.14　安装 Ambari-server 与 MySQL

安装 Ambari-server 与 MySQL，相关操作如下：

```
# yum -y install ambari-server
```

在主节点安装 MySQL，Ambari 使用的默认数据库是 PostgreSQL，用于存储安装元数据，可以使用自己安装的 MySQL 数据库作为 Ambari 元数据库，MySQL 数据库具有良好的扩展性和灵活性。

```
# yum install -y mysql-server
```

设置开机自动启动，命令如下：

```
# chkconfig mysqld on
```

直接启动，命令如下：

```
# service mysqld start
```

通过以下命令查看 MySQL 服务是不是开机自启动。

```
# chkconfig --list | grep mysqld
```

在 MySQL 中分别创建数据库 Ambari、Hive、Oozie 和其相应用户，创建相应的表。代码如下：

```
CREATE DATABASE ambari;
use ambari;
CREATE USER 'ambari'@'%' IDENTIFIED BY 'bigdata';
GRANT ALL PRIVILEGES ON *.* TO 'ambari'@'%';
CREATE USER 'ambari'@'localhost' IDENTIFIED BY 'bigdata';
GRANT ALL PRIVILEGES ON *.* TO 'ambari'@'localhost';
CREATE USER 'ambari'@'master' IDENTIFIED BY 'bigdata';
GRANT ALL PRIVILEGES ON *.* TO 'ambari'@'master';
FLUSH PRIVILEGES;
source /var/lib/ambari-server/resources/Ambari-DDL-MySQL-CREATE.sql
show tables;
use MySQL;
select Host,User,Password from user where user='ambari';
CREATE DATABASE hive;
use hive;
CREATE USER 'hive'@'%' IDENTIFIED BY 'hive';
GRANT ALL PRIVILEGES ON *.* TO 'hive'@'%';
CREATE USER 'hive'@'localhost' IDENTIFIED BY 'hive';
GRANT ALL PRIVILEGES ON *.* TO 'hive'@'localhost';
CREATE USER 'hive'@'master' IDENTIFIED BY 'hive';
GRANT ALL PRIVILEGES ON *.* TO 'hive'@'master';
FLUSH PRIVILEGES;
CREATE DATABASE oozie;
use oozie;
CREATE USER 'oozie'@'%' IDENTIFIED BY 'oozie';
GRANT ALL PRIVILEGES ON *.* TO 'oozie'@'%';
CREATE USER 'oozie'@'localhost' IDENTIFIED BY 'oozie';
GRANT ALL PRIVILEGES ON *.* TO 'oozie'@'localhost';
CREATE USER 'oozie'@'master' IDENTIFIED BY 'oozie';
GRANT ALL PRIVILEGES ON *.* TO 'oozie'@'master';
FLUSH PRIVILEGES;
```

执行成功后，针对 MySQL 数据库再做下面几项工作。

将 mysql-connector-Java.jar 复制到 /usr/share/java 目录下：

```
mkdir /usr/share/java
# mv mysql-connector-java-5.1.32-bin.jar mysql-jdbc-driver.jar
```

将 MySQL-connector-java.jar 复制到 /var/lib/ambari-server/resources 目录下：

```
#cp /usr/share/java/mysql-jdbc-driver.jar /var/lib/ambari-server/resources/mysql-jdbc-driver.jar
```

编辑 ambari.properties：

```
# vi /etc/ambari-server/conf/ambari.properties
```

在配置文件 ambari.properties 中添加如下代码：

```
server.jdbc.driver.path=/usr/share/java/MySQL-connector-java-5.1.32-
bin.jar
```

14.1.15　安装 Ambari

现在开始安装 Ambari，在主节点上进行以下操作：

```
# ambari-server setup
```

下面是配置执行流程，按照提示操作。

（1）提示是否自定义设置，输入 y：

```
Customize user account for ambari-server daemon [y/n] (n)? y
```

（2）ambari-server 账号：

```
Enter user account for ambari-server daemon (root):
```

如果直接回车就是默认选择 root 用户。

如果输入已经创建的用户就会显示：

```
Enter user account for ambari-server daemon (root):ambari
Adjusting ambari-server permissions and ownership...
```

（3）检查防火墙是否关闭。代码如下：

```
Adjusting ambari-server permissions and ownership...
Checking firewall...
WARNING: iptables is running. Confirm the necessary Ambari ports are
accessible. Refer to the Ambari documentation for more details on ports.
OK to continue [y/n] (y)?
```

直接回车

（4）选择设置 JDK，输入 3：

```
Checking JDK...
Do you want to change Oracle JDK [y/n] (n)? y
[1] Oracle JDK 1.8 + Java Cryptography Extension (JCE) Policy Files 8
[2] Oracle JDK 1.7 + Java Cryptography Extension (JCE) Policy Files 7
[3] Custom JDK
==============================================================================
Enter choice (1): 3
```

如果上面选择 3，即自定义 JDK，则需要设置 JAVA_HOME。输入：/usr/java/jdk1.8.0_131

```
WARNING: JDK must be installed on all hosts and JAVA_HOME must be valid on
all hosts.
WARNING: JCE Policy files are required for configuring Kerberos security.
If you plan to use Kerberos,please make sure JCE Unlimited Strength
Jurisdiction Policy Files are valid on all hosts.
Path to JAVA_HOME: /usr/java/jdk1.8.0_131
Validating JDK on Ambari Server...done.
Completing setup...
```

（5）数据库配置，选择 y：

```
Configuring database...
Enter advanced database configuration [y/n] (n)? y
```

（6）选择数据库类型，输入 3：

```
Configuring database...
==============================================================
Choose one of the following options:
[1] - PostgreSQL (Embedded)
[2] - Oracle
[3] - MySQL
[4] - PostgreSQL
[5] - Microsoft SQL Server (Tech Preview)
[6] - SQL Anywhere
==============================================================
Enter choice (3): 3
```

（7）设置数据库的具体配置信息，根据实际情况输入，如果和括号内信息相同，则可以直接回车。如果想重命名，就输入名称。

```
Hostname (localhost):
Port (3306):
Database name (ambari):
Username (ambari):
Enter Database Password (bigdata):
Re-Enter password:
```

14.1.16　安装 Agent 与 Ambari 登录安装

（1）在所有节点安装 Agent，命令如下：

```
# yum install ambari-agent
```

修改 ambari-agent.ini 中要连接的主机名，这里每一个节点都是为了与安装 Ambari 的主机进行连接，因此每个节点都应将 Hostname 配置成主节点所对应的主机机名。代码如下：

```
# vi /etc/ambari-agent/conf/ambari-agent.ini
[server]
hostname=master
url_port=8440
secured_url_port=8441
```

（2）Ambari 登录安装。

从节点启动 Agent，命令如下：

```
# ambari-agent start
```

主节点启动 Server，命令如下：

```
# ambari-server start
```

查看状态，命令如下：

```
# ambari-server status
```

在 Web 页面打开 http://192.168.72.100:8080
账号密码默认是 admin，如图 14.1 所示。

图 14.1　默认 admin 页面

单击 Launch Install Wizard 按钮进入 Ambari 页面，如图 14.2 所示。

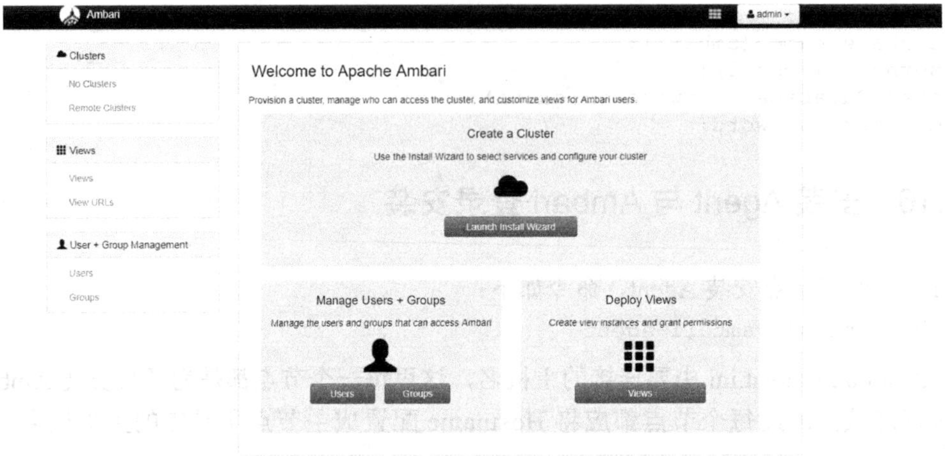

图 14.2　Apache Ambari 页面

给集群命名后，单击 Next 按钮，如图 14.3 所示。
选择 Use Local Repository 并填入本地 YUM 源，如表 14.2 所示。

表 14.2　YUM源

redhat6	HDP-2.6	http://192.168.4.10/HDP/HDP/centos6
redhat6	HDP-UTILS-1.1.0.21	http://192.168.4.10/HDP-UTILS

因为在前面已安装好了 ambari-agent，因而这一步选择第二种方式安装而非上传私钥方式，如图 14.4 所示。

由于 MySQL 是安装在 Master 上的，在此填写数据库所在的机器名，将提前为 Hive 建立的数据库信息填入下表，单击测试连接按钮，如图 14.5 所示。

第 14 章　Ambari 和 CDH

图 14.3　Get Started 页面

图 14.4　Select Version 页面

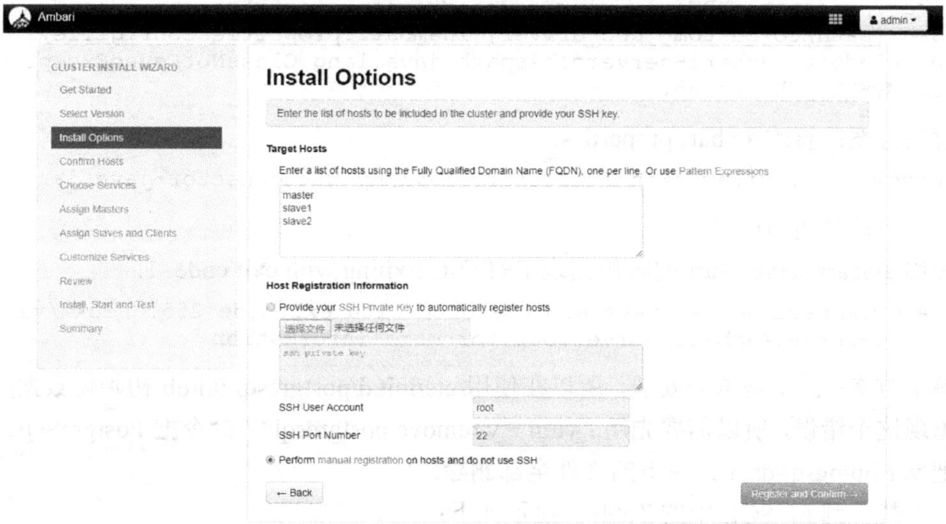

图 14.5　测试连接页面

Hive 数据库的安装，如图 14.6 所示。

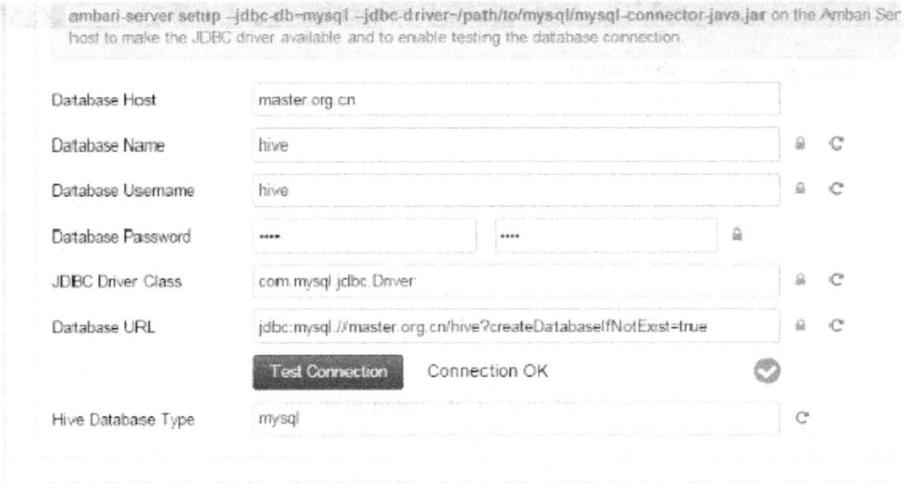

图 14.6 Hive 数据库的安装

14.1.17 安装部署问题解决方案

在安装部署过程中，可能会出现一些问题，在此对这些问题进行了汇总，如果读者遇到类似问题，可以参考解决。

（1）找不到 MySQL 的 JAR 包，显示如下：

```
ERROR[main]DBAccessorImpl:117-Ifyouareusinganon-defaultdatabasefor
AmbariandacustoJDBCdriverjar,youneedtosetproperty"server.jdbc.driver.
path={path/to/custom_jdbc_driver}"inambari.propertiesconfigfile,
toincludeitinambari-serverclasspath.java.lang.ClassNotFoundException:
com.MySQL.jdbc.Driver
```

解决方案：修改 ambari.properties。

```
server.jdbc.driver.path=/usr/share/java/MySQL-connector-java.jar
```

（2）启动时随即退出。

使用 ambari-server start 的时候出现 ERROR: Exiting with exit code -1.

```
REASON:Ambari Server java process died with exitcode 255.Check /var/log/
ambari-server/ambari-server.out for more information
```

解决方案：由于是重新安装，所以在使用/etc/init.d/postgresql initdb 初始化数据库的时候会出现这个错误，所以需要先用"yum‐y remove postgresql*"命令把 PostgreSQL 卸载，然后把/var/lib/pgsql/data 目录下的文件全部删除。

（3）找不到 MySQL 中的 Table，显示如下：

```
Table 'ambari.metainfo' doesn't exist
Error while creating database accessor
java.sql.SQLException: Access denied for user 'ambari'@'master' (using
password: YES)
```

解决方案：在 MySQL 中导入建表脚本：

```
source /var/lib/ambari-server/resources/Ambari-DDL-MySQL-CREATE.sql
```

（4）THP 没有关闭，显示如下：

```
Transparent Huge Pages Issues (1)
The following hosts have Transparent Huge Pages (THP) enabled.THP should
be disabled to avoid potential Hadoop performance issues.
```

解决方案：在/etc/grub.conf 中添加 transparent_hugepage=never。

（5）集群时间不统一，显示如下：

```
Service Issues (1)
The following services should be up Service ntpd or chronyd Not running on
3 hosts
```

解决方案：NTP 服务只在一个节点启动，其他节点关闭 service ntpd stop。

（6）MySQL 版本问题，显示如下：

```
resource_management.core.exceptions.ExecutionFailed:Executionof'/usr/
bin/yum-d0-e0-yinstall accumulo_2_6_0_3_8' returned 1.
You could try using --skip-broken to work around the problem
** Found 1 pre-existing rpmdb problem(s), 'yum check' output follows:
MySQL-community-common-5.7.9-1.el5.x86_64hasmissingrequiresofMySQL=('0',
'5.7.9','1.el5')
```

解决方案：MySQL 版本问题，将 MySQL 卸载后重装。

（7）数据库连接不上。

这里要注意一下 Hive 和 Oozie，如果已经在 MySQL 中创建了元表，选择已经存在的 MySQL，输入 URL 和密码后，单击连接测试总是失败，此时需要先停止 Ambari，执行下面这个命令：

```
ambari-server setup--jdbc-db=MySQL--jdbc-driver=/usr/share/java/MySQL-
connector-java.jar
```

再次连接测试就会成功。

（8）软链接错误。

找到对应文件夹进行删除或重命名。

```
Error:Cannotretrieverepositorymetadata(repomd.xml)forrepository:base.
Pleaseverifyitspath and try again
```

解决方案：使用 yum clean all 命令。

```
yum makecache
```

（9）文件链接存在重复。

安装 HDFS 和 HBASE 的时候出现：

```
/usr/hdp/current/hadoop-client/conf  doesn't exist
```

/etc/Hadoop/conf 文件链接存在是由于/etc/hadoop/conf 和/usr/hdp/current/hadoop-client/conf 目录互相链接，造成死循环，所以要改变其中一个的链接。

```
cd /etc/hadoop
rm -rf conf
ln -s /etc/hadoop/conf.backup /etc/hadoop/conf
```

HBase 也会遇到同样的问题，解决方式同上。

```
cd /etc/hbase
rm -rf conf
ln -s /etc/hbase/conf.backup /etc/hbase/conf
```

ZooKeeper 也会遇到同样的问题，解决方式同上。

```
cd /etc/ZooKeeper
rm -rf conf
ln -s /etc/ZooKeeper/conf.backup /etc/ZooKeeper/conf
```

14.2　CDH 的安装与集群管理

14.1 节中介绍了 Ambari 及其安装，Ambari 是开源免费的。目前业界还有一个用得比较多的工具是 CDH（Cloudera Distribution Hadoop），CDH 是把 Apache 的 Hadoop 开源项目进行了商业化，集成了很多补丁，可以减少大幅的安装工作，可以直接用于生产环境。

14.2.1　什么是 CDH 和 Cloudera Manager 介绍

CDH 是由 Cloudera 维护，在稳定版本的基础上构建的 Apache Hadoop，集成了很多补丁，把 Apache hadoop 开源项目进行商业化的 Hadoop 分布式系统，可直接用于生产环境。

Cloudera Manager 的作用是为了协助管理 CDH 集群，Cloudera Manager 提供了统一的 UI 界面，这样便于系统管理者快速地自动配置和部署 CDH 及相关组件，其中包括 Hadoop 整个生态圈的组件，例如 Hadoop、Hive、HBase、Sqoop 和 ZooKeeper 等。

Cloudera Manager 还提供了各种各样可自定义化的监视诊断及报告功能、集群上统一的日志管理功能等，这些功能可以更加便于企业统一管理和维护自己的数据中心，功能还包括多租户功能、自动恢复功能、高可用容灾部署功能、统一的集群配置管理及实时配置变更功能。另外，Cloudera Manager 被分为免费的版本和功能更加齐备的商用版本，可以大大简化集群中 Hadoop、Hive、Spark 等服务的安装配置管理。

14.2.2　Cloudera Manager 与 Ambari 对比的优势

Cloudera Manager 与 Ambari 各自优点，Ambari 最大的特点是免费，Cloudera Manager 与 Ambari 对比的优势如下：

- 版本更新速度快，划分清晰。
- 易用、稳定。

- 商用版有着更好的技术服务。
- 支持多种安装方式。
- 市场占有率高。

14.2.3 CDH 安装和网卡配置

Hadoop 集群安装时涉及很多组件，一个一个安装配置起来比较麻烦，还要考虑 HA、监控等。使用 Cloudera 可以很简单地部署集群，安装需要的组件，并且可以监控和管理集群。下面开始介绍 CDH 的安装过程。为方便演示下面我们以 3 个节点为例，介绍 CDH 的搭建过程，如表 14.3 所示。

表 14.3 CDH的搭建过程

CDH 集群规划	Master 192.168.4.10
	Slave1 192.168.4.11
	Slave2 192.168.4.12
所用系统	CentOS 6.5
CDH 版本	5.4.0
Cloudera-mananger	5.4.3
JDK 版本	JDK 7.79

配置网卡，在所有节点进行网络配置。

```
# vi /etc/sysconfig/network-scripts/ifcfg-eth0
```

写入：

```
DEVICE= eth0
TYPE =Ethernet
ONBOOT=yes
NM_CONTROLLED=yes
BOOTPROTO=static
IPADDR=192.168.4.10
GATEWAY=192.168.4.1
NATMASK=255.255.255.0
```

14.2.4 修改本机名与定义 DNS 服务器

在所有节点修改主机名和 DNS 服务器：

```
# vi /etc/sysconfig/network
```

把 HOSTNAME 改成集群节点的名字：

```
NETWORKING=yes
HOSTNAME=master
```

定义 DNS 服务器。在所有节点配置域名解析。

```
# vi /etc/resolv.conf
```

写入：

```
nameserver 114.114.114.114
```

接着重启网络：

```
# service network restart
[rootomaster ~]# service network restart
Shutting down interface ethe:           [OK]
Shutting down loopback interface:       [OK]
Bringing up loopback interfacer         [OK]
Bringing up interface eth0: Determining if ip address 192.168 .72.10 is
already in use for device eth0...
                            [OK]
```

可以 ping 一下，测试外网能否连通。

14.2.5 修改 hosts 主机映射关系

在所有节点修改 hosts 文件，实现 IP 和主机映射关系。

```
# vi /etc/hosts
```

写入：

```
192.168.4.10 master
192.168.4.11 slave1
192.168.4.12 slave2
127.0.0.1   localhost locaLhost .localdomain LocaLhost4 LocaLhost4.localdonain4
::1         localhost Localhost.localdomain localhost6 localhost6.localdonain6
192.168.4.10 master
192.168.4.11slavel
192 .168 4 .12 slave2
```

14.2.6 关闭防火墙

在所有节点关闭防火墙，分为以下两种情况。

（1）临时关闭：

```
# service iptables stop
[root@master-]# service iptables stop
iptables:Setting chains  to policy ACCEPT: filter      [OK]
iptables:Flushing  firewall rules:                     [OK]
iptables: Unloading modules :                          [OK]
```

（2）永久关闭，重启后生效：

```
# chkconfig iptables off
[root@master~]# chkconfig iptables off
```

关闭防火墙之后，查看防火墙状态：

```
# service iptables status
[root@master~]# service iptables status
iptables: Firewall is not running.
```

14.2.7 安装 JDK

在所有节点安装 JDK，安装命令和安装过程如下：

```
[root@ambari software]# rpm -ivh jdk-7u79-linux-x64.rpm
Preparing...                ########################[100%]
  1:jdk                     ########################[100%]
Unpacking JAR files...
        rt.jar...
        jsse.jar...
        charsets.jar...
        tools.jar...
        localedata.jar...
        jfxrt.jar...
```

安装完成之后，配置环境变量命令如下：

```
# vi .bash_profile
```

配置文化内容如下：

```
JAVA_HOME=/usr/java/jdk1.7.0_79
JRE_HOME=/usr/java/jdk1.7.0_79/jre
PATH=$PATH:$HOME/bin:$JAVA_HOME/bin
export CLASSPATH=$JAVA_HOME/lib/dt.jar:$JAVA_HOME/lib/tools.jar:JRE_HOME/bin
#.bash_profile
# Get the aliases and functions
if[-f~/.bashrc ]; then
    .-/.bashrc
fi

# User specific environment and startup programs

JAVA HOME=/usr/java/jdk1.7.0 79
JRE_HOME=/usr/java/jdk1.7.0 79/jre
PATH=$PATH: $HOME/bin :$JAVA_HOME/bin
export CLASSPATH=$JAVA HOME/lib/dt.jar:$JAVA_HOME/lib/tools.jar:JRE_HOME/bin

export PATH
```

使配置生效：

```
# source .bash_profile
```

14.2.8 升级 OpenSSL 安全套接层协议版本

很多加密算法中用到了 OpenSSL 这个组件，低版本的 OpenSSL 存在很多漏洞和风险，需要升级到新版本，在所有节点升级 OpenSSL。

```
# rpm -qa | grep openssl
```

结果如下:

```
# openssl-1.0.1e-15.el6.x86_64
```

这个版本比较旧,需要对其进行升级。

升级命令如下:

```
# yum upgrade openssl
```

14.2.9　禁用 SELinux 的强制访问功能

在所有节点关闭 SELinux 命令,因为在 SELinux 强制访问控制体系限制下,进程只能访问那些被指定可以访问的资源,即在它的任务中所需要的文件,不能关闭的原因可能是因为权限冲突。

直接关闭 SELinux 的命令如下:

```
# setenforce 0
```

不立即关闭 SELINUX,需要重启机器后生效的命令如下:

```
# vi /etc/sysconfig/selinux
```

将 SELINUX=enforcing 改为 SELINUX=disabled。

14.2.10　SSH 免密码登录

配置 SSH 免密码登录方式,免密码登录能大大提高工作效率。

主从节点全部生成密钥:

```
# ssh-keygen -t rsa -P ''
```

主节点发送密钥到从节点,命令如下:

```
# scp .ssh/id_rsa.pub root@slave1:~
```

在从节点执行如下命令:

```
# cat id_rsa.pub >> .ssh/authorized_keys
```

在主节点执行如下命令:

```
# cat .ssh/id_rsa.pub >> .ssh/authorized_keys
```

如果没有 wget,则通过 YUM 进行安装。

```
# yum -y install wget
```

14.2.11　同步 NTP 安装

在集群中,如果时间不一致则往往会导致数据混乱,可以通过在所有节点上安装 NTP,用于同步系统时间,也可以通过自己搭一台 NTP 服务器来保持整个集群时间的一致性,

过程如下:

```
# yum install -y ntp
# vi /etc/ntp.conf
restrict 127.0.0.1
restrict-6 ::1
restrict  192.168.127.0 mask 255.255.255.0 nomodify notrap
server  210.72.145.44          #中国授时中心
server  202.112.10.36          #1.cn.pool.ntp.org
server  59.124.196.83          #@0.asia.pool.ntp.org
server  192.168.72.10          #局域网 NTP 本机 IP
server  127.127.1.0            #本地时间
fudge server  127.127.1.0 stratum 10
```

首先,"restrict 192.168.127.0 mask 255.255.255.0 nomodify notrap",意思是从 IP 地址 192.168.217.1-192.168.217.254,默认网关 255.255.255.0 的服务器都可以使用 NTP 服务器进行时间同步。

其次,指定互联网和局域网中作为 NTP 服务器的 IP 地址,在上述配置文件中相关内容如下:

```
server  210.72.145.44
server  202.112.10.36
server  59.124.196.83
server  192.168.72.10
```

另外,当服务器与公用的时间服务器失去联系时以本地时间为客户端提供时间服务,在上述配置文件中相关内容如下:

```
server  127.127.1.0            #本地时间
fudge server  127.127.1.0 stratum 10
```

配置文件修改完成后,保存并退出,然后启动服务。

```
[root@master yum.repos.d]# chkconfig --list ntpd
ntpd            0:off 1:off 2:off 3:off 4:off 5:off 6:off

# chkconfig ntpd on
# service ntpd start
```

启动后,一般需要 5~10 分钟左右的时间才能与外部时间服务器开始同步时间。可以通过命令查询 NTPD 服务情况,查看服务连接和监听如下:

```
[root@master-]# netstat -tlunp | grep ntp
udp    0    0 192.168.72.100:123    0.0.0.0:*         1416/ntpd
udp    0    0 127.0.0.1:123         0.0.0.0:*         1416/ntpd
udp    0    0 0.0.0.0:123           0.0.0.0:*         1416/ntpd
udp    0    0 fe80::20C:29ff:fe28 8fc1:123  :::*      1416/ntpd
udp    0    0 :1:123                :::*              1416/ntpd
udp    0    0 :::123                :::*              1416/ntp
```

重新启动服务。

```
# service ntpd restart
```

将其他节点的时间与 master 同步，在其他所有节点运行命令，就可以将节点间的时间同步了。

```
# ntpdate master
```

14.2.12　安装 MySQL

在主节点安装 MySQL：

```
# yum install -y mysql-server
```

设置开机自动启动 MySQL：

```
# chkconfig mysqld on
```

直接启动 MySQL：

```
# service mysqld start
```

查看 MySQL 服务是否设置为开机自动启动：

```
# chkconfig --list | grep mysqld
```

更改密码：

```
# mysqladmin -u root password '123456'
```

更改密码后进入 MySQL：

```
# mysql -u root -p123456
```

在 MySQL 中分别创建 ambari、hive、oozie 三个数据库表。

```
create database hive DEFAULT CHARSET utf8 COLLATE utf8_general_ci;
create database amon DEFAULT CHARSET utf8 COLLATE utf8_general_ci;
create database oozie default charset utf8 collate utf8_general_ci;
```

授权 root 用户在 mster 节点拥有所有数据库的访问权限。

```
grant all privileges on *.* to 'root'@'master' identified by '123456' with grant option;
flush privileges;
```

14.2.13　安装 Cloudera Manager

在主节点下安装 Cloudera Manager：

Cloudera Manager 下载路径为 http://archive.cloudera.com/cm5/cm/5/。

CDH 下载路径为 http://archive.cloudera.com/cdh5/parcels/。

解压下载好的 Cloudera Manager 至/opt 下，产生两个文件夹：cloudera 和 cm-5.3.0。

14.2.14 添加 MySQL 驱动包和修改 Agent 配置

在主节点,添加 MySQL 驱动包。

MySQL JDBC 驱动下载地址为:

http://download.softagency.net/MySQL/Downloads/Connector-J/

将 MySQL-connector-java-5.1.26-bin.jar 复制至 CM(Cloudera Manager)的 lib 目录如下:

```
# cp /opt/ mysql-connector-java-5.1.26-bin.jar /opt/cm-5.4.3/share/cmf/lib
```

在主节点修改 Agent 配置。

将 agent/config.ini 中的 server_host 改为主节点的主机名。

```
vi /opt/cm-5.4.3/etc/cloudera-scm-agent/config.ini
[General ]
# Hostname of the CM server.
server._host=master
# Port that the CM server is listening on.
server_port=7182
```

14.2.15 初始化 CM5 数据库和创建 cloudera-scm 用户

在主节点初始化 CM5 数据库:

```
/opt/cm-5.4.3/share/cmf/schema/scm_prepare_database.sh mysql cm -h
localhost -u root -p123456 --scm-host localhost scm scm scm
```

其中,scm_prepare_database.sh 表示数据库类型,MySQL 表示数据库名称,localhost 表示 MySQL 数据库主机名,-uroot 表示用 root 身份运行 MySQL,-p123456 表示 MySQL 的 root 密码,--scm-host localhost 表示主机名,scm scm scm 表示数据库名,数据库用户名,数据库密码。

同步 CM-5.43 到其他节点:

```
scp -r /opt/cm-5.4.3 root@slave1:/opt/
scp -r /opt/cm-5.4.3 root@slave2:/opt/
```

在所有节点创建 cloudera-scm 用户:

```
useradd --system--home=/opt/cm-5.4.3/run/cloudera-scm-server/--no-
create-home--shell=/bin/false --comment "Cloudera SCM User" cloudera-scm
```

14.2.16 准备 Parcels

server 和 agent 是用来接收和发送数据的目录,server 端的 parcel-repo 目录会把所有的安装文件全部下载到该目录下,而 agent 也需要安装包,Parcels 就是用来存储指定的安装包。

将下载好的 CDH-xxx.parcels 文件、CDH-xxx.parcels.sha1 文件和 manifest.json 文件移至/opt/cloudera/parcel-repo/中,并将 CDH-xxx.parcels.sha1 重命名为 CDH-xxx.parcels.sha,否则会重新下载。

14.2.17　CDH 的安装配置

在主节点启动 Server：

`# /opt/cm-5.4.3/etc/init.d/cloudera-scm-server start`

所有节点（包括节点）启动 Agent 服务：

`# /opt/cm-5.4.3/etc/init.d/cloudera-scm-agent start`

观察启动服务端的日志，其中查看的日志路径如下：

`# cd /opt/cm-5.4.3/log/cloudera-scm-server`

Cloudera Manager Server 和 Agent 都启动以后，就可以进行 CDH5 的安装配置了。

可以通过浏览器访问主节点的 7180 端口进行测试，默认的用户名和密码均是 admin，如图 14.7 所示。

图 14.7　登录首页

为 CDH 集群安装指定主机，如图 14.8 所示。

图 14.8　为 CDH 群集安装指定主机

选择所装 CDH 的版本号，单击"继续"按钮，如图 14.9 所示。

图 14.9　群集安装

以上步骤如果配置本地 Parcel 包无误，那么图 14.10 中的"已下载"，应该是瞬间就完成了，之后就是耐心等待分配过程，这个时间取决于节点之间的传输速度。

Cloudera 建议将/proc/sys/vm/swappiness 设置为 0，当前设置为 60，使用 sysctl 命令在运行时更改该设置并编辑/etc/sysctl.conf 文件，重启后会保存该设置，如图 14.10 所示。

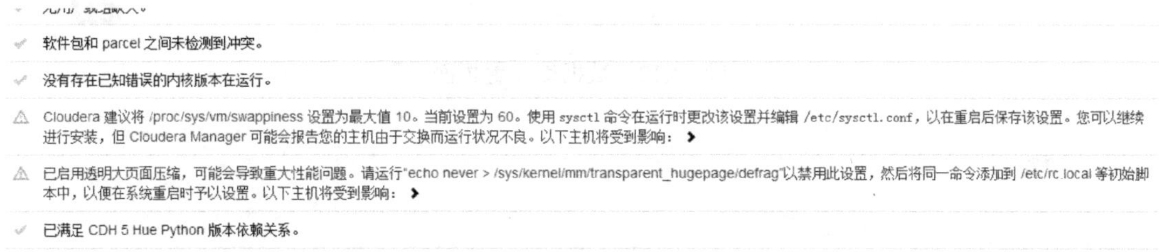

图 14.10　设置建议提示

服务配置通常保持默认就可以了，Cloudera Manager 会根据机器的配置自动进行配置，如果需要特殊调整，可以进行相应设置即可，如图 14.11 和图 14.12 所示。

前面所创建的数据库表选择默认即可，如图 14.13 所示。

至此，CDH 的安装和配置的主要过程就完成了。

图 14.11　自定义角色分配页面

图 14.12　数据库设置页面

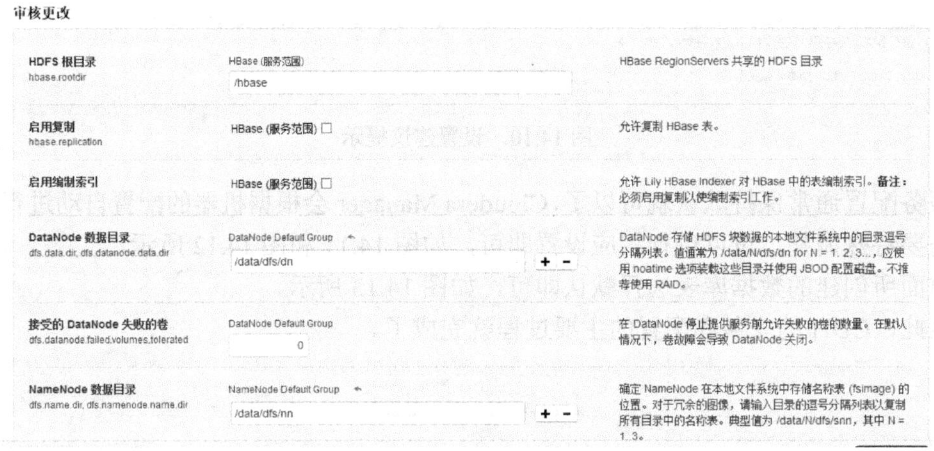

图 14.13　审核更改页面

14.3 小　　结

　　本章主要介绍了什么是 Ambari 及 CDH，以及 Ambari 和 CDH 各自的特点，接着完成了 Ambari 和 CDH 的安装。在安装和配置过程中，读者可以参考官方网站及官方提供的相关手册。

第 15 章 快速且通用的集群计算系统——Spark

Spark 是一个统一的、用于大数据分析处理的、快速且通用的集群计算系统。它开创了不以 MapReduce 为执行引擎的数据处理框架，提供了 Scala、Java、Python 和 R 这 4 种语言的高级 API，以及支持常规执行图的优化引擎。

15.1　Spark 基础知识

Spark 还支持包括用于离线计算的 Spark Core、用于结构化数据处理的 Spark SQL、用于机器学习的 MLlib、用于图形处理的 GraphX 和进行实时流处理的 Spark Streaming 等高级组件，它在项目中通常用于迭代算法和交互式分析。

15.1.1　Spark 的特点

Spark 在性能和通用性上都有显著优势，它是基于内存计算的并行计算框架，这使它的数据处理速度更快，具有高容错性和高可伸缩性。同时 Spark 可以运行在 YARN 上，无缝集成 Hadoop 组件，在已有 Hadoop 集群上使用 Spark。Spark 的特点如图 15.1 所示。

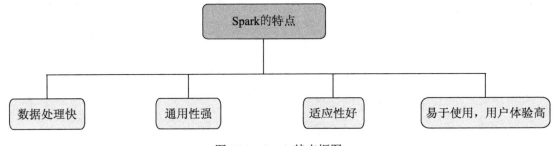

图 15.1　Spark 特点框图

- 数据处理快。Spark 是基于内存的计算框架，数据处理时将中间数据集放到内存中，减少了磁盘 I/O，提升了性能。

- 通用性强。提供了 MLlib、GraphX、Spark Streaming 和 Spark SQL 等多个出色的分析组件，涵盖了机器学习、图形算法、流式计算、SQL 查询和迭代计算等多种功能，组件间无缝、紧密地集成，一站式解决工作流中的问题。
- 适应性好。Spark 具有很强的适应性，能够与 Hadoop 紧密集成，支持 Hadoop 的文件格式，如以 HDFS 为持久层进行数据读写，能以 YARN 作为资源调度器在其上运行，成功实现 Spark 应用程序的计算。
- 易于使用，用户体验高。Spark 提供了 Scala、Java、Python 和 R 这 4 种语言的高级 API 和丰富的内置库，使更多的开发人员能在熟悉的编程语言环境中工作，用简洁的代码进行复杂的数据处理。而且 Scala 和 Python 语言的 REPL（read—eval—print—loop）交互模式使其应用更加灵活。

15.1.2 Spark 和 Hadoop 的比较

Spark 和大多数的数据处理框架不同，它并没有利用 MapReduce 作为计算框架，而是使用自己的分布式集群环境进行并行化计算。它最突出的特点是执行多个计算时，能将作业之间的数据集缓存在跨集群的内存中，因此利用 Spark 对数据集做的任何计算都会非常快，在实际项目中的大规模作业能大大节约时间。

Spark 在内存中存储工作数据集的特点使它的性能超过了 MapReduce 工作流，完美切合了迭代算法的应用要求，这与 MapReduce 每次迭代都生成一个 MapReduce 运行作业，迭代结果在磁盘中写入、读取不同；Spark 程序的迭代过程中，上一次迭代的结果被缓存在内存中，作为下一次迭代的输入内容，极大地提高了运行效率。

Spark 和 MapReduce 的相同点和不同点如下：
- Spark 是基于 MapReduce 的思想而诞生，二者同为分布式并行计算框架。
- MapReduce 进行的是离线数据分析处理，Spark 主要进行实时流式数据的分析处理。
- 在数据处理中，MapReduce 将 Map 结果写入磁盘中，影响整体数据处理速度；Spark 的 DAG 执行引擎，充分利用内存，减少磁盘 I/O，迭代运算效率高。
- MapReduce 只提供了 Map 和 Reduce 两种操作；Spark 有丰富的 API，提供了多种数据集操作类型（如 Transformation 操作中的 map、filter、groupBy、join，以及 Action 操作中的 count 和 collect 等）。

Spark 和 MapReduce 相比其内存消耗较大，因此在大规模数据集离线计算、时效要求不高的项目中，应优先考虑 MapReduce，而在进行数据的在线处理、实时数据计算时，更倾向于选择 Spark。

15.2 弹性分布式数据集 RDD

在实际数据挖掘项目中，通常会在不同计算阶段之间重复用中间数据结果，即上一阶段的输出结果会作为下一阶段的输入，如多种迭代算法和交互式数据挖掘工具的应用等。MapReduce 框架将 Map 后的中间结果写入磁盘，大量磁盘 I/O 拖慢了整体的数据处理速度。RDD（Resilient Distributed Dataset）的出现弥补了 MapReduce 的缺点，很好地满足了基于统一的抽象将结果保存在内存中的需求。Spark 建立在统一的抽象 RDD 上，这使 Spark 的各个组件得以紧密集成，完成数据计算任务。

15.2.1 RDD 的概念

分布式数据集 RDD 是 Spark 最核心的概念，它是在分布式集群节点中跨多个分区存储的一个只读的元素集合，是 Spark 中最基本的数据抽象。每个 RDD 可以分为多个分区，每个分区都是一个数据集片段，同一个 RDD 的不同分区可以保存在集群中不同的节点上，即 RDD 是不可变的、可分区的、里面数据可进行并行计算的、包含多个算子的集合。

RDD 提供了一种抽象的数据架构，根据业务逻辑将现有 RDD 通过转换操作生成新的 RDD，这一系列不同的 RDD 互相依赖实现了管道化，采用惰性调用的方式避免了多次转换过程中的数据同步等待，且中间数据无须保存，直接通过管道从上一操作流入下一操作，减少了数据复制和磁盘 I/O。

15.2.2 RDD 的创建方式

RDD 共有以下 3 种创建方式：
- 使用外部存储系统的数据集（如 HDFS 等文件系统支持的数据集）。
- 通过 Scala 集合或数组以并行化的方式创建 RDD。
- 对现有 RDD 进行转换来创建 RDD。

15.2.3 RDD 的操作

RDD 有转换（Transformation）和动作（Action）两大类操作，转换是加载一个或多个 RDD，从当前的 RDD 转换生成新的目标 RDD，转换是惰性的，它不会立即触发任何数据处理的操作，有延迟加载的特点，主要标记读取位置、要做的操作，但不会真正采取实际行动，而是指定 RDD 之间的相互依赖关系；动作则是指对目标 RDD 执行某个动作，触发 RDD 的计算并对计算结果进行操作（返回给用户或保存在外部存储器中）。

通常我们操作的返回类型判断是转换还是动作：转换操作包括 map、filter、groupBy、join 等，接收 RDD 后返回 RDD 类型；行动操作包括 count、collect 等，接收 RDD 后返回非 RDD，即输出一个值或结果。

15.2.4 RDD 的执行过程

RDD 的执行过程主要包括 RDD 的创建、转换和计算三部分，具体执行过程如图 15.2 所示。

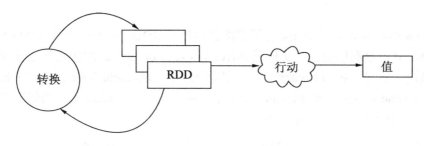

图 15.2 RDD 执行过程图

RDD 的详细执行流程如下：

（1）使用外部存储系统的数据集创建 RDD。

（2）根据业务逻辑，将现有 RDD 通过一系列转换操作生成新的 RDD，每一次产生不同的 RDD 传给下一个转换操作，在行动操作真正计算前，记录下 RDD 的生成轨迹和相互之间的依赖关系。

（3）最后一个 RDD 由行动操作触发真正的计算，并将计算结果输出到外部数据源（返回给用户或保存在外部存储器中）。

下面我们通过一个示例详细讲解 RDD 的工作流程。首先假设有一个数据集 Data，我们要对数据进行计算，输入、计算和输出过程如图 15.3 所示。

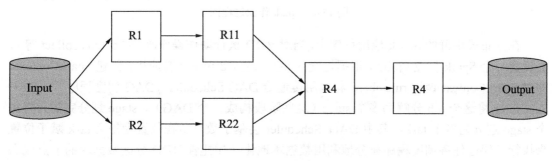

图 15.3 RDD 工作流程图

从外部存储系统的数据集输入数据 Data，创建 R1 和 R2 两个 RDD，经过多次的转换操作后生成了一个新的 RDD，即 R4，此过程中计算一直没有发生，但 RDD 标记了读取位置、要做的操作，Spark 只是记录了 RDD 间的生成轨迹和相互依赖关系，最后一个 RDD 即 R4 的动作操作触发计算时，Spark 才会根据 RDD 之间的依赖关系生成有向无环图 DAG，DAG 描述了 RDD 的依赖关系，也称为"血缘关系（Lineage）"。在一系列的转换和计算结束后，计算结果会输出到外部数据源上。

15.3 Spark 作业运行机制

前两节我们已经介绍了 Spark 的基础概念和核心 RDD，下面来详细地剖析 Spark 作业运行的过程，其中最高层的两个实体是 driver 和 executor，driver 的作用是运行应用程序的 main() 函数，创建 SparkContext，其中运行着 DAGScheduler、TaskScheduler 和 SchedulerBackend 等组件；而 executor 专属于应用，在 Application 运行期间运行并执行应用的任务。Spark 的运行过程如图 15.4 所示。

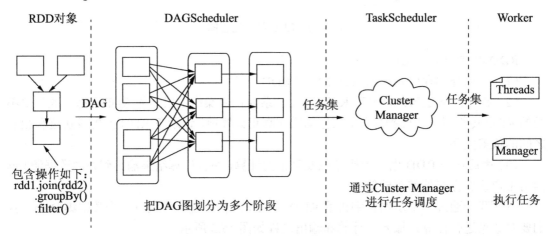

图 15.4　Spark 作业过程图

在分布式集群的 Spark 应用程序上，当对 RDD 执行动作操作时（如 count、collect 等），会提交一个 Spark 作业（job），根据提交的参数设置，driver 托管应用，创建 SparkContext，即对 SparkContext 调用 runJob()，将调用传递给 DAG Scheduler（DAG 调度程序）。DAG Scheduler 将这个 job 分解为多个 stage（这些阶段构成一个 DAG），stage 划分完后，将每个 stage 划分为多个 task，其中 DAG Scheduler 会基于数据所在位置为每个 task 赋予位置来执行，保证任务调度程序充分地利用数据本地化（如托管 RDD 分区数据块的节点或保存 RDD 分区的 executor）。DAG Scheduler 将这个任务集合传给 Task Scheduler，在任务集合发送到 Task Scheduler 之后，Task Scheduler 基于 task 位置考虑的同时构建由 Task 到

Executor 的映射，将 Task 按指定的调度策略分发到 Executor 中执行。在这个调度的过程中，SchedulerBackend 负责提供可用资源，分别对接不同的资源管理系统；无论任务完成或失败，Executor 都向 Driver 发送消息，如果任务失败则 Task Scheduler 将任务重新分配在另一个 Executor 上，在 Executor 完成运行任务后会继续分配其他任务，直到任务集合全部完成。

15.4 运行在 YARN 上的 Spark

我们已经知道 Spark 可以和 Hadoop 紧密集成，而在 YARN 上运行 Spark 的模式恰好提供了与 Hadoop 组件最紧密的集成，它是在我们已部署好的 Hadoop 集群上应用 Spark 最简便的方法。

15.4.1 在 YARN 上运行 Spark

在 Spark 的独立模式中，因为是单独部署到一个集群中，不依赖其他资源管理系统，集群资源调度是 Master 节点负责，只能支持简单的固定资源分配策略，即每个任务固定核数量，每个作业按顺序依次分配资源，资源不够时排队等待，因此通常会遇到一些用户分配不到资源的问题。此时 Spark 就可以将资源调度交给 YARN 负责，YARN 支持动态资源调度，因此能很好地解决这个问题。

我们知道 YARN 是一个资源调度管理系统，它不仅能为 Spark 提供调度服务，还能为其他子系统（如 Hadoop、MapReduce 和 Hive 等）服务，由 YARN 来统一为分布式集群上的计算任务分配资源，提供资源调度，从而有效地避免了资源分配的混乱无序。

15.4.2 Spark 在 YARN 上的两种部署模式

在 YARN 上运行 Spark 时，YARN 的调度模式主要包括 YARN 客户端模式和 YARN 集群模式，下面我们说一下 Spark 的这两种部署模式的含义。
- YARN 集群模式：Spark 程序启动时，YARN 会在集群的某个节点上为它启动一个 Master 进程，然后 Driver 会运行在 Master 进程内部并由这个 Master 进程启动 Driver 程序，客户端提交作业后，不需要等待 Spark 程序运行结束。
- YARN 客户端模式：跟 YARN 集群模式相似的是 Spark 程序启动时，也会启动一个 Master 进程，但 Driver 程序运行在本地而不在这个 Master 进程内部运行，仅仅是利用 Master 来申请资源，直到程序运行结束。

上面我们介绍了 Spark 的两种部署模式的含义，下面说一下二者的区别。

Spark 程序在运行时，在 YARN 集群模式下，Driver 进程在集群中的某个节点上运行，基本不占用本地资源。这种模式适合生产环境的运行方式。

而在 YARN 客户端模式下，Driver 运行在本地，对本地资源会造成一些压力，但它的优点是 Spark 程序在运行过程中可以进行交互，这种模式适合需要交互的计算。

因此，建议具有任何交互式组件的程序都使用 YARN 客户端模式，同时，客户端模式因为任何调试输出都是立即可见的，因此构建 Spark 程序时非常有价值；当用于生成作业时，建议使用 YARN 集群模式，此时整个应用都在集群上运行，更易于保留日志文件以备检查。

15.5 Spark 集群安装

前面介绍了 Spark 相关的基础知识和运行机制，现在我们要动手实践搭建 Spark 集群，然后基于 Spark 集群进行实例讲解。我们在这里搭建的是 Spark 完全分布式集群，用到了三台机器，接下来看一下 Spark 集群的安装步骤。

15.5.1 Spark 安装包的下载

（1）首先访问 Spark 官网 http://spark.apache.org/，如图 15.5 所示。

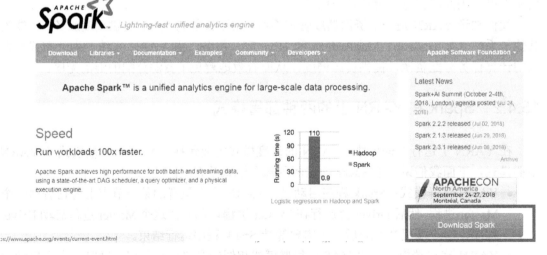

图 15.5 Spark 官网

（2）选择要下载的 Spark 安装包版本及其需要兼容的 Hadoop 版本类型，如图 15.6 所示。

第 15 章 快速且通用的集群计算系统——Spark

图 15.6　Spark 版本选择

（3）在这里我们选择安装 Spark 当前的最新版本 spark-2.3.1，并兼容 2.7 及之后的 Hadoop，然后单击下载按钮，如图 15.7 所示。

图 15.7　安装包

（4）其中，下载网页链接为 https://www.apache.org/dyn/closer.lua/spark/spark-2.3.1/spark-2.3.1-bin- hadoop2.7.tgz，下载镜像内容如 15.8 图所示。

单击下载链接开始下载，之后我们的 Spark 安装包就下载完成了。

图 15.8 下载包链接

15.5.2 Spark 安装环境

因为我们搭建的是 Spark 完全分布式集群，在上传并安装 Spark 安装包前，首先要确认以下 4 点：

（1）一台 Master 和两台 Slave，并已实现 SSH 免密码登录，使我们启动 Spark 时 Master 能通过 SSH 启动远端 Worker。

（2）安装配置好 JDK（这里我使用的是 jdk1.8.0_60）。

（3）Hadoop 分布式集群已搭建完成（启动 Spark 前要先启动 HDFS 和 YARN）。

（4）Scala 已安装并配置好。

因为 Spark 的运行需要 Java 和 Scala 的支持，因此首先需要配置 Java、Scala 运行环境，同时为了实现 Spark 和 Hadoop 的集成，需要基于 Hadoop 分布式集群进行 Spark 的集群部署。最后，因为 Spark 的 Master 和 Worker 需要通过 SSH 进行通信，并利用 SSH 启动远端 Worker，因此必须实现 Master 和 Slave 的 SSH 免密码登录。其中，Java 环境配置、Hadoop 完全分布式搭建和 SSH 免密码设置等关于 Hadoop 的安装和配置方法已经在第 2 章中介绍过，如果未设置，请参考 2.5～2.8 节的内容。

15.5.3 Scala 安装和配置

Spark 的运行需要 Scala 的支持，Scala 语法简洁，同时支持 Spark-Shell，更易于原型设计和交互，接下来我们将介绍 Scala 的安装和配置过程。

1. 安装包下载并解压

首先访问 Scala 官网 https://www.scala-lang.org/download/，单击下载按钮，如图 15.9 所示。

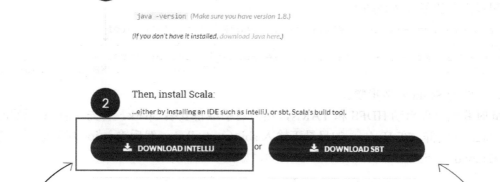

图 15.9 Scala 安装包下载图

其次解压 Scala 到指定目录，这里解压在/opt/software 目录下，Linux 命令如下：

```
[root@master ~]# cd /opt/software
[root@master software]# tar -zxvf scala-2.11.6.tgz
```

2. 配置Scala的环境变量

打开 profile 配置文件，命令如下：

```
[root@master ~]# vi /etc/profile
```

在文件最后添加以下内容：

```
export SCALA_HOME=/opt/software/scala-2.11.6
export PATH=$PATH:$SCALA_HOME/bin

JAVA_HOME=/usr/java/jdk1.8.0_60
export PATH=$PATH:$JAVA_HOME/bin
```

修改好配置文件后关闭文件，为使配置文件立即生效，输入如下命令：

```
[root@master ~]# source /etc/profile
```

接下来我们要检查安装是否成功，输入 scala -version 命令：

```
[root@master ~]# scala -version
Scala code runner version 2.11.6 -- Copyright 2002-2013, LAMP/EPFL
```

当输入命令后显示出以上版本信息内容，则表示已安装成功。

同时 slave1 和 slaver2 节点也需要安装并配置 Scala，将 Scala 的文件和配置文件分别复制到 slave 节点，解压缩安装包，命令如下：

```
[root@master ~]# scp -r /opt/software/scala-2.11.6 root@slave1:/opt/software/scala-2.11.6
[root@master ~]# scp -r /opt/software/scala-2.11.6 root@slave2:/opt/software/scala-2.11.6
```

配置文件的复制命令如下：

```
[root@master ~]# scp /etc/profile root@slave1:/etc/profile
profile                          100% 2021     2.0KB/s   00:00
[root@master ~]# scp /etc/profile root@slave2:/etc/profile
profile                          100% 2021     2.0KB/s   00:00
```

以上即为 Scala 安装步骤。

此时我们可以启动 HDFS 和 YARN，然后着手准备安装 Spark。首先，启动已经配置好的 Hadoop 集群，在 Hadoop 的目录下输入 sbin/start-all.sh，然后输入 jps 观察进程状态，此时 Hadoop 分布式集群和 YARN 都已启动，如图 15.10 所示。

```
[root@master2 hadoop-2.7.0]# sbin/start-all.sh
This script is Deprecated. Instead use start-dfs.sh and start-yarn.sh
Starting namenodes on [master2]
master2: starting namenode, logging to /opt/software/hadoop-2.7.0/logs/hadoop-root-namenode-master2.out
slavenode2: starting datanode, logging to /opt/software/hadoop-2.7.0/logs/hadoop-root-datanode-slavenode2.out
slavenode1: starting datanode, logging to /opt/software/hadoop-2.7.0/logs/hadoop-root-datanode-slavenode1.out
Starting secondary namenodes [master2]
master2: starting secondarynamenode, logging to /opt/software/hadoop-2.7.0/logs/hadoop-root-secondarynamenode-master2.out
starting yarn daemons
starting resourcemanager, logging to /opt/software/hadoop-2.7.0/logs/yarn-root-resourcemanager-master2.out
slavenode1: starting nodemanager, logging to /opt/software/hadoop-2.7.0/logs/yarn-root-nodemanager-slavenode1.out
slavenode2: starting nodemanager, logging to /opt/software/hadoop-2.7.0/logs/yarn-root-nodemanager-slavenode2.out
[root@master2 hadoop-2.7.0]# jps
1536 SecondaryNameNode
1953 Jps
1691 ResourceManager
1343 NameNode
```

图 15.10　Hadoop 和 YARN 成功启动

15.5.4　Spark 分布式集群配置

在 Hadoop 和 YARN 都成功启动的前提下，我们现在开始进行 Spark 分布式集群的搭建，具体步骤如下：

1．上传并解压安装包

将我们在 15.5.1 节中下载好的 spark-2.3.1-bin-hadoop2.7.tgz 安装包上传到 Linux 的 /opt/ software 目录下，并解压安装包到本目录中，命令如下：

```
[root@master ~]# tar -zxvf spark-2.3.1-bin-hadoop2.7.tgz
```

为了简便，修改文件名，命令如下：

```
[root@master ~]# mv spark-2.3.1-bin-hadoop2.7 spark-2.3.1
```

2. Spark集群配置

修改配置文件.bash_profile，输入 vi .bash_profile，修改内容如下：

```
export JAVA_HOME=/usr/java/jdk1.8.0_60
export HADOOP_HOME=/opt/software/hadoop-2.7.0
export SPARK_HOME=/opt/software/spark-2.3.1
export SCALA_HOME=/opt/software/scala-2.11.6
export PATH=$PATH:$JAVA_HOME/bin:$HADOOP_HOME/bin:$SPARK_HOME/bin:
$SCALA_HOME/bin
```

修改配置文件/etc/profile，输入 vi /etc/profile，修改内容如下：

```
export SPARK_HOME=/opt/software/spark-2.3.1
export PATH=$PATH:$SPARK_HOME/bin

export SCALA_HOME=/opt/software/scala-2.11.6
export PATH=$PATH:$SCALA_HOME/bin

JAVA_HOME=/usr/java/jdk1.8.0_60
export PATH=$PATH:$JAVA_HOME/bin
```

进入/opt/software/spark-2.3.1/conf 目录：

```
[root@master ~]# cd /opt/software/spark-2.3.1/conf/
```

复制配置文件 spark-env.sh.template，并将其重命名为 spark-env.sh，命令如下：

```
[root@master conf]# cp spark-env.sh.template spark-env.sh
```

修改 spark-env.sh，在配置文件的末尾添加如下配置（实际中可以根据 spark-env.sh 注释中列出的模式选择不同配置），编辑文件命令如下：

打开文件：

```
vi spark-env.sh
```

添加如下内容：

```
export SPARK_DIST_CLASSPATH=$(/opt/software/hadoop-2.7.0/bin/hadoop classpath)
export JAVA_HOME=/usr/java/jdk1.8.0_60
export SCALA_HOME=/opt/software/scala-2.11.6
export HADOOP_HOME=/opt/software/hadoop-2.7.0
export HADOOP_CONF_DIR=/opt/software/hadoop-2.7.0/etc/hadoop
export SPARK_MASTER_IP=192.158.179.103
export SPARK_WORKER_MEMORY=1g
export SPARK_WORKER_CORES=2
export SPARK_WORKER_INSTANCES=2
export JAVA_HOME=/usr/java/jdk1.7.0_79
```

复制并重命名 slaves.template 文件为 slaves，命令如下：

```
[root@master2 conf]# cp slaves.template slaves
```

打开 slaves 文件，并在文件末尾加入节点地址（前提是 etc/hosts 中已经指定对应地址的主机名称），命令如下：

```
slave1
slave2
```

以上配置 slaves 文件的作用是指定 Worker 节点,在 Spark 启动时会读取此文件,令 Master 和 Worker 通信,启动远端 Worker。

将我们对 Spark 的配置同步到其他节点:

```
[root@master ~]# scp -r /opt/software/spark-2.3.1 root@slave1:/opt/
software/spark-2.3.1
[root@master ~]# scp -r /opt/software/spark-2.3.1 root@slave2:/opt/
software/spark-2.3.1
```

同时将其他节点的配置文件.bash_profile 和/etc/profile 做同样的修改。

3. Spark集群启动

进入/opt/software/spark-2.3.1 目录下,命令如下:

```
cd /opt/software/spark-2.3.1
```

在 Spark 的目录下输入 sbin/start-all.sh 启动 Spark 集群,命令如下:

```
[root@master2 spark-2.3.1]# sbin/start-all.sh
starting org.apache.spark.deploy.master.Master, logging to /opt/software/
spark-2.3.1/logs/spark-root-org.apache.spark.deploy.master.Master-1-
master2.out
slavenode2: starting org.apache.spark.deploy.worker.Worker, logging to /
opt/software/spark-2.3.1/logs/spark-root-org.apache.spark.deploy.worker.
Worker-1-slavenode2.out
slavenode1: starting org.apache.spark.deploy.worker.Worker, logging to /
opt/software/spark-2.3.1/logs/spark-root-org.apache.spark.deploy.worker.
Worker-1-slavenode1.out
slavenode2: starting org.apache.spark.deploy.worker.Worker, logging to /
opt/software/spark-2.3.1/logs/spark-root-org.apache.spark.deploy.worker.
Worker-2-slavenode2.out
slavenode1: starting org.apache.spark.deploy.worker.Worker, logging to /
opt/software/spark-2.3.1/logs/spark-root-org.apache.spark.deploy.worker.
Worker-2-slavenode1.out
```

启动后,为检查 Spark 集群是否启动成功,查看进程命令如下。

主节点启动 Master 进程,命令如下:

```
[root@master2 spark-2.3.1]# jps
7270 ResourceManager
7116 SecondaryNameNode
7773 Jps
6925 NameNode
7695 Master
```

其他节点启动 Worker 进程,命令如下:

```
[root@slavenode1 conf]# jps
3808 Worker
3424 DataNode
3881 Jps
```

```
3770 Worker
3519 NodeManager
```

出现以上结果就表示 Spark 集群搭建成功。

15.6　Spark 实例详解

本章前面介绍了 Spark 的基础知识和作业运行流程，为了让读者更深入地理解 Spark 运行机制，增强 Spark 项目实战经验，提高实践编程能力，下面以网站用户浏览量排名和网站用户归属地客户量统计这两个实例来简单介绍 Spark 的应用。

15.6.1　网站用户浏览次数最多的 URL 统计

现在有一个 IT 教育网站，其中包括 Java、PHP 和 NET 等多个栏目，我们想要根据用户访问的访问日志信息，分别统计出每门学科的前 3 名统计量。

我们已经搜集好网站日志，其中第 1 个字段是访问日期，第 2 个字段是访问的 URL，第 3 个字段是每个栏目都有的一个独立域名，数据如下：

```
20180321101954    http://java.sendto.cn/java/course/javaeeadvanced.shtml
20180321101954    http://java.sendto.cn/java/course/javaee.shtml
20180321101954    http://java.sendto.cn/java/course/android.shtml
20180321101954    http://java.sendto.cn/java/video.shtml
20180321101954    http://java.sendto.cn/java/teacher.shtml
20180321101954    http://java.sendto.cn/java/course/android.shtml
20180321101954    http://php.sendto.cn/php/teacher.shtml
20180321101954    http://net.sendto.cn/net/teacher.shtml
20180321101954    http://java.sendto.cn/java/course/hadoop.shtml
20180321101954    http://java.sendto.cn/java/course/base.shtml
20180321101954    http://net.sendto.cn/net/course.shtml
20180321101954    http://php.sendto.cn/php/teacher.shtml
20180321101954    http://net.sendto.cn/net/video.shtml
20180321101954    http://java.sendto.cn/java/course/base.shtml
20180321101954    http://net.sendto.cn/net/teacher.shtml
20180321101954    http://java.sendto.cn/java/video.shtml
20180321101954    http://java.sendto.cn/java/video.shtml
20180321101954    http://net.sendto.cn/net/video.shtml
20180321101954    http://net.sendto.cn/net/course.shtml
20180321101954    http://java.sendto.cn/java/course/javaee.shtml
20180321101954    http://java.sendto.cn/java/course/android.shtml
20180321101955    http://php.sendto.cn/php/course.shtml
20180321101955    http://net.sendto.cn/net/teacher.shtml
20180321101955    http://php.sendto.cn/php/teacher.shtml
20180321101955    http://java.sendto.cn/java/course/base.shtml
20180321101955    http://net.sendto.cn/net/teacher.shtml
```

```
20180321101955    http://java.sendto.cn/java/course/javaee.shtml
20180321101955    http://php.sendto.cn/php/video.shtml
```

由以上的访问日志可以看出，每个学科的域名是独一无二的，而每个域名类下有不同的访问网址 URL，我们要统计每个域名下排名前 3 的点击量的 URL，可以先将完全相同的 URL 进行分类和统计次数，然后再将每个域名下的 URL 的小类和次数进行累加得到每门学科的所有 URL 类和次数。案例的代码编写如下：

```scala
package com.sendto.urlcount

import java.net.URL

import org.apache.spark.{SparkConf, SparkContext}

/**
 * 根据指定的学科，取出点击量前三名
 */
object AdvUrlCount {

  def main(args: Array[String]) {

    //获取本网站中各学科的域名，用于后面的循环遍历
    val arr = Array("java.sendto.cn", "php.sendto.cn", "net.sendto.cn")

//设置应用名称和节点
    val conf = new SparkConf().setAppName("AdvUrlCount").setMaster("local[2]")
    //创建 SparkContext
val sc = new SparkContext(conf)
    //rdd1 将数据按行切分，先将 URL 的分类和个数统计出来，元组中放的是(URL, 1)
    val rdd1 = sc.textFile("c://sendto.log").map(line => {
      val f = line.split("\t")
      (f(1), 1)
    })
    //根据 URL 的 key 得到每类 URL 访问的次数
    val rdd2 = rdd1.reduceByKey(_ + _)
    //因为我们要取得每门学科的访问次数，提取域名
    val rdd3 = rdd2.map(t => {
      val url = t._1
      val host = new URL(url).getHost
      (host, url, t._2)    //这里得到的是(域名,url,访问次数)
    })

    //对所有学科循环遍历，根据科目将访问次数前名的数据提取出来
    for (ins <- arr) {
      对每门学科的域名进行拦截，符合某域名类下的 URL 都拦截出来并根据次数进行排序
      val rdd = rdd3.filter(_._1 == ins)
      val result= rdd.sortBy(_._3, false).take(3)
      //在这里可以通过 JDBC 向数据库中存储数据
```

```
        //ID，学院，URL，次数， 访问日期
        println(result.toBuffer)
    }
    sc.stop()
  }
}
```

15.6.2 用户地域定位实例

某运营商通过网关拿到了自己的客户访问记录，包括时间、IP、访问域名、访问资源和用户相关浏览器和系统信息，现在要根据客户的这些数据定位客户的地域信息，得到客户在各省的分布情况。

客户访问记录如下：

20090121000132095572000|125.213.100.123|show.51.com|/shoplist.php?phpfile=shoplist2.php
20090121000132124542000|117.101.215.133|www.jiayuan.com|/19245971|Mozilla/4.0 (compatib
20090121000132406516000|117.101.222.68|gg.xiaonei.com|/view.jsp?p=389|Mozilla/4.0 (comp
20090121000132581311000|115.120.36.118|tj.tt98.com|/tj.htm|Mozilla/4.0 (compatible; MSI
20090121000132864647000|123.197.64.247|cul.sohu.com|/20071227/n254338813_22.shtml|Mozil20090121000133296729000|222.55.57.176|down.chinaz.com|/|Mozilla/4.0 (compatible; MSIE 6
20090121000133331104000|123.197.66.93|www.pkwutai.cn|/down/downLoad-id-45383.html|Mozil
20090121000133446262000|115.120.12.157|v.ifeng.com|/live/|Mozilla/4.0 (compatible; MSIE
20090121000133456256000|115.120.7.240|cqbbs.soufun.com|/3110502342~1~2118/23004348_2300
20090121000133586141000|117.101.219.241|12.zgwow.com|/launcher/index.htm|Mozilla/4.0(co

在已知客户 IP 的情况下，想要定位客户的归属地，首先需要知道每个省市不同运营商的 IP 段是多少，根据我们取得的客户 IP 是否包含在某个省运营商的 IP 段内，来判断客户是否属于某个省，最后累加每个省的人数。

假设我们已经收集到了每个省市的各大运营商的 IP 段和相关地域等信息，其中，IP 段已经按升序排好，收集数据 operatorIP.txt 如下：

1.0.1.0|1.0.3.255|16777472|16778239|亚洲|中国|福建|福州||电信|350100|China|CN|119.306239|26.075302
1.0.8.0|1.0.15.255|16779264|16781311|亚洲|中国|广东|广州||电信|440100|China|CN|113.280637|23.125178
1.0.32.0|1.0.63.255|16785408|16793599|亚洲|中国|广东|广州||电信|440100|China|CN|113.280637|23.125178
1.1.0.0|1.1.0.255|16842752|16843007|亚洲|中国|福建|福州||电信|350100|China

```
|CN|119.306239|26.075302
1.1.2.0|1.1.7.255|16843264|16844799|亚洲|中国|福建|福州||电信|350100|China
|CN|119.306239|26.075302
1.1.8.0|1.1.63.255|16844800|16859135|亚洲|中国|广东|广州||电信|440100|China
|CN|113.280637|23.125178
1.2.0.0|1.2.1.255|16908288|16908799|亚洲|中国|福建|福州||电信|350100|China
|CN|119.306239|26.075302
1.2.2.0|1.2.2.255|16908800|16909055|亚洲|中国|北京|北京|海淀|北龙中网|110108
|China|CN|116.29812|39.95931
1.2.4.0|1.2.4.255|16909312|16909567|亚洲|中国|北京|北京||中国互联网信息中心
|110100|China|CN|116.405285|39.904989
1.2.5.0|1.2.7.255|16909568|16910335|亚洲|中国|福建|福州||电信|350100|China
|CN|119.306239|26.075302
```

在已知客户 IP 和相关地域对应 IP 的规则库之后，就可以进行代码的编写，将所有客户的归属地求出并累加各归属地个数。具体代码如下：

```scala
package com.sendto.iplocation

import org.apache.spark.{SparkConf, SparkContext}

object IPLocation {

    //这里定义了一个 IP 地址的转换方法，将所有字符串类型的 IP 转换为 Long 类型，
    再进行比较
  def ip2Long(ip: String): Long = {
    val fragments = ip.split("[.]")
    var ipNum = 0L
    for (i <- 0 until fragments.length){
      ipNum =  fragments(i).toLong | ipNum << 8L
    }
    ipNum
  }
    //利用二分法来寻找客户 IP 属于的 IP 段
    //这里传入的是 long 类型的 IP 和提取的对应规则，即(开始 PI 段,结束 IP 段,所在省)
  def binarySearch(lines: Array[(String, String, String)], ip: Long) : Int = {
    var low = 0
    var high = lines.length - 1
    while (low <= high) {
      val middle = (low + high) / 2
//先将客户 IP 和所有已排序 IP 段的中间 IP 段进行比较，如果恰好在这个 IP 段中间，
则返回这个 IP 段
      if ((ip >= lines(middle)._1.toLong) && (ip <= lines(middle)._2.toLong))
        return middle
      //如果不在中间 IP 段，客户 IP<中间 IP 段的开始值，则在上半段，那么就在上半段继续
      实行二分法进行匹配
```

```scala
      if (ip < lines(middle)._1.toLong)
        high = middle - 1
      else {
        low = middle + 1
      }
    }
    -1
}

def main(args: Array[String]) {
    //创建 SparkContext
  val conf = new SparkConf().setMaster("local[2]").setAppName("IpLocation")
  val sc = new SparkContext(conf)
  //读取规则库数据,把有价值的、可以进行比较的映射规则提取出来
  val ipRulesRdd = sc.textFile("c://operatorIp.txt").map(line =>{
    val fields = line.split("\\|")
    val start_num = fields(2)
    val end_num = fields(3)
    val province = fields(6)
    (start_num, end_num, province)    //将 IP 段的起始、结尾和归属省提取并返回
  })
  //整合全部的 IP 映射规则
  val ipRulesArrary = ipRulesRdd.collect()

  //广播规则:将数据广播出去,即将 IP 映射规则广播给所有属于这个任务的 Executor
  val ipRulesBroadcast = sc.broadcast(ipRulesArrary)

  //加载要处理的数据
  val ipsRDD = sc.textFile("c://access_log").map(line => {
    val fields = line.split("\\|")
    fields(1)
  })
  //调用我们之前编写的函数,将 IP 转换为 long 类型,并根据映射规则找到所有 IP 所属 IP
  //段的索引,再将对应的索引从映射规则中提取出来,得到每个 IP 对应的省信息,再根据省信
  //息进行累加求出每个省分别多少客户
  val result = ipsRDD.map(ip => {
    val ipNum = ip2Long(ip)
    val index = binarySearch(ipRulesBroadcast.value, ipNum)
    val info = ipRulesBroadcast.value(index)
    //info 内容为(IP 的起始 Num, IP 的结束 Num,省份名)
    // //将每个省的返回规则个数进行累加,得到每个省的客户个数
    info
  }).map(t => (t._3, 1)).reduceByKey(_+_)

  })

  println(result.collect().toBuffer)
```

```
    sc.stop()

  }
}
```

最终打印结果如下:

ArrayBuffer(陕西,1824),(河北,383),(云南,126),(重庆,868),(北京,1535)

以上就是我们求得的客户地区和客户个数。

15.7 小　　结

本章首先介绍了 Spark 的基础知识,包括核心概念和运行机制;然后讲解了 Spark 分布式集群的安装步骤,使读者能够独立地进行 Spark 平台搭建;最后通过两个 Spark 实例加强读者对并行计算框架的整体认知,从而更深入地理解 Spark 项目的思考方式和问题解决方法。

第 3 篇
Hadoop 项目案例实战

▶▶ 第 16 章　基于电商产品的大数据业务分析系统实战

▶▶ 第 17 章　用户画像分析实战

▶▶ 第 18 章　基于个性化的视频推荐系统实战

▶▶ 第 19 章　电信离网用户挽留实战

第 16 章 基于电商产品的大数据业务分析系统实战

前面章节中我们已经介绍了大数据的相关技术,包括 HDFS、MapReduce、Hive 和 Flume 等技术。这些技术到底如何使用呢?本章我们会基于一个电商产品的大数据业务分析系统案例把这些知识串联起来,使读者对相关知识有整体的理解。

掌握这些案例,对于读者学习大数据技术起到举足轻重的作用,因此建议读者把这些程序在自己的环境中运行起来。关于环境的搭建,可以参考相关章节。

16.1 项目背景、实现目标和项目需求

随着互联网的快速发展,越来越多的行业都在尝试从网站挖掘来访客户的潜在价值,比如用户在打开网页后浏览了哪个页面,进行了什么操作,哪些用户来自于山东,哪些用户来自于北京,哪些用户购买了哪些商品,哪些商品容易销售等,进而挖掘客户的潜在需求。比如当用户多次浏览了 A 类商品,则说明用户对 A 类商品有购买意愿,如果有大批用户来自山东地区,则说明山东地区具有巨大的市场潜力。

说到这里,不得不提到天猫"双 11"。据报道,2017 年天猫"双 11"交易额超过 2016 年的 1 207 亿元人民币,再次创下新纪录。其实,不仅是天猫,很多电商后台都会有日志收集和分析系统,会记录用户浏览页面的相关操作和购买行为,并在后台以日志形式存在,如用户的搜索、收藏、加入购物车、购买和评论等都对商家有很大的参考价值。比如,用户通过各类浏览器访问的网站最后会保存在磁盘上的 logs 目录下,如图 16.1 所示。

本章,我们将介绍一个简化版本的商城日志分析项目,使读者不仅可以掌握大数据项目的整体流程,而且可以同时掌握大数据中的核心技术应用。

首先我们从电商的参与者着手,这样便于通过不同的维度进行分析,而电商的参与者主要包括买方、卖方和快递三大类。接着可以对不同的用户群体进行不同维度的分析,包括基于买家的分析、基于商品的分析,以及买家购买了哪些商品的分析,如买家的消费额度等。

第 16 章 基于电商产品的大数据业务分析系统实战

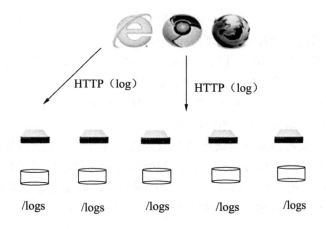

图 16.1 用户通过各类浏览器访问的网站保存在磁盘上的 logs 目录下

16.2 功能与流程

本案例由数据采集、数据存储、数据处理和数据可视化 4 部分组成，具体流程如图 16.2 所示。

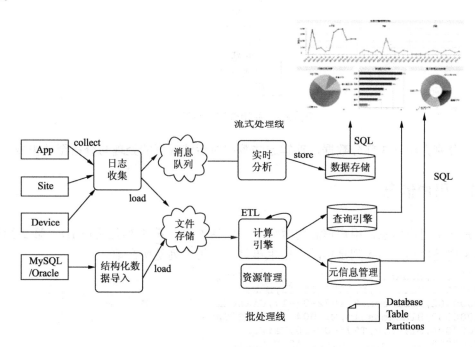

图 16.2 大数据处理流程

开发大数据项目的第一项工作往往是采集原始数据，在这个项目中，原始数据包括 3 部分，分别是用户信息、商品信息和购买记录。

- 用户信息包括：用户 ID、性别、年龄（出生年月）、所在地。
- 商品信息包括：商家 ID、类别、商品的商家。
- 购买记录包括：订单号、用户 ID、商品 ID、交易时间、价格、发货城市、收货城市、来源网站、快递单号、快递公司。

这里强调一点，读者需要和传统的关系型数据库区分开，这些信息初始是以 log 的形式生成，并以文本的样式存储在磁盘上，比如电商后台会把用户登录时的 IP 写在磁盘上的 log 文件中。其中，用户数据和商品数据量相对较少，而交易记录数据往往非常多，大型的商务网站需要将这些数据存放在大数据平台 HDFS 上，在后面会逐步讨论。原始数据的关系如图 16.3 所示。

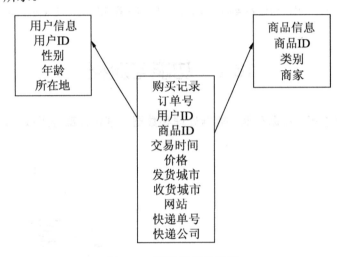

图 16.3 原始数据关系图

下面分别介绍这 3 部分数据，理解这些数据对后续数据的处理与分析至关重要。

16.2.1 用户信息

首先介绍用户信息，用户信息内容如下，包括 5 列，第 1 列是用户 ID，第 2 列是姓名，第 3 列是性别，第 4 列是年龄（出生年月），第 5 列是所在地。

```
00000000,Mcdaniel,F,1980-04-12,BeiJing
00000001,Harrell,F,1989-05-12,HuNan
00000002,Robinson,F,1972-07-17,JiangXi
00000003,Blankenship,M,2004-05-11,SiChuan
00000004,Miles,M,1975-09-20,TaiWan
00000005,Hughes,M,1997-01-12,SiChuan
00000006,Hale,F,2001-01-27,YunNan
00000007,Jackson,M,1987-03-05,XinJiang
```

```
00000008,Brooks,M,1990-03-20,SiChuan
00000009,Johnston,F,1985-02-19,GuangXi
00000010,Smith,F,1977-02-06,ShangHai
```

16.2.2 商品信息

商品信息内容如下，包括3列，第1列是商家ID、第2列是类别、第3列是商家，比如第一条记录的商家ID是00000000，类别是Sports，品牌（商家）是PUMA。

```
00000000,sports,PUMA
00000001,clothes,ZARA
00000002,food,NIULANSHAN
00000003,television,SAMSUNG
00000004,computer,SONY
00000005,sports,ANTA
00000006,sports,KAPPA
00000007,cosmetic,MEIFUBAO
00000008,sports,NB
00000009,computer,LENOVO
00000010,telephone,HTC
```

16.2.3 购买记录

购买记录描述了用户购买了哪些商品，数据内容如下代码。第1列是订单号，第2列是用户ID，第3列是商品ID，第4列是交易时间，后续依次是价格、发货城市、收货城市、来源网站、快递单号和快递公司。

```
0000000000,00000468,00000030,1494235629,347,HaiNan,LiaoNing,TIANMAO,
869905775627268,YUANTONG,186.160.56.200,tr
0000000001,00000026,00000448,1494235629,339,YunNan,HeBei,TIANMAO,
675929469467,SHENTONG,97.126.176.12,sid
0000000002,00000543,00000081,1494235629,332,SiChuan,TaiWan,TIANMAOCHAOSHI,
4010167066854172,SHENTONG,20.130.157.1,byn
0000000003,00000397,00000217,1494235629,669,YunNan,GuangXi,TIANMAO,
342636875611434,SHUNFENG,135.165.86.149,cv
0000000004,00000543,00000342,1494235629,472,HeBei,GuiZhou,TAOBAO,
371933698610006,SHUNFENG,41.249.58.179,bs
0000000005,00000088,00000177,1494235629,834,XiangGang,NeiMengGu,TAOBAO,
6011254950459740,SHENTONG,237.177.13.163,eu
0000000006,00000451,00000486,1494235629,309,NeiMengGu,ShangHai,TAOBAO,
180032521959143,EMS,58.150.39.28,gl
0000000007,00000515,00000926,1494235629,758,NeiMengGu,HuNan,TAOBAO,
4214464510259775,ZHONGTONG,162.67.8.3,lg
0000000008,00000409,00000678,1494235629,978,ShanXi3,Aomen,TAOBAO,
4174495333403156,YUNDA,178.30.137.244,ig
0000000009,00000000,00000162,1494235629,73,HeNan,HeNan,TIANMAO,
4029121613570020,EMS,184.240.228.28,sd
0000000010,00000320,00000197,1494235629,627,ZheJiang,HuNan,TIANMAOCHAOSHI,
30055208479673,ZHONGTONG,21.203.205.227,hsb
```

16.3 数据收集

前面讲到，原始数据需要通过数据采集来完成，而数据采集通常分为两种方式，第一种方式是从手机 App、网站和其他设备上通过日志收集工具进行收集，这里的日志收集工具可以使用 Flume；第二种方式是从现有的系统中进行收集，如 MySQL/Oracle 导入，导入的工具可以使用 Sqoop。

针对本例来说，下一步就需要进行数据收集了，前面也提到由于购买记录数据量庞大，需要放到大数据存储平台 HDFS 上进行存储，将数据从本地传到 HDFS 上的工具可以用 Flume。关于 Flume 的内容，读者可以参考相关章节，下面我们只讲解在该例中如何使用 Flume。

16.3.1 Flume 的配置文件

首先，Flume 存放在/opt/software/flume/apache-flume-1.8.0-bin 路径下，需要在 Flume 的 conf 文件夹下创建配置文件 logAnalysis.properties 进行 Flume 的相关配置。配置文件内容如下：

```
#Source 从 Client 收集数据,传递给 Channel,这里 source 的名字是 logSource
mylogAgent.sources = logSource
# Channel 连接 Sources 和 Sinks ，Channel 可以理解为一个队列,在这里 Channel 的名字设置为 fileChannel
mylogAgent.channels = fileChannel
# Sink 从 Channel 收集数据,这里的名字是 hdfsSink,将从 Channel 收集过来的数据最终放到 HDFS 上
mylogAgent.sinks = hdfsSink
#将 Source 的属性定义为 Exec,使用 Exec 的时候需要指定 Shell 命令来对日志进行读取
mylogAgent.sources.logSource.type = spooldir
mylogAgent.sources.logSource.spoolDir = /opt/data/loganalysis/records/
#将 fileChannel 绑定到 Source 上
mylogAgent.sources.logSource.channels = fileChannel
#定义 Sink 的类型
mylogAgent.sinks.hdfsSink.type = hdfs
#将收集到的数据放在 HDFS 上并按照特定的格式进行显示
mylogAgent.sinks.hdfsSink.hdfs.path = hdfs:// master:9000/flume/record/%Y-%m-%d/%H%M
# 设置 HDFS 文件的前缀。默认是 FlumeData,这里设置前缀为 record_log
mylogAgent.sinks.hdfsSink.hdfs.filePrefix=transaction_log
#产生新文件的间隔,默认是 30（秒）,0 表示不以时间间隔为准
mylogAgent.sinks.hdfsSink.hdfs.rollInterval= 120
#event 达到特定大小时再产生一个新文件,默认是 10,0 表示不以 event 数目为准
mylogAgent.sinks.hdfsSink.hdfs.rollCount= 1000
#文件到达特定大小时重新产生一个新文件,默认是 1024bytes,0 表示不以文件大小为准
```

```
mylogAgent.sinks.hdfsSink.hdfs.rollSize= 0
#类似于"四舍五入",下面3行代码的意思是每10分钟生成一个文件
mylogAgent.sinks.hdfsSink.hdfs.round = true
mylogAgent.sinks.hdfsSink.hdfs.roundValue = 10
mylogAgent.sinks.hdfsSink.hdfs.roundUnit = minute

# hdfs.fileType 用于控制文件类型,这里设置为 DataStream
mylogAgent.sinks.hdfsSink.hdfs.fileType = DataStream
mylogAgent.sinks.hdfsSink.hdfs.useLocalTimeStamp = true
#将 FileChannel 绑定到 Sink 上
mylogAgent.sinks.hdfsSink.channel = fileChannel
# Each channel's type is defined.
#FileChannel 的类型设置为 file
mylogAgent.channels.fileChannel.type = file
# 存放检查点目录, checkpointDir 是一个目录
mylogAgent.channels.fileChannel.checkpointDir=
/opt/software/flume/apache-flume-1.8.0-bin/dataCheckpointDir
#存放数据的目录
mylogAgent.channels.fileChannel.dataDirs=
/opt/software/flume/apache-flume-1.8.0-bin /dataDir
```

16.3.2 启动 Flume

Flume 配置文件创建完后,就可以启动 Flume 了。根据前面的配置,Flume 启动完成后会检测/opt/data/loganalysis/records/文件夹,并把其中的文件上传到 HDFS 上。上传到 HDFS 上的主要原因是 HDFS 是分布式文件系统,在 HDFS 上可以存放大量的数据,所以不需要担心数据量过大而无法存储的问题。下面是 Flume 的启动命令:

```
./flume-ng agent  --conf-file ../conf/flume-conf-logAnalysis.properties
--name logAgent -Dflume.root.logger=INFO,console
```

其中,flume-ng agent 代表启动 Flume 代理;--conf-file 指定配置文件为 logAnalysis. properties;--name 指定代理的名字是 logAgent,-Dflume.root.logger 是设置日志输出级别和显示方式。Flume 启动完成后,可以在后台控制台看到以下信息:

```
18/04/27 09:34:00 INFO node.PollingPropertiesFileConfigurationProvider:
Configuration provider starting
18/04/27 09:34:00 INFO node.PollingPropertiesFileConfigurationProvider:
Reloading configuration file:../conf/flume-conf-logAnalysis.properties
18/04/27 09:34:00 INFO conf.FlumeConfiguration: Processing:hdfsSink
```

16.3.3 查看采集后的文件

启动 Flume 之后,就可以访问 http://ip:50070 了,其中 IP 是 NameNode 的地址。由于配置了监测/opt/data/loganalysis/records/文件夹,所以 Flume 采集文件夹中的文件并放到 HDFS 上。关于 HDFS 上的存放路径配置是 mylogAgent.sinks.hdfsSink.hdfs.path = hdfs:// master:9000/flume/record/%Y-%m-%d/%H%M ,其中,master 是 NameNode 节点的机器名,

9000 是端口，由于存放路径是/%Y-%m-%d 的格式标识，所以最终以年-月-日格式显示，如图 16.4 所示。

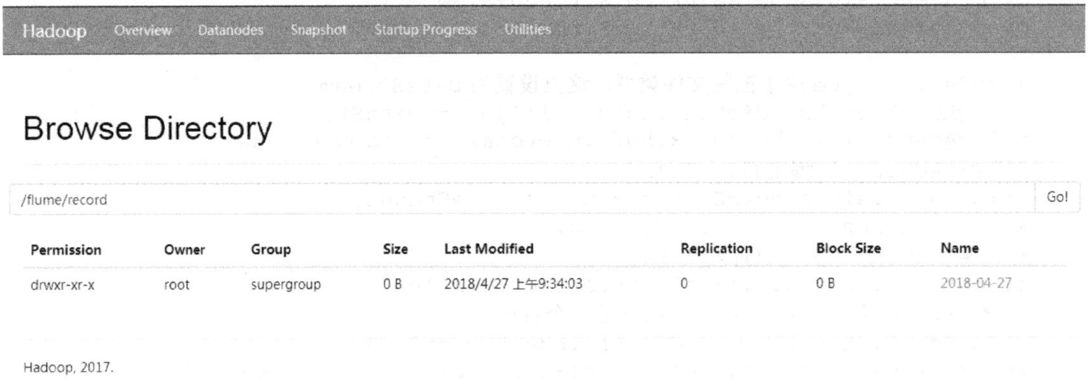

图 16.4　基于浏览器查看 HDFS 上的文件

单击 2018-04-27 链接，可以看到如图 16.5 所示的信息。其中，0930 对应%H%M，表示是 9 点 30 分生成的文件。

图 16.5　显示 0930 文件

继续点开 0930 文件，可以看到有一个 transaction_log.1524193717651 文件。这就是 Flume 存放在 HDFS 上的文件，在这里读者需要注意的是，这个文件是存放在了集群上，而不是在某台特定的计算机上。

16.3.4　通过后台命令查看文件

查看生成的文件有两种方式，第一种是通过浏览器查看，另外一种是通过后台 HDFS 命令查看，命令如下：

```
hdfs dfs -ls /flume/record/2018-04-27/0930
```

其中，/flume/record/2018-04-27/0930 代表存放文件的路径。

当在后台执行查询命令时，可以列出相关文件，执行命令和结果如下：

```
[root@master records]# hdfs dfs -ls /flume/record/2018-04-27/0930
Found 5 items
-rw-r--r--1rootsupergroup11100282018-04-2709:34/flume/record/2018-04-27/0930/transaction_log.1524792841496
-rw-r--r--1rootsupergroup11102242018-04-2709:34/flume/record/2018-04-27/0930/transaction_log.1524792841497
-rw-r--r--1rootsupergroup11107752018-04-2709:34/flume/record/2018-04-27/0930/transaction_log.1524792841498
-rw-r--r--1rootsupergroup11110632018-04-2709:34/flume/record/2018-04-27/0930/transaction_log.1524792841499
-rw-r--r--1rootsupergroup111882018-04-2709:34/flume/record/2018-04-27/0930/transaction_log.1524792841500.tmp
```

扩展名是.tmp 代表还没有生成完毕的临时文件。

16.3.5　查看文件内容

前面我们通过第 2 种方式查看了 HDFS 上的文件，接着可以查看文件的内容，相关命令是 hdfs dfs -cat ，比如想要查看/flume/record/2018-04-27/0930/transaction_log.1524792841496 文件，可以执行以下命令：

```
hdfs dfs -cat  /flume/record/2018-04-27/0930/transaction_log.1524792841496 | tail -10
```

结果如下：

```
[root@masterrecords]#hdfs dfs -cat /flume/record/2018-04-27/0930/transaction_log.1524792841496| tail -10
0000009989,00000324,00000157,1494237443,344,ChongQing,BeiJing,JUHUASUAN,4189333890504,SHENTONG,12.153.117.42,fil
0000009990,00000256,00000077,1494237443,208,GuiZhou,GuiZhou,TIANMAO,4816864308189,YUANTONG,118.110.192.76,os
0000009991,00000331,00000295,1494237443,761,JiLin,TaiWan,TIANMAO,340787335790904,SHENTONG,208.109.80.9,uk
0000009992,00000340,00000427,1494237443,564,GuangDong,Aomen,TIANMAOCHAOSHI,30292252741340,YUNDA,87.248.43.49,ast
0000009993,00000092,00000422,1494237443,791,AnHui,JiangXi,JUHUASUAN,180094320378750,EMS,184.28.246.104,dv
0000009994,00000402,00000264,1494237443,848,HeNan,LiaoNing,JUHUASUAN,869985226241000,SHUNFENG,48.126.5.253,se
0000009995,00000497,00000561,1494237444,130,ShanXi3,SiChuan,TAOBAO,4259501077678107,ZHONGTONG,12.172.169.147,xh
0000009996,00000401,00000618,1494237444,443,JiangSu,ShanXi3,TIANMAO,210097518565199,YUNDA,241.98.145.191,mt
0000009997,00000166,00000895,1494237444,131,NingXia,JiangSu,JUHUASUAN,6011606123756890,YUNDA,241.226.176.91,ro
0000009998,00000553,00000061,1494237444,21,HeNan,SiChuan,JUHUASUAN,375999018058735,EMS,205.110.143.124,dz
```

由于数据量比较大，为了只显示一部分数据，在"|"之后加上了 tail -10，其作用是

显示最后 10 条数据。接着将用户信息 user.list 和商品信息 brand.list 上传到 HDFS 的相应目录下。

16.3.6　上传 user.list 文件

上传 user.list 文件。代码如下：

```
hdfs dfs -put user.list /flume/user
```

执行完成后，在浏览器中查看相应结果，如图 16.6 所示。

图 16.6　浏览器中查看 user.list 文件

16.3.7　上传 brand.list 目录

上传 brand.list 目录。代码如下：

```
hdfs dfs -put brand.list /flume/brand
```

结果如图 16.7 所示。

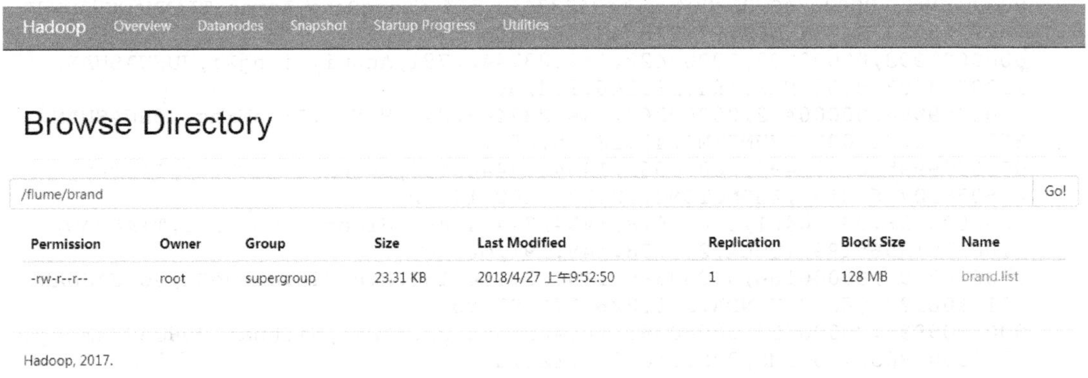

图 16.7　在浏览器中查看 brand.list 文件

通常，用户信息和商品信息是存放在关系型数据库中，在这里将这两类数据上传到 HDFS 上的目的是为了后续和交易数据进行整合分析。

简要回顾，目前我们已经完成了数据的收集工作，总体逻辑是通过 Flume 将生成的 log 文件上传到 HDFS 上，本例中的 log 文件可以是各类数据，如用户操作行为、运行环境、访问 IP 等，只要是想记录下来的信息，都可以放入 log 日志，除了进行数据分析，也方便后续的查询稽核。

16.4 数据预处理

前面我们将收集的数据上传到了 HDFS 上，下面就需要完成数据的清洗工作了。所谓数据清洗，主要是对各种"脏"数据（不符合要求）进行相应处理，一是为了解决数据的质量问题，二是使数据更加便于做挖掘与分析。不同的目的，清洗方式和规则也不一样，包括但不局限于"数据完整性""数据唯一性""数据合法性""数据一致性"等。

通过查看 HDFS 上的文件内容，我们发现某些数据需要进行处理，比如第 4 列的交易时间是从 1970 年到现在的秒数，而我们更习惯看到类似年-月-日的样式，如 2018-08-08 23:39。

```
0000044213,00000363,00000252,1494243770,41,GuangXi,YunNan,TIANMAOCHAOSHI,
4631586785877033,ZHONGTONG,240.26.213.17,unm
```

下面就可以使用 MapReduce 来完成上述格式的转化，将类似 1494243770 这样的长整数样式换成 2018-08-08 23:39 这样的年月日时分秒样式。在此创建 MapReduce 的核心类 DateUtilMapper.java。代码如下：

```java
package com.mr.etl;

import java.io.IOException;
import java.text.SimpleDateFormat;

import org.apache.hadoop.io.LongWritable;
import org.apache.hadoop.io.Text;
import org.apache.hadoop.mapreduce.Mapper;

public class DateUtilMapper extends Mapper<LongWritable, Text, Text, Text>{
    protected void map(LongWritable key, Text value,
            Context context)
            throws IOException, InterruptedException {
        String[] words =value.toString().split(",");
        
        String content = "";
        
        for(int i = 0;i<words.length-2;i++){
            
            //如果是第 4 列，则进行数据处理
```

```
            if(i==3){
                //按照 yyyy-MM-dd HH:mm:ss 格式对数据处理
                SimpleDateFormat sdf = new SimpleDateFormat("yyyy-MM-dd HH:mm:ss");
                content = content+ sdf.format(Long.parseLong(words[3]+"000"))+"\t";

                //对于不是第 4 列的数据，直接进行字符串连接
            }else if(i==words.length-3){

                    content=content+words[i]+"";

            }else{
                //处理完后数据换行
                content=content+words[i]+"\t";
            }
        }
        context.write(new Text(content), new Text(""));

    }
}
```

以上代码是数据处理的核心代码，其中最核心的逻辑是当遇到第 4 列时，则将呈现格式转化成类似 2018-08-08 08:08 这样的年月日时分秒，其他情况则直接将字符串进行连接。

与 Mapper 对应的是 Reducer，以下是 DateUtilReducer 的核心代码。

```
DateUtilReducer.java
package com.mr.etl;

import java.io.IOException;

import org.apache.hadoop.io.IntWritable;
import org.apache.hadoop.io.Text;
import org.apache.hadoop.mapreduce.Reducer;

public class DateUtilReducer extends Reducer<Text, Text, Text, Text>{

    protected void reduce(Text arg0, Iterable<Text> arg1,
            Context arg2)
            throws IOException, InterruptedException {

        arg2.write(arg0, new Text(""));
    }
}
```

这里需要说明的是，由于核心代码在 Mapper 中已经处理完成，所以在 Reducer 中没有复杂的逻辑处理。

另外，Mapper 和 Reducer 需要通过一个主函数进行调用，我们通过 RunJob.java 作为主函数来调用 Mapper 和 Reducer，相关代码如下：

RunJob.java
package com.mr.etl;

import org.apache.hadoop.conf.Configuration;
import org.apache.hadoop.fs.FileSystem;
import org.apache.hadoop.fs.Path;
import org.apache.hadoop.io.Text;
import org.apache.hadoop.mapreduce.Job;
import org.apache.hadoop.mapreduce.lib.input.FileInputFormat;
import org.apache.hadoop.mapreduce.lib.output.FileOutputFormat;

public class RunJob {

    public static void main(String[] args) {

        Configuration conf =new Configuration();

        conf.set("fs.defaultFS", "hdfs://localhost:9000");

        try {
            FileSystem fs =FileSystem.get(conf);
            Job job =Job.getInstance(conf,"wc");

            job.setJarByClass(RunJob.class);

            job.setMapperClass(DateUtilMapper.class);
            job.setReducerClass(DateUtilReducer.class);

            job.setMapOutputKeyClass(Text.class);
            job.setMapOutputValueClass(Text.class);

            FileInputFormat.addInputPath(job, new Path("/flume/record/2018-04-27/0930"));

            Path output =new Path("/opt/logs/record_dimension/");
            if(fs.exists(output)){
                fs.delete(output, true);
            }
            FileOutputFormat.setOutputPath(job, output);

            boolean f= job.waitForCompletion(true);
            if(f){
                System.out.println("job Complete");
            }
        } catch (Exception e) {
            e.printStackTrace();
        }
    }
}
```

在以上代码中：
```

FileInputFormat.addInputPath(job,newPath("/flume/record/2018-04-27/0930")) 的作用是设置了 HDFS 上的数据源。Path output =new Path("/opt/logs/record_dimension/")的作用是设置了数据处理后文件的存放路径。

启动数据清洗程序，在 Mapper、Reducer 和 RunJob 主函数编写完毕后，就可以上传到服务器并启动主函数完成数据的清洗工作了，调用命令如下：

hadoop jar dateutils.jar com.mr.etl.RunJob

通过运行以上命令，最终会在主函数中会调用 Mapper 和 Reducer 来完成数据的转化和清洗工作。主函数运行完后，会出现以下成功提示：

```
18/04/27 10:02:37 INFO mapreduce.Job: Counters: 35
    File System Counters
        FILE: Number of bytes read=194404854
        FILE: Number of bytes written=213063644
        FILE: Number of read operations=0
        FILE: Number of large read operations=0
        FILE: Number of write operations=0
        HDFS: Number of bytes read=20929407
        HDFS: Number of bytes written=4577913
        HDFS: Number of read operations=67
        HDFS: Number of large read operations=0
        HDFS: Number of write operations=14
    Map-Reduce Framework
        Map input records=44224
        Map output records=44224
        Map output bytes=4577982
        Map output materialized bytes=4666460
        Input split bytes=725
        Combine input records=0
        Combine output records=0
        Reduce input groups=44223
        Reduce shuffle bytes=4666460
        Reduce input records=44224
        Reduce output records=44223
        Spilled Records=88448
        Shuffled Maps =5
        Failed Shuffles=0
        Merged Map outputs=5
        GC time elapsed (ms)=300
        Total committed heap usage (bytes)=1670381568
    Shuffle Errors
        BAD_ID=0
        CONNECTION=0
        IO_ERROR=0
        WRONG_LENGTH=0
        WRONG_MAP=0
        WRONG_REDUCE=0
    File Input Format Counters
        Bytes Read=4911177
    File Output Format Counters
        Bytes Written=4577913
job Complete
```

通常情况下，当看到 job Complete 时，则表示数据已经处理完成。如果我们想查看处理后的相关文件，则可以在后台通过相关命令列出文件夹/opt/logs/record_dimension/下的文件了，执行命令和结果如下：

```
[root@master software]# hdfs dfs -ls /opt/logs/record_dimension/
Found 2 items
-rw-r--r--   1 root supergroup  02018-04-2710:02/opt/logs/record_dimension/_SUCCESS
-rw-r--r--   1 root supergroup45779132018-04-2710:02/opt/logs/record_dimension/part-r-00000
```

然后可以查看 part-r-00000 文件内容：

```
[root@master software]# hdfs dfs -cat /opt/logs/record_dimension/part-r-00000 | tail -10
0000044213   00000363    00000252    2017-05-08  19:42:50    41   GuangXi    YunNan     TIANMAOCHAOSHI    4631586785877033    ZHONGTONG
0000044214   00000264    00000337    2017-05-08  19:42:50    974  HeNan      XiZang     TAOBAO            3112965475218387    ZHONGTONG
0000044215   00000347    00000835    2017-05-08  19:42:51    458  ZheJiang   AnHui      JUHUASUAN         869957068104422     EMS
0000044216   00000189    00000610    2017-05-08  19:42:51    883  XiangGang  SiChuan    JUHUASUAN         180093303671603     SHENTONG
0000044217   00000570    00000217    2017-05-08  19:42:53    409  GuangXi    ZheJiang   TIANMAOCHAOSHI    30031673061320      SHENTONG
0000044218   00000454    00000541    2017-05-08  19:42:53    95   GanSu      ShanDong   TIANMAOCHAOSHI    3096474337270330    YUANTONG
0000044219   00000453    00000094    2017-05-08  19:42:53    734  ShanXi1    ShanXi1    JUHUASUAN         180077602041071     YUNDA
0000044220   00000043    00000314    2017-05-08  19:42:53    208  ChongQing  JiangXi    JUHUASUAN         5544770908298714    SHENTONG
0000044221   00000200    00000479    2017-05-08  19:42:53    187  ShanXi3    ShangHai   JUHUASUAN         869935332633749     EMS
0000044222   00000116    00000916    2017-05-08  19:42:53    437  NingXia    FuJian
```

从上面的结果中可以看到，第 4 列的数据已经变成了年-月-日样式，说明数据已经转化完成。

16.5　数据分析——创建外部表

到目前为止，我们已经将数据上传到了 HDFS 上，完成了数据收集、处理与清洗工作，那么下一步就可以进行数据分析了。在此，我们可以通过 Hive 对其进行分析，关于 Hive 的介绍与安装配置，请读者参考相应章节。

看过本书 Hive 章节的读者应该知道，Hive 的 HQL 操作都是基于"表"的操作，所以为了通过 Hive 完成数据的分析，我们需要创建 3 张表，分别对应前面处理后的 HDFS 上的 3 个文件 userinfo、brandinfo 和 recordsinfo。在此我们创建的是外部表，关于外部表和内部表的区别，读者可以查看相关章节。

创建 user 外部表的相关语句如下：

```
create external table if not exists userinfo(
 userid STRING,
 username STRING,
 gender STRING,
 birthdate DATE,
 province STRING
)ROW FORMAT DELIMITED
 FIELDS TERMINATED BY ','
 location 'hdfs://localhost:9000/flume/user/';
```

在 Hive 中执行以上代码，提示 OK 字样，则代表创建成功。

```
 hive> create external table if not exists userinfo(
 > userid STRING,
 > username STRING,
 > gender STRING,
 > birthdate DATE,
 > province STRING
 > )ROW FORMAT DELIMITED
 > FIELDS TERMINATED BY ','
 > location 'hdfs://localhost:9000/flume/user/';
OK
Time taken: 0.608 seconds
```

在以上创建表的代码中，localhost 是 NameNode 的本机地址；FIELDS TERMINATED BY ','代表每个字段以逗号","隔开；location 'hdfs://localhost:9000/flume/user/' 代表 HDFS 上面的数据位置，通过 Location 的设置，最终会将表和文件 user.list 中的数据完成映射，由于之前将数据上传到了 /flume/user/ 下面，所以这里的 Location 就指向了 /flume/user/。

完成了表的创建，也完成了表和文件的映射关系，接下来就可以查看数据了，为了避免显示数据过多，在这里通过 limit 10 命令来限制显示条数为 10 条，执行命令及结果如下：

```
hive> select * from userinfo limit 10;
OK
00000000    Mcdaniel     F    1980-04-12    BeiJing
00000001    Harrell  F    1989-05-12    HuNan
00000002    Robinson     F    1972-07-17    JiangXi
00000003    Blankenship M    2004-05-11    SiChuan
00000004    Miles    M    1975-09-20    TaiWan
00000005    Hughes   M    1997-01-12    SiChuan
00000006    Hale     F    2001-01-27    YunNan
00000007    Jackson  M    1987-03-05    XinJiang
00000008    Brooks   M    1990-03-20    SiChuan
00000009    Johnston     F    1985-02-19    GuangXi
Time taken: 0.091 seconds, Fetched: 10 row(s)
```

为了完成后续的数据分析，只有 userinfo 表是不够的，还需要再创建 brandinfo 表和 recordsinfo 表。为了确认表创建完成后和相应数据进行了关联，我们在创建完表后，均进行简单的查询来确保数据映射没有问题。

首先创建外部表 brandinfo。

```
create external table if not exists brandinfo (
```

```
 brandid STRING,
 category STRING,
 brandname STRING
)ROW FORMAT DELIMITED
 FIELDS TERMINATED BY ','
 location 'hdfs://localhost:9000/flume/brand/';
```

表创建完成后,通过 Hive 查看 brandinfo 表中的数据。

```
hive> select * from brandinfo limit 10;
OK
00000000    sports    PUMA
00000001    clothes   ZARA
00000002    food      NIULANSHAN
00000003    television  SAMSUNG
00000004    computer    SONY
00000005    sports    ANTA
00000006    sports    KAPPA
00000007    cosmetic    MEIFUBAO
00000008    sports    NB
00000009    computer    LENOVO
Time taken: 0.087 seconds, Fetched: 10 row(s)
```

接着创建外部表 recordsinfo。

```
create external table if not exists recordsinfo(
 recordid STRING,
 userid STRING,
 brandid STRING,
 trancation_date TIMESTAMP,
 price INT,
 from_province STRING,
 to_province STRING,
 website STRING,
 express_number STRING,
 express_company STRING
)
 ROW FORMAT DELIMITED
 FIELDS TERMINATED BY ','
 location 'hdfs://localhost:9000/flume/record/2018-04-27/0930/';
```

表创建完成后,通过 Hive 查看 recordsinfo 表中的数据。

```
hive> select * from recordsinfo limit 10;
OK
0000044222  00000116    00000916    NULL    437 NingXia FuJian  TAOBAO
86997   NULL
0000000000  00000468    00000030    NULL    347 HaiNan  LiaoNing
TIANMAO 869905775627268 YUANTONG
0000000001  00000026    00000448    NULL    339 YunNan  HeBei   TIANMAO
675929469467    SHENTONG
0000000002  00000543    00000081    NULL    332 SiChuan TaiWan
TIANMAOCHAOSHI  4010167066854172    SHENTONG
0000000003  00000397    00000217    NULL    669 YunNan  GuangXi TIANMAO
342636875611434 SHUNFENG
0000000004  00000543    00000342    NULL    472 HeBei   GuiZhou TAOBAO
```

```
371933698610006  SHUNFENG
0000000005    00000088    00000177    NULL    834 XianqGanq    NeiMengGu
TAOBAO    6011254950459740    SHENTONG
0000000006    00000451    00000486    NULL    309 NeiMengGu    ShangHai
TAOBAO    180032521959143    EMS
0000000007    00000515    00000926    NULL    758 NeiMengGu    HuNan
TAOBAO    4214464510259775    ZHONGTONG
0000000008    00000409    00000678    NULL    978 ShanXi3 Aomen    TAOBAO
4174495333403156    YUNDA
Time taken: 0.139 seconds, Fetched: 10 row(s)
```

目前为止，我们已经将 3 张 Hive 表创建完成，而且可以分别从 3 张表中查询数据，接下来就可以通过建立模型进行数据分析了。

16.6 建 立 模 型

在建立模型之前，要先了解有哪些需求，我们现在主要有 3 个需求，分别是查询各年龄段用户消费总额、查询各品牌销售总额，以及查询各省份消费总额。

16.6.1 各年龄段用户消费总额

首先，读者需要考虑两个关键词，一是"年龄"，二是"总额"。我们知道"年龄"数据在 userinfo 表中，"总额"数据在 recordsinfo 表中，因此需要将 userinfo 表和 recordsinfo 表进行关联，以下是 HQL 语句：

```
select cast(DATEDIFF(CURRENT_DATE,birthdate)/365 as int) as age,sum(price)
as totalprice from recordsinfo join userinfo on recordsinfo.userid=
userinfo.userid group by cast(DATEDIFF(CURRENT_DATE,birthdate)/365 as int)
order by totalPrice desc
```

其中，DATEDIFF()函数是用于计算当前时间和 birthdate 字段的差别天数，cast()函数是将 DATEDIFF 得到的结果转化成 int 类型。

在 Hive 中执行以下命令：

```
hive> select cast(DATEDIFF(CURRENT_DATE,birthdate)/365 as int) as age,
sum(price) as totalprice from recordsinfo
    > join userinfo on recordsinfo.userid=userinfo.userid group by
    > cast(DATEDIFF(CURRENT_DATE,birthdate)/365 as int) order by totalPrice
      desc
```

执行完毕后，会输出以下结果：

```
2018-04-27 10:10:35,983 Stage-3 map = 100%,  reduce = 100%
Ended Job = job_local1359080733_0002
MapReduce Jobs Launched:
Stage-Stage-2:  HDFS Read: 9950780 HDFS Write: 0 SUCCESS
```

```
Stage-Stage-3:  HDFS Read: 9950780 HDFS Write: 0 SUCCESS
Total MapReduce CPU Time Spent: 0 msec
OK
39      707080
13      679280
1       679279
41      649885
44      635443
24      595125
14      589813
16      587366
9       586603
...
```

代码中的上半部分是 MapReduce 执行的描述，下半部分是年龄和销量的统计，左边是年龄，右边是销量。

16.6.2 查询各品牌销售总额

与第一个需求的分析逻辑类似，"品牌"数据在 brandinfo 表中，"销售"数据在 recordsinfo 表中，所以需要将 brandinfo 表和 recordsinfo 表进行关联，查询各品牌销售额度的 HQL 语句如下：

```
select brandname,sum(price) as totalprice from recordsinfo join
brandinfo on recordsinfo.brandid=brandinfo.brandid group by
brandinfo.brandname order by totalPrice desc
```

在 Hive 中执行以下命令。

```
hive> select brandname,sum(price) as totalprice from recordsinfo join
    > brandinfo on recordsinfo.brandid=brandinfo.brandid group by
    > brandinfo.brandname order by totalPrice desc
```

执行结束后，出现 OK 字样，则代表关联成功，查询结果左列是品牌名信息，右列是销量信息。

```
2018-04-27 10:14:16,509 Stage-3 map = 100%, reduce = 100%
Ended Job = job_local873791642_0004
MapReduce Jobs Launched:
Stage-Stage-2:  HDFS Read: 19773134 HDFS Write: 0 SUCCESS
Stage-Stage-3:  HDFS Read: 19773134 HDFS Write: 0 SUCCESS
Total MapReduce CPU Time Spent: 0 msec
OK
SAMSUNG 742665
TCL 657428
HLA 599171
DELL    566776
ASUS    560944
SONY    545402
APPLE   520915
MOTOROLA        474754
DHC 472575
```

16.6.3 查询各省份消费总额

查询省份的消费总额逻辑和前两项类似，用户身份信息在 userinfo 表中，消费信息在 recordsinfo 中，所以将 userinfo 和 recordsinfo 进行关联，HQL 语句如下：

```
select province,sum(price) as totalprice from recordsinfo
join userinfo on recordsinfo.userid=userinfo.userid
group by userinfo.province order by totalPrice desc
```

在 Hive 中执行以下命令：

```
hive> select province,sum(price) as totalprice from recordsinfo
    > join userinfo on recordsinfo.userid=userinfo.userid
    > group by userinfo.province order by totalPrice desc
```

得到相应结果，左边是省份信息，右边是消费额信息，并且消费额按照从高到低顺序进行了排序。

```
2018-04-27 10:15:47,751 Stage-3 map = 100%,  reduce = 100%
Ended Job = job_local1940751870_0006
MapReduce Jobs Launched:
Stage-Stage-2:  HDFS Read: 29595488 HDFS Write: 0 SUCCESS
Stage-Stage-3:  HDFS Read: 29595488 HDFS Write: 0 SUCCESS
Total MapReduce CPU Time Spent: 0 msec
OK
SiChuan 1322949
TaiWan  992801
NeiMengGu    914987
QingHai 846420
JiLin   820037
HuNan   767805
HeNan   758084
XinJiang    749827
TianJin 735897
```

根据以上的分析和运行 HQL，我们得到了查询结果，为了避免重复查询和完成后续的数据可视化，我们可以将查询结果插入一张临时表中，并将临时表中的数据通过 Sqoop 导入 MySQL 关系型数据库中。

16.6.4 使用 Sqoop 将数据导入 MySQL 数据库

为了将数据存到 MySQL 中，我们需要 3 个步骤，对应每一个分析结果创建一张 Hive 内部表和 MySQL 表，最终将数据导入 MySQL 中进行存储。步骤为：第一步，在 Hive 中创建内部表；第二步，将分析结果插入临时表中；第 3 步，通过 Sqoop 将 Hive 内部表中的数据再导入 MySQL 数据库中。在此我们只讨论"查询各省份消费总额"需求，另外两个与之类似，读者可以自行完成。

（1）在 Hive 上创建内部表，并通过 Location 设置其在 HDFS 上面的位置。

```
create table if not exists t_statisticsbyprovince (
    province  STRING,
    totalspending  float
    )ROW FORMAT DELIMITED
    FIELDS TERMINATED BY ','
location 'hdfs://localhost:9000/flume/result/province/';
```

上面建表语句中的 Location 指向了 hdfs://localhost:9000/flume/result/province/，所以最终插入的数据会存放到/flume/result/province /下面。

（2）在 MySQL 中创建一张对应的表，用来与 Hive 中的表对应，便于通过 Sqoop 完成数据从 Hive 导入 MySQL 对应的表中。

```
mysql> create table tbl_statisticsbyprovince(statisticsid int not null
auto_increment, province varchar(20),totalspending float,primary key
(statisticsid));
    Query OK, 0 rows affected (0.01 sec)
```

（3）将查询结果插入 Hive 内部表中。

```
insert into t_statisticsbyprovince5
select province,sum(price) as totalSpending from recordsinfo
join userinfo on recordsinfo.userid=userinfo.userid
group by userinfo.province order by totalSpending desc
```

执行完后，就可以查看在 t_statisticsbyprovince 中的数据了。

```
hive> select * from t_statisticsbyprovince;
OK
SiChuan 1322949.0
TaiWan  992801.0
NeiMengGu   914987.0
QingHai 846420.0
JiLin   820037.0
HuNan   767805.0
HeNan   758084.0
```

目前，数据已经插入 hive 的内部表中了，接着可以使用 Sqoop 将数据导入 MySQL 数据库中。

（4）通过 Sqoop 将 Hive 数据导入 MySQL 数据库中。

```
sqoop export --connect jdbc:mysql://localhost:3306/loganlysis --username
root --password root --table tbl_statisticsbyprovince4 --columns province,
totalspending --export-dir hdfs://localhost:9000/flume/result/province/
000000_0 -m 1
```

运行结果如下：

```
18/04/27 10:38:26 INFO mapreduce.Job:  map 100% reduce 0%
18/04/27 10:38:26 INFO mapreduce.Job: Job job_local527674974_0001
completed successfully
18/04/27 10:38:26 INFO mapreduce.Job: Counters: 20
    File System Counters
```

```
        FILE: Number of bytes read=18029260
        FILE: Number of bytes written=18531346
        FILE: Number of read operations=0
        FILE: Number of large read operations=0
        FILE: Number of write operations=0
        HDFS: Number of bytes read=582
        HDFS: Number of bytes written=0
        HDFS: Number of read operations=9
        HDFS: Number of large read operations=0
        HDFS: Number of write operations=0
    Map-Reduce Framework
        Map input records=34
        Map output records=34
        Input split bytes=133
        Spilled Records=0
        Failed Shuffles=0
        Merged Map outputs=0
        GC time elapsed (ms)=0
        Total committed heap usage (bytes)=150470656
    File Input Format Counters
        Bytes Read=0
    File Output Format Counters
        Bytes Written=0
18/04/27 10:38:26 INFO mapreduce.ExportJobBase: Transferred 582 bytes in
2.7284 seconds (213.3157 bytes/sec)
18/04/27 10:38:26 INFO mapreduce.ExportJobBase: Exported 34 records.
```

（5）查看 MySQL 中的数据。

通过以下查询结果可以看出，Hive 中的数据已经通过 Sqoop 导入 MySQL 数据库中。

```
MySQL> select * from tbl_statisticsbyprovince4;
+--------------+--------------+---------------+
| statisticsid | province     | totalspending |
+--------------+--------------+---------------+
|            1 | SiChuan      |       1322950 |
|            2 | TaiWan       |        992801 |
|            3 | NeiMengGu    |        914987 |
|            4 | QingHai      |        846420 |
|            5 | JiLin        |        820037 |
|            6 | HuNan        |        767805 |
|            7 | HeNan        |        758084 |
|            8 | XinJiang     |        749827 |
```

16.7　数据可视化

按照大数据的整体流程，接下来就要完成数据可视化了，关于数据可视化，可以使用 ECharts，网址为 http://echarts.baidu.com，官网实例如图 16.8 所示。

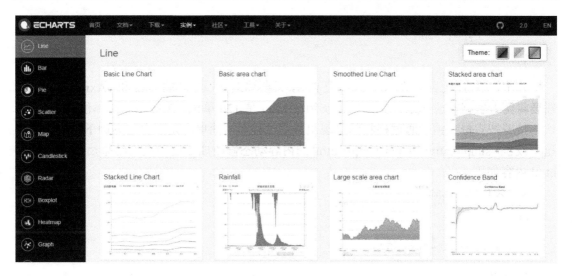

图 16.8　官网实例

在官网中有很多案例，包括 Line（线图）、Bar（柱状图）、Pie（饼状图）等常用的图形。下面我们找到柱状图案例，如图 16.9 所示。

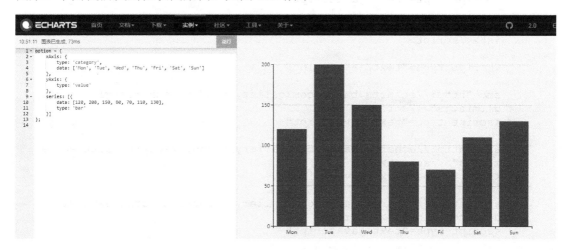

图 16.9　柱状图案例

只要改变左边的数据，右边的图形就会发生改变。代码如下：

```
option = {
    xAxis: {
        type: 'category',
        data: ['Mon', 'Tue', 'Wed', 'Thu', 'Fri', 'Sat', 'Sun']
    },
    yAxis: {
        type: 'value'
    },
    series: [{
```

```
            data: [120, 200, 150, 80, 70, 110, 130],
            type: 'bar'
        }]
    };
```

其中，xAxis 代表 x 轴，yAxis 代表 y 轴，只要修改 xAxis 中的 data 和 yAxis 中的 data，需要将两者数据对应起来，比如以上实例中 Mon 是 120 单位，Tue 是 200 单位等。

接着把省份和销量通过柱状图呈现出来，过程非常简单，先把官方案例下载下来，这里下载 bar-simple.html，将 bar-simple.html 的内容动态修改成上述的查询结果后，bar-simple.html 内容如下：

```
<!DOCTYPE html>
<html style="height: 100%">
<head>

<meta charset="utf-8">

</head>
<body style="height: 100%; margin: 0">

    <div id="container" style="height: 100%"></div>

    <script type="text/javascript"

src="http://echarts.baidu.com/gallery/vendors/echarts/echarts.min.js"></script>

    <script type="text/javascript"

src="http://echarts.baidu.com/gallery/vendors/echarts-gl/echarts-gl.min.js"></script>
    <script type="text/javascript"

src="http://echarts.baidu.com/gallery/vendors/echarts-stat/ecStat.min.js"></script>
    <script type="text/javascript"

src="http://echarts.baidu.com/gallery/vendors/echarts/extension/dataTool.min.js"></script>
    <script type="text/javascript"

src="http://echarts.baidu.com/gallery/vendors/echarts/map/js/china.js"></script>
    <script type="text/javascript"

src="http://echarts.baidu.com/gallery/vendors/echarts/map/js/world.js"></script>
    <script type="text/javascript"

src="http://api.map.baidu.com/api?v=2.0&ak=ZUONbpqGBsYGXNIYHicvbAbM"></script>
    <script type="text/javascript"

src="http://echarts.baidu.com/gallery/vendors/echarts/extension/bmap.
```

```
        min.js"></script>
        <script type="text/javascript"
        src="http://echarts.baidu.com/gallery/vendors/simplex.js"></script>
        <script type="text/javascript">
var dom = document.getElementById("container");
var myChart = echarts.init(dom);
var app = {};
option = null;
option = {
    xAxis: {
  type: 'category',
 data: ['SiChuan', 'TaiWan', 'NeiMengGu', 'QingHai', 'JiLin', 'HuNan', 'HeNan']
    },
    yAxis: {
        type: 'value'
    },
    series: [{
        data: [1322950, 992801, 914987, 846420, 820037, 767805, 758084],
        type: 'bar'
    }]
};
;
if (option && typeof option === "object") {
    myChart.setOption(option, true);
}
        </script>
</body>
</html>
```

运行上述文件,可以直观地看出每个省份的销售额度,如图 16.10 所示。

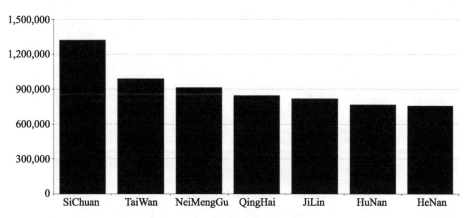

图 16.10 每个省份的销售额度

以上是一个简化的过程，实际工作中可以和前后端技术整合起来，此处不再赘述。

16.8 小　　结

本章通过一个完整的案例，完成了从数据采集、数据清洗与转化，到数据分析，再到最终的数据展现，使读者对大数据的项目开发有个整体认知。需要注意的是，本章中的商业案例进行了脱敏和简化，读者可以进行进一步研究。

第 17 章 用户画像分析实战

当前企业获得用户的成本越来越高,所以也越发关注用户的长期价值。本章介绍一个基于用户画像的案例,着重讲解如何构建一个具体的用户画像,并应用到实际项目当中。

本章主要内容涉及项目背景和项目分析过程,同时会基于开发过程及具体逻辑代码进行业务分析,最终完成用户画像的构建。本章案例的用户画像主要包括性别和年龄段。

17.1 项目背景

随着智能手机的普及,各种手机 App 的用户也越来越多。可是由于用户需求的多样化及竞争的加剧,App 目标市场的细分也变得越来越重要,精准营销和优化用户体验在企业战略发展中的意义越发重大。

根据用户的信息和行为动作,通过一些标签把用户描绘出来,这里描绘的标签就是用户画像。通过对用户画像的分析,可以判断用户的兴趣和偏好,从而实现在产品研发中以需求为导向,实现精准营销。

用户画像分为很多维度。在本项目中的用户画像主要分为两个维度,第一个是性别维度,第二个是年龄段维度,同时以"手机号"作为每个用户的唯一标识,如图 17.1 所示。

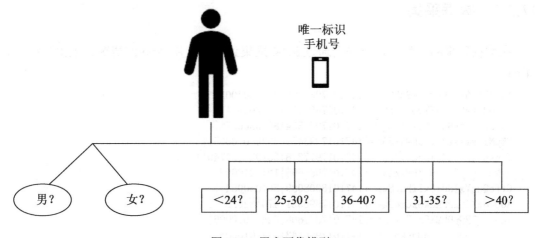

图 17.1 用户画像模型

17.2 项目目标与项目开发过程

本项目主要实现以下目标：
- 分析手机用户使用 App 的行为，进行用户画像的构建，包括用户性别和年龄属性。
- 将用户画像进行实时存储，以手机号为 Key，以年龄和性别为 Value 的形式存入 HBase 中。

项目开发过程主要分为数据采集、数据预处理、模型构建、数据分析和数据存储，如图 17.2 所示。

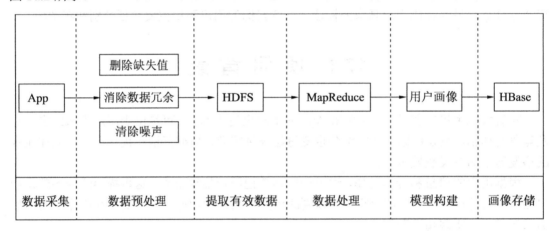

图 17.2 项目开发流程图

17.2.1 数据采集

在数据采集的过程中，首先会通过 SDK 采集到用户使用 App 的情况。采集数据格式如下：

```
NXxY3tn5XsuFcyzEw8qP8g==|1471017931598|23|330400||
+/rmMLtMV+s+gXTDoOaoxQ==|1471017991701|0|220499||
pRxqXdxHty8oF2NGI/1tNg==|1471017987354|80|050057||
H5pNI1xm1wt3Pyf9E9jY5A==|1471017993459|3338|010005||
TiasbUQxuUzZjXSFuW770w==|1471017977218|33797|151740||
qxNdo1FZp8i/nmaZPf+ZGw==|1471017996695|101|010005||
GtfeQucYxTd0i6xIiFdw8w==|1471018008089|56|050054||
AOpF8zX4LzPPKdkGzCmc0Q==|1471017996321|470|320291||
wApjYCPRGpDb2XPhiEzz1A==|1471017971808|3241|010006||
KGbfFXK0Hu6LIjhQjhkM0g==|1471017996443|342|010005||
EshDtEikJHum4g1jv/G1+Q==|1471017987273|170|070106||
```

在程序处理过程中会通过"|"进行分割,分割后得到多列,我们只提取其中几列即可,包括第 0、1、2、3 列,其中第 0 列记录了编码后的手机号,第 1 列记录了打开 App 的时间,第 2 列记录了 App 的使用时长,第 3 列记录了打开 App 的 AppID。

17.2.2 数据预处理

由于有些数据不满足条件,需要进行一定的预处理操作,这里的数据预处理需要完成以下两个任务。

- 删除缺失值,用户信息中唯一标识手机号的缺失数据可以剔除,减少分析过程中不必要的处理。
- 消除噪声,减少处理量,对于用户闪退 App 的数据,计时太短,忽略不计。

17.2.3 模型构建

在构建用户画像时,需要用到权重基础信息表和用户画像表。

1. 权重基础信息表

为了计算用户画像,也就是用户年龄和性别概率,事先对每个 App 下用户的性别和年龄计算权重进行设定。如下面的 AppTab 表,以第一条数据为例,对使用 QQ 软件的用户来说,第一列表示软件 AppID 为 10001,第 2 列是软件名称,第 3 列是男性权重,第 4 列为女性权重,第 5、6、7、8、9 列分别为年龄段<24 岁、25~30 岁、31~35 岁、36~40 岁、>40 岁的计算权重。AppTab 表中有大量的数据,在此仅取出前 10 条为例。

AppTab 表如下:

```
10001|QQ|0.001|0.001|0|0.2|0.3|0.2|0.3
10002|飞信|0.001|0.001|0|0.2|0.3|0.2|0.3
10003|MSN|0.001|0.001|0|0.2|0.3|0.2|0.3
10004|阿里旺旺|0.001|0.001|0|0.2|0.3|0.2|0.3
10005|微信|0.001|0.001|0|0.2|0.3|0.2|0.3
10006|陌陌|0.001|0.001|0|0.2|0.3|0.2|0.3
10007|米聊|0.001|0.001|0|0.2|0.3|0.2|0.3
10008|飞聊|0.001|0.001|0|0.2|0.3|0.2|0.3
10009|来往|0.001|0.001|0|0.2|0.3|0.2|0.3
10010|连我|0.001|0.001|0|0.2|0.3|0.2|0.3
```

2. 用户画像表

用户画像表的作用是存储手机号与所对应用户计算后的性别权重和年龄段权重,通过手机号对应到某个用户,这样就可以知道某个用户的性别概率和年龄段概率。该表每天会计算汇总一次,由当天凌晨统计前一天数据,用户画像表如表 17.1 所示。

表 17.1 用户画像表

字段名	类型	长度	说明
PhoneNumber	byte	16	手机号
man	float	16	男性权重
woman	float	16	女性权重
Age1	float	16	年龄段1权重
Age2	float	16	年龄段2权重
Age3	float	16	年龄段3权重
Age4	float	16	年龄段4权重
Age5	float	16	年龄段5权重

17.2.4 数据分析

本节将介绍数据分析。首先，用户画像是标签的集合，这里用年龄和性别标签来构建用户模型，但是怎么知道用户划分到哪个标签呢？首先我们针对 App 分类设计一个标签库，如表 17.2 所示。

表 17.2 针对App分类设计标签库

AppID	App名称	男性权重	女性权重	年龄段1	年龄段2	年龄段3	年龄段4	年龄段5
10001	QQ	0.001	0.001	0	0.2	0.3	0.2	0.3
10002	飞信	0.001	0.001	0	0.2	0.3	0.2	0.3
10003	MSN	0.001	0.001	0	0.2	0.3	0.2	0.3
10004	阿里旺旺	0.001	0.001	0	0.2	0.3	0.2	0.3
10005	微信	0.001	0.001	0	0.2	0.3	0.2	0.3
10006	陌陌	0.001	0.001	0	0.2	0.3	0.2	0.3
10007	米聊	0.001	0.001	0	0.2	0.3	0.2	0.3
……								

以上定义了 App 对应的男女权重和年龄权重，我们可以根据标签库中的基础数据和用户对 App 的使用情况，对某个用户进行画像，某用户的性别和年龄概率计算方法如下。

1. 性别概率计算方法

```
sum = (oldman + oldwoman + (appman + appwoman)*使用时长);
newman = (oldman + appman*times) / sum;
newwoman = (oldwoman + appwoman*times) / sum;
```

下面公式中各属性的意义。

- **appman**：标签库中设定打开某 App 的男性权重；
- **appwoman**：标签库中设定打开某 App 的女性权重；
- **oldman**：用户当前画像的男性权重（在分析用户数据前，首先为用户赋予初始化性别和年龄权重作为用户当前画像，这个画像会在计算用户新画像时取出，和标签库中设定的权重一起计算用户新画像，新画像在下一次计算时就成为了当前画像，这个当前画像是不断更新的）；
- oldwoman：用户当前画像的女性权重；
- newman：计算后用户新画像的男性权重；
- newwoman：计算后用户新画像的女性权重。

关于性别权重的计算，是根据用户已存在画像的性别权重、App 的标签库中的性别权重和使用时长计算得到的，即一个用户的 App 使用信息在他自身特征信息上的累加和更新。

2．年龄概率计算方法

```
sum = (age1 + age2 + age3 + age4 + age5 ) + (pAge1 + pAge2 + pAge3 + pAge4
+ pAge5 )*times;
nage1 = (pAge1*times + age1) / sum;
nage2 = (pAge2*times + age2) / sum;
nage3 = (pAge3*times + age3) / sum;
nage4 = (pAge4*times + age4) / sum;
nage5 = (pAge5*times + age5) / sum;
```

下面是公式中各元素的意义。
- age：用户画像中的年龄权重；
- pAge：用户打开 App 对应的年龄计算权重；
- nage：用户画像的年龄新权重。

用户性别权重和某一年龄段权重的计算由用户当前性别年龄段权重、标签库中 App 的性别年龄权重和使用时长获得。如果用户没有当前权重，则使用用户首次打开某个 App 时在标签库中设置的权重作为初始权重。

17.3　核心代码解读

通过前面两节内容的介绍，我们构建了用户画像模型，并分析了用户画像计算方法，接下来将利用大数据技术对用户的 App 使用信息进行处理和计算，从代码层面深度剖析用户画像的过程。

17.3.1　项目流程介绍

项目流程主要分为以下 3 部分：

（1）将用户使用的初始信息上传至 HDFS 文件系统中。

（2）MapReduce 的部分主要分为两个任务，第一个任务是基于存储在 HDFS 中的 App 使用数据，将用户和其使用总次数、总时长提取出来。MapReduce 的第二个任务是将所有用户一天中的使用信息根据手机号进行分别汇集，逐个提取用户所使用的 AppID，再根据 AppID 结合权重表分别计算此用户的性别概率和年龄概率，这样就可以得到用户的新画像。

（3）最后将结果以手机号为 Key、用户画像特征属性为 Value 的形式存储至 HBase 中。

17.3.2　核心类的解读

核心代码共包含 9 个类、3 个 .xml 配置文件和 1 个 .properties 文件。项目结构如图 17.3 所示。

图 17.3　项目结构图

接下来对项目文件的作用进行简要概述。
- core-site.xml：定义 HDFS 的端口和 Hadoop 的临时存储路径。
- hdfs-site.xml：定义 Secondary NameNode 的基本信息。
- UserProfile.properties：定义了从原始数据提取的列号和连接 HBase 的地址。
- UserProfile.java：用户画像的实体类，包含性别、年龄等实例变量。
- LoadConfig.java：从 UserProfile.properties 中读取配置信息。
- ReadFile.java：从 HDFS 中读取文件。
- ReadFromHdfs.java：从 HDFS 中读取数据后，提取 App 对应的性别和年龄权重放入 Map 中，用于后面用户画像中性别和年龄概率的计算。

- TextArrayWritable.java：自定义的一个用来存放字符数组的类。
- UserProfileMapReduce.java：提取原始数据中的有用信息，进行处理和合并，在主方法中定义 MapReduce 的任务。
- UserProfileMapReduce2.java：继续提取 UserProfileMapReduce.java 中定义的任务一输出的数据，归集每个用户的使用信息，结合 ReadFromHdfs.java 中得到的 App 对应的用户年龄和性别计算权重，最终得到用户画像。
- UserProfilePutInHbaseMap.java：将通过 MapReduce 任务得到的用户画像信息提取出来。
- UserProfilePutInHbaseReduce.java：将用户画像存储至 HBase 中。

以上是整个项目的结构说明，下面对每个类分别详细介绍。

17.3.3 core-site.xml 配置文件

core-site.xml 和 hdfs-site.xml 是配置 Hadoop 的文件，在 core-site.xml 中定义了 HDFS 的入口 hdfs://192.168.179.96:9000 和 Hadoop 的节点临时存放路径。代码如下：

```
<?xml version="1.0" encoding="UTF-8"?>
<?xml-stylesheet type="text/xsl" href="configuration.xsl"?>
<configuration>
    <property>
        <name>fs.defaultFS</name>
        <value>hdfs://192.168.179.96:9000</value>
    </property>
    <property>
        <name>hadoop.tmp.dir</name>
        <value>/opt/software/hadoop-2.5.1</value>
    </property>
</configuration>
```

17.3.4 hdfs-site.xml 配置文件

hdfs-site.xml 文件中定义了 Secondary NameNode 的信息。代码如下：

```
<?xml version="1.0" encoding="UTF-8"?>
<?xml-stylesheet type="text/xsl" href="configuration.xsl"?>
<configuration>
    <property>
        <name>dfs.namenode.secondary.http-address</name>
        <value>192.168.179.97:50090</value>
    </property>
    <property>
        <name>dfs.namenode.secondary.https-address</name>
        <value>192.168.179.97:50091</value>
    </property>
</configuration>
```

17.3.5 UserProfile.properties 配置文件

properties 文件中配置了需要读取的列号和 HBase 的连接信息等。代码如下：

```
Delimiter=\\|
Date=11
phoneID=0
appID=15
frequency=1
duration=12
consite=hbase.ZooKeeper.quorum
hbaseip=192.168.179.96,192.168.179.97,192.168.179.98
overtime=dfs.socket.timeout
time=180000
tableProfile=user_profile
```

17.3.6 LoadConfig.java：读取配置信息

读取 UserProfile.properties 文件，利用其中配置的需要读取数据的列号从用户信息中提取有用信息，并读取 HBase 的连接信息等进行数据存储。代码如下：

```java
package util;
import java.io.IOException;
import java.io.InputStream;
import java.util.Properties;
public class LoadConfig {
    static Properties properties;
    static{
        properties = new Properties();
        //可以获取资源文件
    InputStream inStream = Thread.currentThread().getContextClassLoader().
    getResourceAsStream("./resource/UserProfile.properties");
        try {
            properties.load(inStream);
        } catch (IOException e) {
            e.printStackTrace();
        }
    }
    //UserProfile
    public String Delimiter = properties.getProperty("Delimiter");
    public String Date = properties.getProperty("Date");
    public String phoneID = properties.getProperty("phoneID");
    public String appID = properties.getProperty("appID");
    public String frequency = properties.getProperty("frequency");
    public String duration = properties.getProperty("duration");
    //Hbase
    public String consite = properties.getProperty("consite");
    public String hbaseip = properties.getProperty("hbaseip");
```

```java
    public String overtime = properties.getProperty("overtime");
    public String time = properties.getProperty("time");
    public String tableProfile = properties.getProperty("tableProfile");
}
```

17.3.7　ReadFile.java：读取文件

从 HDFS 上读取 appTab.txt 文件，从而得到 App 对应的性别和年龄计算权重。代码如下：

```java
package util;
import java.io.BufferedReader;
import java.io.InputStreamReader;
import org.apache.hadoop.conf.Configuration;
import org.apache.hadoop.fs.FSDataInputStream;
import org.apache.hadoop.fs.FileSystem;
import org.apache.hadoop.fs.Path;
/** 读取文件 **/
public class ReadFile {
    public static BufferedReader fileReader(String fileName) throws
    Exception {
        Configuration conf = new Configuration();
        FileSystem fs = FileSystem.get(conf);
        FSDataInputStream input = fs.open(new Path(fileName));
        BufferedReader br = new BufferedReader(new InputStreamReader
        (input));
        return br;
    }
}
```

17.3.8　ReadFromHdfs.java：提取信息

我们在上面的 ReadFile.java 文件中读取了 HDFS 上的文件，现在将读取的文件中 App 对应的 AppID 和年龄、性别计算权重提取出来。代码如下：

```java
package util;

import java.io.BufferedReader;
import java.util.HashMap;
import java.util.Map;

public class ReadFromHdfs {
    public static String appTable = "/data/appTab.txt";
    private static Map<String, String[]> appMap = new HashMap<String,
    String[]>();
    static {
        try {
            StringBuffer stringBuffer = new StringBuffer();
            String line = null;
            BufferedReader appWeight = ReadFile.fileReader(appTable);
```

```java
            while ((line = appWeight.readLine()) != null) {
                String[] appArray = line.split("\\|");
                // appName
                stringBuffer.append(appArray[1]).append(",");
                //性别权重
                stringBuffer.append(appArray[2]).append(",").append
                (appArray[3]).append(",");
                //年龄段权重  stringBuffer.append(appArray[4]).append (",").
                append(appArray[5]).append(",").append(appArray[6]).
                append(",");
                stringBuffer.append(appArray[7]).append(",").append
                (appArray[8]);
                //sb="QQ, 0.001, 0.001, 0, 0.2, 0.3, 0.2, 0.3"
                String[] appToValueArray = stringBuffer.toString().
                split(",");
                appMap.put(appArray[0], appToValueArray);//appMap=
                {10001,[QQ, 0.001, 0.001, 0, 0.2, 0.3, 0.2, 0.3]}
                stringBuffer.delete(0, stringBuffer.length());
                                     //把加载缓冲区的 sb 内容删除掉
            }
        } catch (Exception e) {
            e.printStackTrace();
        }
    }
    public static Map<String, String[]> getAppMap() {
        return appMap;
    }
    public static void setAppMap(Map<String, String[]> appMap) {
        ReadFromHdfs.appMap = appMap;
    }
}
```

以上数据处理后得到{"appId",[appName，男性权重，女性权重，年龄 1 权重，年龄 2 权重，年龄 3 权重，年龄 4 权重，年龄 5 权重，]}的数据形式，将其存入 AppMap 中，这个 AppMap 的 appId 用于匹配后面 UserProfileMapReduce2 类中提取的用户信息的 appId，从而利用用户信息和初始化权重进行用户画像。

17.3.9　UserProfile.java：创建用户画像

UserProfile.java 文件中定义了一个用户画像的实体类，包括画像的性别和年龄权重等属性，并且定义了用户画像中性别和年龄权重的初始化和计算方法。代码如下：

```java
package userprofile;
import java.math.BigDecimal;
public class UserProfile {
    //用户画像
     //属性
    private String openTime;
    private String phoneNum;
    private double man;
    private double woman;
```

```java
    private double age1;
    private double age2;
    private double age3;
    private double age4;
    private double age5;

    //重写toString()方法
    public String toString() {
        StringBuffer stringBuffer = new StringBuffer();
        stringBuffer.append(openTime).append("|");
        stringBuffer.append(phoneNum).append("|");
        stringBuffer.append(new BigDecimal(man).setScale(3, 4).
            doubleValue()).append("|");
        stringBuffer.append(new BigDecimal(woman).setScale(3, 4).
            doubleValue()).append("|");
        stringBuffer.append(new BigDecimal(age1).setScale(3, 4).
            doubleValue()).append("|");
        stringBuffer.append(new BigDecimal(age2).setScale(3, 4).
            doubleValue()).append("|");
        stringBuffer.append(new BigDecimal(age3).setScale(3, 4).
            doubleValue()).append("|");
        stringBuffer.append(new BigDecimal(age4).setScale(3, 4).
            doubleValue()).append("|");
        stringBuffer.append(new BigDecimal(age5).setScale(3, 4).
            doubleValue()).append("|");
        return stringBuffer.toString();
    }
    /** 性别融合 */
    public void mixGender(double man2, double woman2, long times) {
        double sum = (this.man + this.woman + (man2 + woman2) * times);
        if(sum != 0){
            this.man = (this.man + man2 * times) / sum;
            this.woman = (this.woman + woman2 * times) / sum;
        }
    }

    /** 年龄段融合 */
    public void mixAge(double nAge1, double nAge2, double nAge3, double nAge4, double nAge5, long times) {
        double sum = (age1 + age2 + age3 + age4 + age5 )  // 之前的App的
            + (nAge1 + nAge2 + nAge3 + nAge4 + nAge5 ) * times;// 当前的App的
        if(sum != 0){
            this.age1 = (nAge1 * times + age1) / sum;
            this.age2 = (nAge2 * times + age2) / sum;
            this.age3 = (nAge3 * times + age3) / sum;
            this.age4 = (nAge4 * times + age4) / sum;
            this.age5 = (nAge5 * times + age5) / sum;
        }
    }
    /** 初始化男女概率 */
    public void initGender(float man, float woman) {
    float sum = man + woman;
        if(sum != 0){
        this.man = man/ sum;
        this.woman = woman / sum;
```

```java
        }
    }
    /** 初始化年龄段概率 */
    public void initAge(float nAge1, float nAge2, float nAge3, float nAge4,
    float nAge5) {
        float sum = nAge1 + nAge2 + nAge3 + nAge4 + nAge5;
        if(sum != 0){
            this.age1 = nAge1 / sum;
            this.age2 = nAge2 / sum;
            this.age3 = nAge3 / sum;
            this.age4 = nAge4 / sum;
            this.age5 = nAge5 / sum;
        }
    }
    // setter and getter method
    public String getOpenTime() {
        return openTime;
    }

    public void setOpenTime(String openTime) {
        this.openTime = openTime;
    }
    public String getPhoneNum() {
        return phoneNum;
    }
    public void setPhoneNum(String phoneNum) {
        this.phoneNum = phoneNum;
    }
    public double getMan() {
        return man;
    }
    public void setMan(double man) {
        this.man = man;
    }
    public double getWoman() {
        return woman;
    }
    public void setWoman(double woman) {
        this.woman = woman;
    }
    public double getAge1() {
        return age1;
    }
    public double getAge2() {
        return age2;
    }
    public double getAge3() {
        return age3;
    }
    public double getAge4() {
        return age4;
    }
    public double getAge5() {
        return age5;
    }
```

17.3.10　TextArrayWritable.java：字符串处理工具类

因为进行用户画像时需要提取多个属性值，数据处理和归集时用到了多个字符串，因此我们定义了一个字符数组的类 TextArrayWritable.java。代码如下：

```java
package util;
import java.util.ArrayList;
import org.apache.hadoop.io.ArrayWritable;
import org.apache.hadoop.io.Text;
public class TextArrayWritable extends ArrayWritable {
    public TextArrayWritable() {
            super(Text.class);
    }
     public TextArrayWritable(String[] strings) {
        super(Text.class);
        Text[] texts = new Text[strings.length];
        for (int i = 0; i < strings.length; i++) {
            texts[i] = new Text(strings[i]);
        }
        set(texts);
    }
     public TextArrayWritable(ArrayList<String> strings) {
        super(Text.class);
        Text[] texts = new Text[strings.size()];
        int i = 0;
        for (String str : strings) {
            texts[i] = new Text(str);
            i++;
        }
        set(texts);
    }
    public ArrayList<String> toArrayList(String[] writables) {
        ArrayList<String> arraylist = new ArrayList<String>();
        for (String writable : writables) {
            arraylist.add(writable.toString());
        }
        return arraylist;
    }
    public ArrayList<String> toArrayList() {
        return toArrayList(super.toStrings());
    }
}
```

17.3.11　MapReduce 任务 1：UserProfileMapReduce.java

这里定义了 MapReduce 的第一个任务，Mapper 用于数据的分割和提取，Reducer 用于数据的合并。具体任务过程如下：

（1）Mapper 首先进行数据的提取处理，即从用户数据中提取出时间、手机号、AppID、用户使用次数和用户使用时长信息。代码如下：

```java
package userprofilemr;
import java.io.IOException;
import java.text.SimpleDateFormat;
import org.apache.hadoop.conf.Configuration;
import org.apache.hadoop.fs.Path;
import org.apache.hadoop.hbase.mapreduce.TableMapReduceUtil;
import org.apache.hadoop.hbase.mapreduce.TableOutputFormat;
import org.apache.hadoop.io.LongWritable;
import org.apache.hadoop.io.Text;
import org.apache.hadoop.mapreduce.Job;
import org.apache.hadoop.mapreduce.Mapper;
import org.apache.hadoop.mapreduce.Reducer;
import org.apache.hadoop.mapreduce.lib.input.FileInputFormat;
import org.apache.hadoop.mapreduce.lib.input.TextInputFormat;
import org.apache.hadoop.mapreduce.lib.output.FileOutputFormat;
import org.apache.hadoop.mapreduce.lib.output.TextOutputFormat;
import userprofilehbase.UserProfilePutInHbaseMap;
import userprofilehbase.UserProfilePutInHbaseReduce;
import userprofilemr.UserProfileMapReduce2.MyMap2;
import userprofilemr.UserProfileMapReduce2.MyReduce2;
import util.LoadConfig;
import util.TextArrayWritable;

public class UserProfileMapReduce {
    public static LoadConfig config = new LoadConfig();

    public static class MyMap extends Mapper<LongWritable, Text, Text,
TextArrayWritable> {
        Text k = new Text();    //参数为读取数据的下标偏移量,可序列化字符串,
        可序列化的字符数组

    public void map(LongWritable key, Text value, Context context)
        throws IOException, InterruptedException {

        String dataLine = value.toString();
            //即通过 Separator="|"分割
        String[] dataArray = dataLine.split(config.Delimiter);
        //将手机号和 appd 的 ID 号进行拼接,作为标识用户的 uniquekey,其中第 0 列为手
        机号,第 15 列是 appID
        String uniqueKey = dataArray[Integer.parseInt(config.phoneID)]
            + dataArray[Integer.parseInt(config.appID)];

        String[] val = new String[5];
            String currentTime = dataArray[Integer.parseInt(config.
            Date)];//Date=11
            SimpleDateFormat sdf = new SimpleDateFormat("yyyyMMdd");
            val[0] = sdf.format(Long.parseLong(currentTime));   //时间
            val[1] = dataArray[Integer.parseInt(config.phoneID)];
                                                                //手机号
            val[2] = dataArray[Integer.parseInt(config.appID)];//AppID
            val[3] = "1";                                       //计数
            val[4] = dataArray[Integer.parseInt(config.duration)];
                                                                //使用时长
            k.set(uniqueKey);
```

```
        context.write(k, new TextArrayWritable(val));
        //k=1512005776610004   val=[20180602,15120057766,10004,1,33]
    }
}
```

（2）Reducer 将用户使用某 App 的使用次数和使用时长进行合并，将合并后的信息以字符串形式写入文本。代码如下：

```
public static class MyReduce extends Reducer<Text, TextArrayWritable,
Text, Text> {
    Text v = new Text();
    public void reduce(Text key, Iterable<TextArrayWritable> values,
        Context context) throws IOException, InterruptedException {
        long sum = 0;
        int frequency = 0;
        String[] res = new String[5];
        boolean flg = true;
        //k=1512005776610004   values=[20180602,15120057766,10004,1,12]
        for (TextArrayWritable t : values) {
            String[] vals = t.toStrings();
            if (flg) {
                res = vals;
            }
            if (vals[3] != null) {                    //计数不为空
                frequency = frequency + 1;
            }
            if (vals[4] != null) {                    //使用时长
                sum += Long.valueOf(vals[4]);
            }
        }
        res[3] = String.valueOf(frequency);
        res[4] = String.valueOf(sum);
        StringBuffer sb = new StringBuffer();
        sb.append(res[0]).append("|");                //时间
        sb.append(res[1]).append("|");                //手机号
        sb.append(res[2]).append("|");                //AppID
        sb.append(res[3]).append("|");                //计数
        sb.append(res[4]);                            //使用时长
        v.set(sb.toString());
          //20180603|qxNdolFZp8i/nmaZPf+ZGw==|10005|6|136
            context.write(null, v);

        }
    }
```

以上 UserProfileMapReduce 的第一个任务就完成了，即提取出精炼用户信息文本后，进行用户使用 App 次数和时长的合并，得到如 key=1512005776610004   value=[20180602,15120057766,10004,1,33]的用户数据，这条信息的意义为 key=手机号+appId，value=[打开日期,用户手机号+appId,打开次数,使用此 app 总时长]。

（3）主方法：设置了关于 HDFS 的连接 URL、要读取的原始数据存放目录和处理后的输出路径，并且定义了关于 HBase 的连接，并指明了 MapReduce 的任务中 Mapper 和

Reducer 过程的类。

```java
public static void main(String[] args) throws Exception {
    Configuration conf = new Configuration();
        conf.set("fs.defaultFS", "hdfs://192.168.179.96:9000");
        Job job1 = Job.getInstance(conf, "UserProfileMapReduceJob1");
                                                        //任务名称

        job1.setJarByClass(UserProfileMapReduce.class);   //主方法
        job1.setMapperClass(MyMap.class);                 //Mapper 方法
        job1.setReducerClass(MyReduce.class);             //Reducer 方法
            //k=1512005776610004  values=[20180602,15120057766,10004,
            1,12]
        job1.setMapOutputKeyClass(Text.class);       //Map 输出的 Key 类型
        job1.setMapOutputValueClass(TextArrayWritable.class);
                                              //Map 输出的 Value 类型
            //values="20180602|15120057766|10004|1|12"
        job1.setOutputKeyClass(Text.class);
        job1.setOutputValueClass(Text.class);
        job1.setInputFormatClass(TextInputFormat.class);
        job1.setOutputFormatClass(TextOutputFormat.class);

        FileInputFormat.addInputPath(job1, new Path("/data/userdata.
        txt"));                                          //输入路径
        FileOutputFormat.setOutputPath(job1, new Path("/data/out.
        txt"));                                          //输出路径
    //输出数据"20180603|qxNdolFZp8i/nmaZPf+ZGw==|10005|6|136,.....
    "意义：小明在 2018 年 6 月 3 日打开了微信 6 次，共 136 分钟
    Boolean state1 = job1.waitForCompletion(true);
        System.out.println("job1 执行成功！！！");
        if (state1) {
            conf = new Configuration();

           conf.set("fs.defaultFS", "hdfs://192.168.179.96:9000");
           Job job2 = Job.getInstance(conf, "UserProfileMapReduceJob2");

           job2.setJarByClass(UserProfileMapReduce.class);
           job2.setMapperClass(MyMap2.class);
           job2.setReducerClass(MyReduce2.class);
        job2.setMapOutputKeyClass(Text.class);
           job2.setMapOutputValueClass(Text.class);
           job2.setOutputKeyClass(Text.class);
           job2.setOutputValueClass(Text.class);
           job2.setInputFormatClass(TextInputFormat.class);
           job2.setOutputFormatClass(TextOutputFormat.class);

           FileInputFormat.addInputPath(job2, new Path("/data/out.
           txt"));                                       //输入路径
           FileOutputFormat.setOutputPath(job2, new Path("/data/out2.
           txt"));                                       //输出路径
             Boolean state2 = job2.waitForCompletion(true);
    System.out.println("job2 执行成功！！！");
        if (state2) {
```

```
            conf = new Configuration();
            //设置ZooKeeper
            conf.set(config.consite, config.hbaseip);//hbase 的 ip
            //设置 HBase 表名称
            conf.set(TableOutputFormat.OUTPUT_TABLE, config.tableProfile);
            //将该值改大，防止 HBase 超时退出
            conf.set(config.overtime, config.time);
            Job job3 = Job.getInstance(conf,"UserProfilePutInHbase");
            job3.setJarByClass(UserProfileMapReduce.class);
            TableMapReduceUtil.addDependencyJars(job3);
            FileInputFormat.setInputPaths(job3, new Path(args[2]));
            job3.setMapperClass(UserProfilePutInHbaseMap.class);
            job3.setMapOutputKeyClass(Text.class);
                 job3.setMapOutputValueClass(Text.class);

              job3.setReducerClass(UserProfilePutInHbaseReduce.class);
            job3.setOutputFormatClass(TableOutputFormat.class);
            job3.waitForCompletion(true);
        }
            }
        }
    }
```

17.3.12　MapReduce 任务 2：UserProfileMapReduce2.java

　　UserProfileMapReduce2 主要是对每个用户的所有数据进行归集，根据这些数据对用户进行画像。具体过程如下：

　　（1）Mapper 处理精简后的用户信息得到多条"手机号和 App 使用信息"，用于后续合并每个手机用户的多条 App 使用信息。代码如下：

```
package userprofilemr;
import java.io.IOException;
import java.util.HashMap;
import java.util.Map;
import java.util.Set;
import org.apache.hadoop.io.LongWritable;
import org.apache.hadoop.io.Text;
import org.apache.hadoop.mapreduce.Mapper;
import org.apache.hadoop.mapreduce.Reducer;
import userprofile.UserProfile;
import util.ReadFromHdfs;
public class UserProfileMapReduce2 {
    public static class MyMap2 extends Mapper<LongWritable, Text, Text,
    Text> {
        Text k = new Text();
        //20180603|qxNdolFZp8i/nmaZPf+ZGw==|10005|6|136
        public void map(LongWritable key, Text value, Context context)
        throws IOException, InterruptedException {

                String dataLine = value.toString();
                String[] dataArray = dataLine.split("\\|");
```

```java
                    String newkey = dataArray[1] ; // phone=15120057766
                    k.set(newkey);
                    context.write(k, value);
                                    //合并同一个手机号的某个日期的使用次数和时长
        }
    }
```

（2）Reducer 将 Mapper 输出信息中的 AppID 取出，与 ReadFromHdfs 类中获取的 AppMap 中的 AppID 进行匹配，匹配成功则从 AppMap 中提取相应 App 的性别和年龄计算权重来进行用户画像。此外定义一个用户画像集合 UserProfileMap，用于存放用户手机号和最新画像。代码如下：

```java
    public static class MyReduce2 extends Reducer<Text, Text, Text, Text> {
        Map<String, String[]> appMap = ReadFromHdfs.getAppMap();

    Text v = new Text();
        public void reduce(Text key, Iterable<Text> values, Context context)
        throws IOException, InterruptedException {

            //appMap={10001,[QQ, 0.001, 0.001, 0, 0.2, 0.3, 0.2, 0.3]}
        Map<String, UserProfile> UserProfileMap = new HashMap<String, UserProfile>();

            //返回值是个只存放 Key 值的 Set 集合，即 AppID 的集合
        Set<String> keySet = UserProfileMap.keySet();
            String keyPhone = null;
            for (Text t : values) {
     String[] dataArray = t.toString().split("\\|");
     keyPhone = dataArray[1];                    //用户 MDN
            String appID = dataArray[2];                //AppID
            //根据 AppID 获取对应的标签信息
              if (appID.length() > 0) {                 //AppID 不能为空
        if (appMap.get(appID) == null) {
                continue;
                }
                String favourite = appMap.get(appID)[2];
    //例：appMap={10001,[QQ, 0.001, 0.001, 0, 0.2, 0.3, 0.2, 0.3]}, ...
            float man = Float.parseFloat(appMap.get(appID)[1]);
            float woman = Float.parseFloat(appMap.get(appID)[2]);
            float age1 = Float.parseFloat(appMap.get(appID)[3]);
            float age2 = Float.parseFloat(appMap.get(appID)[4]);
            float age3 = Float.parseFloat(appMap.get(appID)[5]);
            float age4 = Float.parseFloat(appMap.get(appID)[6]);
            float age5 = Float.parseFloat(appMap.get(appID)[7]);
            long times = Long.parseLong(dataArray[4]);//使用时长
```

从用户数据中提取出手机号，如果这个手机号在用户画像集合中存在，即某用户在这一天打开了另外的 App，并已为此用户画过像，则取出这个画像，并利用 App 的新权重和之前画像的性别、年龄权重对用户重新画像。代码如下：

```java
    if (UserProfileMap.containsKey(keyPhone)==true) {
//如果 Map 中,所有的 Key 中只要有 keyMDN 的值,就为 true,即此用户有画像
//画像过的手机号如果再次出现在数据中,即某用户在这一天打开了另外的 App,则将之前画过
的像(即权重)结合新数据来分析用户画像
        UserProfile userProfile = UserProfileMap.get(keyPhone);
                                         //取出原来画像的性别和年龄权重
        userProfile.mixGender(man, woman, times);   //性别权重
        userProfile.mixAge(age1, age2, age3, age4, age5, times);
                                         //年龄段权重
    } else {
//如果这个手机号没有画像(用户第一次出现),则取初始化性别、年龄权重作为此用户画像
        UserProfileMap.put(keyPhone, createProfileData(dataArray,
            favourite,man, woman, age1, age2, age3, age4, age5, times));
    }
    }
    }
    for (String keys : keySet) {              //把每个用户的画像取出
        v.set(UserProfileMap.get(keys).toString());
        context.write(null, v);
    }
    }
    }
// 创建画像数据
private static UserProfile createProfileData(String[] dataArray,
        String favourite,                 //兴趣爱好
        float man, float woman,           //性别
        float age1, float age2, float age3, float age4, float age5,
                                         //年龄
        long times) {
    UserProfile userProfile = new UserProfile();
    userProfile.setOpenTime(dataArray[0]);
    userProfile.setPhoneNum(dataArray[1]);
    //初始化
    userProfile.initAge(age1, age2, age3, age4, age5);
    userProfile.initGender(man, woman);
    return userProfile;
}
}
```

以上我们就得到根据用户的 App 使用情况来描绘的用户画像了,下面来介绍关于用户画像的存储问题。

17.3.13 UserProfilePutInHbaseMap.java:提取用户画像

我们创建的这个用户画像模型刚好符合 HBase 中行键和列族数据的 Key-Value 存储规则,且考虑到存在用户量大,以及用户 App 使用信息数据多的问题,可以利用 HBase 以 Key-Value 的方式进行实时存储。具体过程如下所述。

通过 Mapper 将用户信息的手机号提取出来作为 Key，转换为手机号和用户信息的数据对形式并输出。代码如下：

```
package userprofilehbase;
import java.io.IOException;
import org.apache.hadoop.io.LongWritable;
import org.apache.hadoop.io.Text;
import org.apache.hadoop.mapreduce.Mapper;
public class UserProfilePutInHbaseMap extends Mapper<LongWritable, Text, Text, Text>{
    Text k2 = new Text();
    Text v2 = new Text();
    @Override
    protected void map(LongWritable key, Text value, Mapper<LongWritable, Text, Text, Text>.Context context)
            throws IOException, InterruptedException {
        String line = value.toString();
        String[] splited = line.split("\\|");
        k2.set(splited[1]);
        v2.set(line);
        context.write(k2, v2);
    }
}
```

这样我们就得到了用户手机号和用户信息的一组画像数据，下面将数据存入 HBase。

17.3.14　UserProfilePutInHbaseReduce：存储用户画像

通过 Reducer 将用户画像写入 HBase，将画像的手机号、性别和年龄存储至列族 profile 的 phone、man、woman、age1…age5 列。

```
package userprofilehbase;
import java.io.IOException;
import org.apache.hadoop.hbase.client.Durability;
import org.apache.hadoop.hbase.client.Mutation;
import org.apache.hadoop.hbase.client.Put;
import org.apache.hadoop.hbase.mapreduce.TableReducer;
import org.apache.hadoop.hbase.util.Bytes;
import org.apache.hadoop.io.NullWritable;
import org.apache.hadoop.io.Text;
import org.apache.hadoop.mapreduce.Reducer;
public class UserProfilePutInHbaseReduce extends TableReducer<Text, Text, NullWritable>{
    @SuppressWarnings("deprecation")
    @Override
    protected void reduce(Text k2, Iterable<Text> val,Reducer<Text, Text, NullWritable,Mutation>.Context context)
            throws IOException, InterruptedException {
```

```java
        for (Text v2 : val) {
            String[] splited = v2.toString().split("\\|");
            //rowkey
            if(k2.toString().length()!=0){
            Put put = new Put(Bytes.toBytes(k2.toString()));
            //跳过写入 Hlog，提高写入速度
            put.setDurability(Durability.SKIP_WAL);
    put.add(Bytes.toBytes("profile"), Bytes.toBytes("phoneNum"), Bytes.
    toBytes(splited[1]));
    put.add(Bytes.toBytes("profile"), Bytes.toBytes("man"), Bytes.toBytes
    (splited[2]));
    put.add(Bytes.toBytes("profile"), Bytes.toBytes("woman"), Bytes.toBytes
    (splited[3]));
    put.add(Bytes.toBytes("profile"), Bytes.toBytes("age1"), Bytes.toBytes
    (splited[4]));

        }
    }
}
```

17.4 项目部署

项目部署主要分为环境搭建和核心代码运行两部分，具体过程如下：

（1）搭建 Hadoop 环境，将经过预处理的原始数据上传至 HDFS 文件系统中，包括 userdata.txt 和 appTab.txt。

（2）运行 UserProfileMapReduce 中的主方法，运行结果如下：

```
20180813|+/rmMLtMV+s+gXTDoOaoxQ==|0.5|0.5|0.1|0.2|0.2|0.2|0.3|
20180813|+0F2tNRRntB4ErNVJFatmA==|0.5|0.5|0.1|0.2|0.2|0.2|0.3|
20180813|+1/YWq4baqEa47r+55EghA==|0.5|0.5|0.116|0.2|0.2|0.2|0.284|
20180813|+2SJcwT8h0TSsxVd2pBMjA==|0.508|0.492|0.1|0.2|0.2|0.2|0.3|
20180813|+2YYprC2jdMfQntMAn8lmA==|0.4|0.6|0.0|0.1|0.3|0.3|0.3|
20180813|+2k9y/1uapW4IW6LUq0LGQ==|0.7|0.3|0.0|0.2|0.3|0.2|0.3|
20180813|+3M/cSZerfq+XDC9kMWWcA==|0.5|0.5|0.1|0.2|0.2|0.2|0.3|
20180813|+3NZW71R8R8nL78M3JB8tw==|0.5|0.5|0.1|0.2|0.2|0.2|0.3|
20180813|+40FbnopHa5sY/CzDDf8cg==|0.5|0.5|0.0|0.2|0.3|0.2|0.3|
20180813|+5k9calLrKDGvHyzvO0pvg==|0.5|0.5|0.1|0.2|0.2|0.2|0.3|
20180813|+5wU5NlU1iJUZ5sMN91DUw==|0.5|0.5|0.2|0.2|0.2|0.2|0.2|
20180813|+69Ng7CRkKTr2hbYtgeiAw==|0.8|0.2|0.0|0.2|0.3|0.2|0.3|
20180813|+6YeD7QUw8Hy6qN7Dxk7vQ==|0.8|0.2|0.0|0.2|0.3|0.2|0.3|
20180813|+844+iumk6xvhTO2fvpSaQ==|0.5|0.5|0.2|0.2|0.2|0.2|0.2|
20180813|+8qfBkXQ+sS/G4fRs1AcQQ==|0.5|0.5|0.1|0.2|0.2|0.2|0.3|
……
```

从上述结果中，可以分析得到用户的画像信息，比如 20180813|+2k9y/1uap W4IW6LUq0LGQ==|0.7|0.3|0.0|0.2|0.3|0.2|0.3|，其中 "==" 左边是用于标识用户的手机信息，右边是各个数据

的权重。从这条记录中可以看出，man 的概率是 0.7，woman 的概率是 0.3，这样在后续进行产品推荐的时候，就可以对此用户进行男性商品的个性化推荐。

17.5 小　　结

本章根据已知的用户 App 使用信息和 App 计算权重信息，利用年龄和性别计算公式建立了用户画像，并通过用户行为分析实现对手机 App 用户的特征描绘，这对于商家要进行精准营销和个性化推荐有重要作用。

第 18 章 基于个性化的视频推荐系统实战

用户的喜好对于商家来说已经成为了一项重要信息,用户倾向的推荐也越来越重要。本章将基于一个用户推荐系统案例,介绍如何通过协同过滤算法为已知历史行为的用户推荐他可能喜欢的视频。这其中涉及推荐系统核心算法、项目架构、模型构建和项目代码分析的详细过程,致力于为视频观看者提供更具针对性的视频推荐。

18.1 项目背景

随着互联网的快速发展,我们已经进入了信息爆炸的时代,各种信息充斥在我们的生活中,在这种情况下,信息的提供者和消费者都遇到了很大的挑战。对于信息提供者来说,由于不知道消费者的具体需求,所以可能会给消费者提供了其并不需要的信息,不但没有给消费者带来帮助,反而变成了对消费者的信息骚扰。

在这种情况下,如果能够根据消费者的历史行为推断出其所倾向的信息,然后再有针对性地发送信息,这对于消费者来说是非常有价值的,这也是推荐系统产生的根源。

推荐系统和搜索引擎有一定的区别,比如推荐系统是根据用户的历史行为数据进行推荐,属于主动式推荐。而搜索引擎是根据用户输入的关键字进行推荐,属于被动式推荐。

18.2 项目目标与推荐系统简介

该项目的主要目标是介绍如何基于用户的历史行为完成视频的推荐。首先会介绍推荐系统的核心思想和算法,接着介绍 Mahout 并通过实例演示 Mahout 的应用,最后讲解 Lambda 架构和视频推荐案例。

为了使读者更容易理解本章的推荐系统,本节首先对推荐系统的基础知识进行总体的介绍与说明。

18.2.1 推荐系统的分类

下面我们介绍几个关于推荐的例子,其中推荐系统的分类如图 18.1 所示。

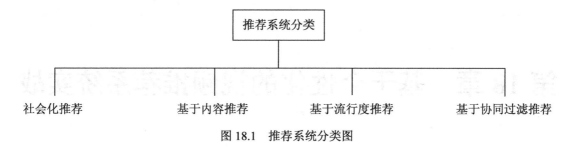

图 18.1　推荐系统分类图

- 社会化推荐：例如，向自己的亲朋好友咨询，请他们推荐好看的视频，这就是社会化推荐。
- 基于内容推荐：如果打开百度进行搜索，根据搜索返回的结果，选择自己喜欢的视频，这就是基于内容的推荐。
- 基于流行度推荐：例如，打开近期视频热度排行榜，根据目前受欢迎的程度来选择视频，这就属于基于流行度推荐。
- 基于协同过滤的推荐：根据用户以往看过的视频，找到类似的视频推荐给用户，这就是基于协同过滤的推荐。

在本例中，主要介绍基于协同过滤的推荐，这种推荐也是企业中用得比较多的推荐方法。

18.2.2　推荐模型的构建流程

推荐模型的构建流程如图 18.2 所示，共分为 4 部分，分别是数据获取、特征构建、机器学习算法应用和输出预测。

图 18.2　推荐模型的构建流程

其中，数据获取是指得到历史数据，如用户购买商品的记录；特征构建是构建一个关系矩阵，如通过矩阵表示用户与物品或者物品与物品之间的关系。下面是一个特征提取的案例，行代表用户，列代表物品 item，1 和 0.25 代表权重。例如，用户购买了某样商品，则权重为 1，如果是浏览了商品则权重为 0.25，具体如表 18.1 所示。

表 18.1 用户浏览权重表

Items\Users	1	2	3	4	5	6	...	n
1	1	—	0.25	—	—	1	—	—
2	—	—	—	—	—	—	—	—
3	1	—	0.25	—	—	—	—	—
4	—	0.25	—	1	—	—	—	—
5	—	—	—	—	—	0.25	—	—
6	—	—	1	—	—	—	—	—
...	—	—	—	—	1	—	—	—
m	—	1	—	—	0.25	0.25	—	—

18.2.3 推荐系统核心算法

下面介绍推荐系统常用的协同过滤算法。协同过滤算法是基于用户行为的推荐,这个行为可以是过去对商品的浏览、购买和评分。其中,协同过滤算法的基本逻辑是:"和你有类似爱好的人所喜欢的物品,也是你所喜欢的"或者"和你所喜欢的物品类似的物品,你也是喜欢的"。基于上面的基本逻辑,协同过滤算法分为两类:User-Based 和 Item-Based。

其中,Used-Based 类是基于用户的推荐,通过不同用户对商品的评分,计算用户之间的相似度。Item-Based 是基于不同商品的推荐,通过用户对不同商品的评分,计算商品间的相似度。

无论是基于用户的推荐还是基于商品的推荐,都要计算相似度。计算相似度有多种方法,常用的算法有:Jaccard 相似度、余弦相似度、对数似然相似度和皮尔逊相似度。具体计算方法如下:

1. Jaccard相似度

定义:有两个集合 a、b,Jaccard 相似度定义为 a、b 交集的大小与 a、b 并集大小的比值,Jaccard 值越大说明相似度越高。

公式 Jaccard Coefficient:

$$J(a,b) = \frac{|a \cap b|}{|a \cup b|} = \frac{|a \cap b|}{|a|+|b|-|a \cap b|}$$

举例:假设有集合 a,b,计算它们的 Jaccard 相似度,如图 18.3 所示。

$$J(a,b) = \frac{(1+1)}{(1+1+1)+(1+1+1)-(1+1)}$$

当 a 和 b 都为空时,$jaccard(a,b)=1$。

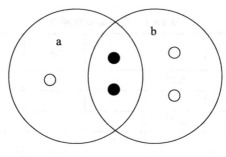

图 18.3 a、b 集合

2. 余弦相似度

定义：已知向量 *a*、*b*，余弦相似度是通过计算两个向量之间的余弦值来比较这两个向量的相似度，余弦值越大说明越相似。

公式 Cosine similarity：

$$\cos\theta = \frac{\boldsymbol{a}\cdot\boldsymbol{b}}{\|\boldsymbol{a}\|\|\boldsymbol{b}\|}$$

举例：向量 *a*、*b* 的坐标分别为 (1, 0.5)，(1, 1)，计算 *a*、*b* 向量的余弦相似度为：

$$\cos\theta = \frac{x_1 x_2 + y_1 y_2}{\sqrt{x_1^2 + y_1^2} \times \sqrt{x_2^2 + y_2^2}} = \frac{1\times 1 + 0.5\times 1}{\sqrt{1^2 + 0.5^2} \times \sqrt{1^2 + 1^2}} \approx 0.95$$

3. 对数似然相似度

定义：假设已知 A、B 两个用户对不同商品的偏好情况，通过创建偏好数据的矩阵，利用矩阵熵、行熵和列熵计算对数似然相似度。

公式：Similarity=2*(matrixEntropy−rowEntropy−columnEntropy)

其中，rowEntropy 为行熵，columnEntropy 为列熵，matrixEntropy 为矩阵熵，N 为总商品数量。

4. 皮尔逊相似度

定义：已知 *x*，*y* 两个 *n* 维向量，皮尔逊相似度通过两个向量之间的协方差和标准差的比值求得。其中，协方差是描述 *x*，*y* 相关程度的量，方差越大，标准差越大，说明这个变量的相似度越低。

公式 Pearson Correlation：

$$corr(\boldsymbol{a},\boldsymbol{b}) = \frac{\sum_{i=1}^{n}(x_i - \overline{x})(y_i - \overline{y})}{\sqrt{\sum_{i=1}^{n}(x_i - \overline{x})^2}\sqrt{\sum_{i=1}^{n}(y_i - \overline{y})^2}}$$

本章在进行用户推荐时使用的是皮尔逊相似度，下面先看一个实例。

已知用户购买商品的评分表（见表 18.2），我们通过这个评分表来计算商品与商品间的相似度，以及用户和用户间的相似度（评分为 0 时表示用户未购买此商品）。

表 18.2 用户购买商品评分表

Items \ Users	1	2	3	4	5	6
1	5	4	5	5	3	3
2	0	1	1	0	2	3
3	5	0	4	0	3	5
4	5	5	4	4	4	5
5	1	3	2	1	0	0
6	2	0	2	1	1	0

我们在这里只是简单介绍相似度的计算方法，下面列出 user 相似度和 Item 相似度的计算过程。

已知 user1 和 user3 对商品评分的均值：$\bar{u}_1 = 3, \bar{u}_3 = 3$

user1,3 相似度 $corr(u_1, u_3)$：

$$corr(u_1, u_3) = \frac{\sum_{i=1}^{6}(u_{1i} - \bar{u}_1)(u_{3i} - \bar{u}_3)}{\sqrt{\sum_{i=1}^{6}(u_{1i} - \bar{u}_1)^2}\sqrt{\sum_{i=1}^{6}(u_{3i} - \bar{u}_3)^2}}$$

$$= \frac{(5-3)\times(2)+(-3)\times(1-3)+(5-3)\times(4-3)+(5-3)\times(4-3)+(1-3)\times(2-3)+(2-3)\times(2-3)}{\sqrt{2^2+3^2+2^2+2^2+2^2+1^2}\sqrt{2^2+2^2+1^2+1^2+1^2+1^2}}$$

已知所有用户对 Item1 和 Item4 商品评分的均值：$\bar{I}_1 = 5.83, \bar{I}_4 = 4.5$

Item1,4 相似度 $corr(I_1, I_4)$：

$$corr(I_1, I_4) = \frac{\sum_{i=1}^{6}(I_{1i} - \bar{I}_1)(I_{4i} - \bar{I}_4)}{\sqrt{\sum_{i=1}^{6}(I_{1i} - \bar{I}_1)^2}\sqrt{\sum_{i=1}^{6}(I_{4i} - \bar{I}_4)^2}}$$

$$= \frac{(-0.83)\times(0.5)+(-1.83)\times(0.5)+(-0.83)\times(-0.5)+(-0.83)\times(-0.5)+(-2.83)\times(-0.5)+(-2.83)\times(0.5)}{\sqrt{0.83^2+1.83^2+0.83^2+0.83^2+2.83^2+2.83^2}\sqrt{0.5^2+0.5^2+0.5^2+0.5^2+0.5^2+0.5^2}}$$

≈ 0.41

如果使用基于 User-Based 的协同过滤算法，那么我们需要求出所有用户间的皮尔逊相似度，然后通过 K 最近邻分类算法找到和被推荐用户最相似的 K 个用户，将他们评分高且被推荐用户未购买过的商品推荐给这些用户。

如果使用基于 Item-Based 的协同过滤算法，我们需要求出所有商品间的皮尔逊相似度，然后找到和被推荐用户购买过（评分高）的商品相似度最大的商品，推荐给这些用户。

在开发推荐系统项目时，为了完成推荐，通常会选择成熟的框架，比如 Mahout、SparkMLib 或者 scikit-learn 等，下面基于 Mahout 框架进行介绍。

18.2.4 如何基于 Mahout 框架完成商品推荐

Mahout 框架源自于 Apache 的开源框架，它是基于机器学习算法的实现。目前，Mahout 框架主要侧重于推荐引擎、聚类及分类，在使用 Mahout 框架时，只需要按照特定的步骤即可完成推荐系统主流程的构建。基于 Mahout 框架构建推荐系统的主要步骤如下：

（1）将原始数据映射到 Mahout 框架定义的 DataModule 中。
（2）调用相似度组件、推荐组件。
（3）计算排名估计值。
（4）评估推荐结果。

为了完成以上步骤，Mahout 框架提供了相应的接口如下。

- DataModel Interface：把原始数据映射为 Mahout 兼容的格式。
- UserSimilarity Interface ：用于计算用户间的相似度。
- ItemSimilarity Interface：用于计算物品间的相似度。
- UserNeighborhood Interface：定义用户或者物品间的"邻近"值。
- Recommender Interface：实现了具体的推荐算法，完成物品推荐。
- 下面基于 Mahout 框架提供的 User-Based 协同过滤算法进行简单的商品推荐实例介绍，其中用到的 Mahout 的相关类包括：
- 使用 FileDataModel 读入本地数据文件。
- 使用 PearsonCorrelationSimilarity 计算皮尔逊相似度。
- 基于 GenericUserBasedRecommener 构建基于用户的推荐。

18.2.5 基于 Mahout 框架的商品推荐实例

为了使读者对 Mahout 框架的商品的具体应用有更加深刻的认识，下面基于 Mahout 框架完成商品的推荐，需求是根据用户给商品的评分，计算用户和用户之间对商品的喜好相似度，完成推荐。

该项目的数据源存储在一个 csv 文件中，结构如下所示，其中第 1 列代表用户，第 2 列代表商品编号，第 3 列代表用户给商品的评分。

```
1,101,3.0
1,102,4.0
1,103,5.0
1,104,3.0
1,108,2.0
1,109,5.0
1,110,1.0
2,102,4.0
2,101,3.0
2,102,4.0
```

```
2,108,3.0
2,102,4.0
2,101,5.0
2,104,4.0
3,102,5.0
3,108,4.0
3,107,4.0
3,105,3.0
3,102,2.0
```

接着基于 Mahout 框架完成商品推荐，步骤如下：

（1）通过 FileDataModel 读取文件。

```
DataModel model = new FileDataModel(new File("d:/data.csv"));
```

（2）基于 Pearson 计算用户间的相似度。

```
UserSimilarity similarity = new PearsonCorrelationSimilarity(model);
```

（3）找到前 5 个最近的"邻居"。

```
UserNeighborhood neighborhood = new NearestNUserNeighborhood(5, similarity, model);
```

（4）构建推荐模型。

```
Recommender recommender = new GenericUserBasedRecommender(model, neighborhood, similarity);
```

（5）完成推荐，recommend 代表给第 2 号用户，推荐 3 个商品。

```
List<RecommendedItem> recommendedItems = recommender.recommend(2, 3);
```

完整代码如下：

```java
package com.sendto;

import org.apache.mahout.cf.taste.impl.model.file.*;
import org.apache.mahout.cf.taste.impl.similarity.*;
import org.apache.mahout.cf.taste.impl.neighborhood.*;
import org.apache.mahout.cf.taste.impl.recommender.*;
import org.apache.mahout.cf.taste.model.DataModel;
import org.apache.mahout.cf.taste.similarity.*;
import org.apache.mahout.cf.taste.neighborhood.*;
import org.apache.mahout.cf.taste.recommender.*;

import java.io.File;
import java.util.List;

public class RecommenderIntro {
    private RecommenderIntro() {
    }

    public static void main(String[] args) throws Exception{
    //通过 FileDataModel 读取原始文件
        DataModel model = new FileDataModel(new File("d:/data.csv"));
        UserSimilarity similarity = new PearsonCorrelationSimilarity(model);
        UserNeighborhood neighborhood = new NearestNUserNeighborhood(5,
        similarity, model);
```

```
Recommender recommender = new GenericUserBasedRecommender(model,
neighborhood, similarity);
 //给第 2 个用户推荐 3 个商品
List<RecommendedItem> recommendedItems = recommender.recommend(2, 3);

for (RecommendedItem recommendedItem: recommendedItems){
   System.out.println(recommendedItem);
}
    }
}
```

运行结果如下，表示给第 2 个用户推荐了 103、105 和 107 号商品，推荐分数分别为 5.0、4.3 和 4.0。

```
RecommendedItem[item:103, value:5.0]
RecommendedItem[item:105, value:4.3333335]
RecommendedItem[item:107, value:4.0]
```

18.3　推荐系统项目架构

前面介绍了推荐系统的基本实现方式，但在企业中开发推荐系统时，往往需要基于特定的架构。目前应用比较广泛的是基于 Lambda 架构。Lambda 架构由 Twitter 工程师 Nathan Marz 提出，它提供了离线数据和实时数据的混合平台，如图 18.4 所示。

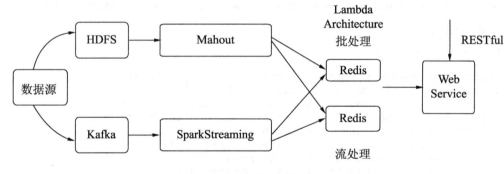

图 18.4　项目架构图

在 Lambda 架构中，主要包括离线处理层、实时处理层和服务层。在离线处理部分将产生的数据存储到 HDFS 上，通过 MapReduce 进行离线处理，MapReduce 将处理完后的模型存储到 Redis 中。在实时处理部分 Spark Streaming 基于 Kafka 实时得到数据并从 Redis 中获得推荐结果。服务层通过 Restful 完成推荐结果的呈现，将结果推荐给用户。

下面分别介绍 Lambda 架构中的离线处理部分和在线处理部分。

1．离线处理部分

离线处理部分把存放在 HDFS 上的数据通过 Mahout 得到推荐模型，并将模型存入

Redis 中,如图 18.5 所示。

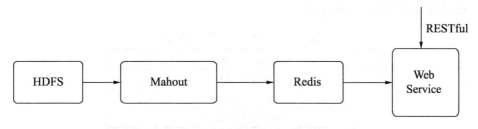

图 18.5　离线处理部分架构图

2. 在线处理部分

在线处理部分中,SparkStreaming 从 Kafka 中得到用户行为,并从数据模型中取出推荐列表,实时完成商品推荐,如图 18.6 所示。

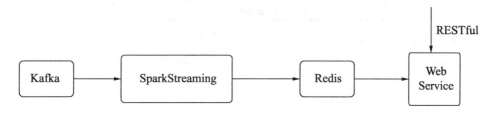

图 18.6　在线处理部分架构图

在该项目中,比如编号为 998 的用户点击了某个商品,SparkStreaming 会从数据模型中取出推荐列表,并把该用户没有看过的视频推荐给他。

18.4　推荐系统模型构建

前面介绍了推荐系统的架构,我们接着介绍推荐系统的原始数据格式和推荐模型,以及最终推荐结果的呈现。下面先介绍原始数据。

该推荐系统中所用到的原始数据是"用户给视频打分数据",数据格式如下:

```
userId,videoId, score
111,    400 , 4
112,    401 , 5
113,    400 , 3
……
```

其中,第 1 列代表用户编号,第 2 列代表视频编号,第 3 列代表用户对视频的评分。

首先,这些数据会通过 Flume 被采集到 HDFS 上。这些原始数据经过 Mahout 的计算得到推荐模型后,会放入 Redis 中,训练得到的推荐模型主要有两类,分别是"视频与视频相关度表"和"用户与视频得分表"。其中,"视频与视频相关度表"主要记录了视频与视频之间的相似程度,如表 18.3 所示。

表 18.3 视频与视频相关度表

Rec_Item_Similarity
Item_ID
Similar_Item
Similarity_Score

"用户与视频得分表"记录了用户给视频的评分情况,如表 18.4 所示。

表 18.4 用户与视频得分表

Rec_User_Item_Base
User_ID
Item_ID
Recommendation_Score

除了模型构建,还需要获得用户行为,在该项目中,用户的行为封装在 NewEvent 类中,NewEvent 类中主要记录用户看了哪部视频。代码如下:

```
public class NewEvent{
   private long userId;
   private long itemId;
   ……
}
```

有了用户行为后,结合推荐模型就可以进行推荐了。主体处理逻辑是 Kafka 会把 NewEvent 类传给 SparkStreaming,SparkStreaming 结合 NewEvent 和存放在 Redis 中的推荐模型完成推荐过程。

推荐格式如下:

```
Key: RUI:998
Value: "[{\"id\":556,\"s\":0.96209},{\"id\":9,\"s\":0.95146},{\"id\":192,\"s\":0.93146},{\"id\":927,\"s\":0.91130}]"
```

其中,998 是被推荐的用户 ID,Value 中是为此用户推荐的视频,分别为:556,表示它与 998 用户看过的视频相似度为 0.96209;9,表示它的相似度为 0.95146;192,表示它的相似度为 0.93146;927,表示它的相似度为 0.91130。

18.5 核心代码

实现用户推荐的核心代码共包括 3 部分,分别为离线部分、在线部分和公共部分,分别对应 offline、online 和 util 这 3 个包,下面我们详细介绍这 3 部分代码的作用和实现过程。

18.5.1 公共部分

公共部分主要分为 Config 类、RedisTool 类和 VideoSimilarity 类 3 个类，它们的主要作用和实现过程如下。

（1）Config.java：定义 Redis 和 Kafka 的服务器和端口等连接参数。代码如下：

```java
package com.sendto.util;

public final class Config{
    public static final String REDIS_SERVER = "192.168.179.96";
    public static final String KAFKA_SERVER = "192.168.179.96";
    public static final String KAFKA_PATH = KAFKA_SERVER + ":9092";
    public static final String KAFKA_TOPICS= "recommend";
}
```

（2）RedisTool.java：设置 Redis 连接池的参数，并进行连接池的初始化，定义了获取 Jedis 实例和释放 Jedis 资源的方法。代码如下：

```java
package com.sendto.util;

import redis.clients.jedis.Jedis;
import redis.clients.jedis.JedisPool;
import redis.clients.jedis.JedisPoolConfig;

public final class RedisTool {
    //Redis 连接设置：服务器 IP、端口号、访问密码
    private static String ADDRESS = Config.REDIS_SERVER;
    private static int PORT = 6379;
    private static String PASSWORD = "123456";
    //在 borrow 一个 Jedis 实例时，设置是否提前进行 validate 操作；如果为 true，
    //则得到的 Jedis 实例均是可用的
    private static boolean borrowTest = true;

    private static int TIMEOUT = 10000;
    private static JedisPool jedisPool = null;

    //初始化 Redis 连接池
    static {
      try {
          JedisPoolConfig config = new JedisPoolConfig();

          config.setTestOnBorrow(borrowTest);
          jedisPool = new JedisPool(config, ADDRESS,PORT,TIMEOUT,PASSWORD);
      } catch (Exception e) {
          e.printStackTrace();
        }
    }
    //获取 Jedis 实例
    public synchronized static Jedis getJedis() {
        try {
            if (jedisPool != null) {
```

```
                Jedis jedisSource = jedisPool.getResource();
                return jedisSource;
            } else {
                return null;
            }
        } catch (Exception e) {
            e.printStackTrace();
            return null;
        }
    }
    // 释放 Jedis 资源
    public static void returnResource(final Jedis jedisSource) {
        if (jedisSource != null) {
            jedisPool.returnResource(jedisSource);
        }
    }
}
```

（3）VideoSimilarity.java：是定义了视频和视频相似度的类，可以通过 set 和 get 方法获取视频相似度。代码如下：

```
package com.sendto.util;

public class VideoSimilarity implements Comparable<VideoSimilarity> {
    private long videoId; //videoID
    private Double similarity; //similarity
    public VideoSimilarity() {
        this.videoId = -1;
        this.similarity = 0d;
    }
    public VideoSimilarity(long videoId, Double similarity) {
        this.videoId = videoId;
        this.similarity = similarity;
    }
    public long getId() {
        return videoId;
    }
    public void setId(long videoId) {
        this.videoId = videoId;
    }
    public Double getS(){
        return similarity;
    }
    public void setS(Double similarity) {
        this.similarity = similarity;
    }
    public boolean equals(Object obj) {
        if (!(obj instanceof VideoSimilarity))
            return false;
        if (obj == this)
            return true;
        // Double 类型的数据不应该直接比较
        return this.videoId == ((VideoSimilarity) obj).videoId && this.
         similarity == ((VideoSimilarity) obj).similarity;
    }
```

```java
    public int hashCode(){
        return (int)(videoId + similarity);
    }
    public int compareTo(VideoSimilarity obj) {
        if(this.similarity > obj.similarity) {
            return 1;
        } else if(this.similarity < obj.similarity) {
            return -1;
        }
        return 0;
    }
    @Override
    public String toString() {
        return "id:" + videoId + ",similarity:" + similarity;
    }
}
```

18.5.2 离线部分

离线部分主要分为 HDFSDataModel.java、UserItemSimilarityToRedis.java、ItemSimilarityToRedis.java 和 RecommendTable.java 4 个类,这 4 个类的主要作用如下。

- HDFSDataModel.java:数据的采集和转换。
- ItemSimilarityToRedis.java:得到视频-视频相似度表。
- UserItemSimilarityToRedis.java:得到用户-视频信息表。
- RecommendTable.java:结合视频-视频和用户-视频表,对用户进行视频推荐。

具体实现过程见下方代码。

(1) HDFSDataModel.java:从 HDFS 中将用户对视频的评分数据提取出来,并进行格式的转换后存入系统临时缓存目录的 data.txt 中。代码如下:

```java
package com.sendto.offline;

import org.apache.commons.io.Charsets;
import org.apache.hadoop.conf.Configuration;
import org.apache.hadoop.fs.FileSystem;
import org.apache.hadoop.fs.Path;
import org.apache.mahout.cf.taste.impl.model.GenericDataModel;
import org.apache.mahout.cf.taste.impl.model.file.FileDataModel;
import java.io.*;
import java.util.regex.Pattern;

public class HDFSDataModel extends FileDataModel {
    private static final String colDelimiter = "::";

//定为字符串的正则表达式必须首先被编译为 Pattern 的实例,然后将得到的模式用于创建 matcher 对象进行匹配
    private static final Pattern colDelimiterPattern = Pattern.compile(colDelimiter);

    public HDFSDataModel(Configuration config, String hdfsPath) throws
```

```java
IOException {
  this(config, new Path(hdfsPath));
}

public HDFSDataModel(Configuration config, Path hdfsPath) throws
IOException {
  super(storeHdfsFileToLocal(config, hdfsPath, colDelimiter));
}

private static File storeHdfsFileToLocal(Configuration config, Path path,
String colDelimiter) {
  //用逗号","代替"::"存入系统的临时缓存目录中
  File readFile = new File(new File(System.getProperty("java.io.tmpdir")),
  "data.txt");
  if (readFile.exists()) {
    readFile.delete();
  }
  try{
    Writer writer = new OutputStreamWriter(new FileOutputStream(readFile),
    Charsets.UTF_8);

    FileSystem fileSystem = path.getFileSystem(config);
    BufferedReader br = new BufferedReader(new InputStreamReader
    (fileSystem.open(path)));
    String oneLine = br.readLine();
    while (oneLine != null) {
      int lastDelimiterIndex = oneLine.lastIndexOf(colDelimiter);
      if ( lastDelimiterIndex < 0) {
        throw new IOException("Unexpected input format on line: " +
        oneLine);
      }
      String subStr = oneLine.substring(0, lastDelimiterIndex);
      String replacedLine = colDelimiterPattern.matcher(subStr).
      replaceAll(",");
      writer.write(replacedLine);
      writer.write('\n');
      oneLine = br.readLine();
    }
  } catch(IOException e){
    e.printStackTrace();
  }
  return readFile;
}
```

（2）UserItemSimilarityToRedis.java：获取 Jedis 实例，对于格式处理后的数据 DataModel，我们将其中的用户 ID 和用户看过的视频信息取出，存入 Redis 中。代码如下：

```java
package com.sendto.offline;

import com.alibaba.fastjson.JSON;
import com.sendto.util.RedisTool;

import org.apache.mahout.cf.taste.common.TasteException;
import org.apache.mahout.cf.taste.impl.common.FastIDSet;
```

```java
import org.apache.mahout.cf.taste.impl.common.LongPrimitiveIterator;
import org.apache.mahout.cf.taste.model.DataModel;
import redis.clients.jedis.Jedis;

import java.util.concurrent.CountDownLatch;
import java.util.concurrent.Executors;
public class UserItemSimilarityToRedis {
  private DataModel dataModel = null;
  private Jedis jedis = null;
  private CountDownLatch countLatch = new CountDownLatch(1);

  public UserItemSimilarityToRedis(DataModel dataModel) {
    this.dataModel = dataModel;
    this.jedis = RedisTool.getJedis();
  }

  public void redisStorage() {
    Executors.newSingleThreadExecutor().submit(
        new Runnable() {
          public void run() {
            realization();
            countLatch.countDown();
          }
        }
    );
  }

  private void realization() {
    try {
      LongPrimitiveIterator userIds = dataModel.getUserIDs();
      while(userIds.hasNext()) {
        long userID = userIds.nextLong();
        FastIDSet idSet = dataModel.getItemIDsFromUser(userID);
        String key = "UI:" + userID;
        String videos = JSON.toJSONString(idSet.toArray());
        jedis.set(key, videos);
        System.out.println("Stored User:" + key);
      }
    } catch(TasteException te) {
      te.printStackTrace();
    }
  }

  public void waitUtilDone() throws InterruptedException {
    countLatch.await();
  }
}
```

（3）ItemSimilarityToRedis.java：计算相似度的代码将在主函数所在的类 RecommendTable.java 中实现，即代码（4）中实现。现在假设相似度已经计算出来，我们将每个 VideoID（即视频 ID）和它相似的多个"VideoID+相似度"提取出来，以 VideoID 为 Key，"相似 Vedio+相似度"json 串为 Value 存入 Redis 表中，即得到视频与视频相似度表。代码如下：

```java
package com.sendto.offline;

import java.io.IOException;
import com.alibaba.fastjson.JSON;
import com.sendto.util.RedisTool;
import com.sendto.util.VideoSimilarity;

import org.apache.mahout.cf.taste.similarity.precompute.SimilarItem;
import org.apache.mahout.cf.taste.similarity.precompute.SimilarItems;
import org.apache.mahout.cf.taste.similarity.precompute.SimilarItemsWriter;
import redis.clients.jedis.Jedis;

public class ItemSimilarityToRedis implements SimilarItemsWriter {
  private long itemCount = 0;
  private Jedis jedis = null;
  public void open() throws IOException {
    jedis = RedisTool.getJedis();
  }

  public void add(SimilarItems similarItems) throws IOException {
    VideoSimilarity[] videos = new VideoSimilarity[similarItems.
    numSimilarItems()];
    int counter = 0;
    for (SimilarItem item: similarItems.getSimilarItems()) {
      videos [counter] = new VideoSimilarity(item.getItemID(), item.
      getSimilarity());
      counter++;
    }
    String key = "II:" + similarItems.getItemID();
    String videoItems = JSON.toJSONString(videos);
    jedis.set(key, videoItems);
    itemCount++;
    if(itemCount % 100 == 0) {
      System.out.println("Store " + key + " to redis, total:" + itemCount);
    }
  }

  public void close() throws IOException {
    jedis.close();
  }
}
```

（4）RecommendTable.java：实现主函数，首先定义 HDFS 的文件读取路径；然后分别生成"用户-视频信息表"和"视频-视频最大似然相似度表"；最后利用 item-based 推荐器来多线程地为用户推荐 5 部视频。代码如下：

```java
package com.sendto.offline;

import org.apache.hadoop.fs.Path;
import org.apache.mahout.cf.taste.impl.recommender.GenericItemBased
Recommender;
import org.apache.mahout.cf.taste.impl.similarity.LogLikelihood
Similarity;
import org.apache.mahout.cf.taste.impl.similarity.precompute.
```

```java
MultithreadedBatchItemSimilarities;
import org.apache.mahout.cf.taste.model.DataModel;
import org.apache.mahout.cf.taste.recommender.ItemBasedRecommender;
import org.apache.mahout.cf.taste.similarity.precompute.BatchItemSimilarities;
import org.apache.hadoop.conf.Configuration;
import javax.security.auth.login.AppConfigurationEntry;
import java.io.File;

public final class RecommendTable {

  private RecommendTable() {}

  public static void main(String[] args) throws Exception {

    Configuration config = new Configuration();
    config.set("fs.defaultFS","hdfs://192.168.179.96:9000");
    DataModel dataModel = new HDFSDataModel(config,new Path("/data/ratings.dat"));

    UserItemSimilarityToRedis userItemSimilarity = new UserItemSimilarityToRedis(dataModel);
    userItemSimilarity.redisStorage();

    ItemBasedRecommender recommender = new GenericItemBasedRecommender(dataModel,
        new LogLikelihoodSimilarity(dataModel));

    BatchItemSimilarities theadDeal = new MultithreadedBatchItemSimilarities(recommender, 5);

    int similarNum = theadDeal.computeItemSimilarities(Runtime.getRuntime().availableProcessors(), 1,new ItemSimilarityToRedis());
    userItemSimilarity.waitUtilDone();
  }
}
```

18.5.3 在线部分

在线部分主要分为 UserItemEvent.java、UserItemProducer.java 和 OnlineRecommender.java 这 3 个类，其主要作用如下。

- UserItemEvent.java：存储用户点击视频行为的类。
- UserItemProducer.java：将用户和点击行为发送到 Kafka 中的 Topic 中。
- OnlineRecommender.java：消费 Kafka 中的数据，在线将用户历史点击数据提取出来，结合离线部分计算出的视频相似度，找出和用户点击过的视频相类似的视频推荐给用户。

下面详细介绍在线部分各类的作用和实现过程。

（1）UserItemEvent.java：存储了用户点击行为的类，对用户点击数据进行初始化。

```java
package com.sendto.online;

public class UserItemEvent {
    private long userId;
    private long itemId;

    public UserItemEvent() {
      this.userId = -1L;
      this.itemId = -1L;
    }

    public UserItemEvent(long userId, long itemId) {
      this.userId = userId;
      this.itemId = itemId;
    }

    public long getUserId() {
      return userId;
    }

    public void setUserId(long userId) {
      this.userId = userId;
    }

    public long getItemId() {
      return itemId;
    }

    public void setItemId(long itemId) {
      this.itemId = itemId;
    }
}
```

（2）UserItemProducer.java：定义 Kafka 的 Topic，将用户和点击数据利用 Producer 发送到 Kafka 的 Topic 中。

```java
package com.sendto.online;

import com.alibaba.fastjson.JSON;
import util.Parameter;
import org.apache.log4j.Logger;
import java.util.Properties;
import kafka.javaapi.producer.Producer;
import kafka.producer.KeyedMessage;
import kafka.producer.ProducerConfig;

public class UserItemProducer implements Runnable {
  private static final Logger logger = Logger.getLogger(UserItemProducer.
    class);
  private final String topic;

  public UserItemProducer(String topic) {
    this.topic = topic;
  }
```

```java
    static UserItemEvent[] userItemEvents = new UserItemEvent[]{
        new UserItemEvent(1000000L, 123L),
        new UserItemEvent(1000001L, 111L),
        new UserItemEvent(1000002L, 500L),
        new UserItemEvent(1000003L, 278L),
        new UserItemEvent(1000004L, 681L),
    };

    public void run() {
      Properties props = new Properties();
      props.put("metadata.broker.list", Parameter.KAFKA_PATH);
      props.put("serializer.class", "kafka.serializer.StringEncoder");
      props.put("producer.type", "async");
      ProducerConfig config = new ProducerConfig(props);
      Producer<Integer, String> producer = null;
      try {
        System.out.println("Producing messages");
        producer = new Producer<>(config);
        for (UserItemEvent event : userItemEvents) {
          String eventAsStr = JSON.toJSONString(event);
          producer.send(new KeyedMessage<Integer, String>(
              this.topic, eventAsStr));
          System.out.println("Sending messages:" + eventAsStr);

        }
        System.out.println("Done sending messages");
      } catch (Exception exception) {
        logger.fatal("Error while producing messages", exception);
        logger.trace(null, exception);
        System.err.println("Error while producing messages: " + exception);
      } finally {
        if (producer != null) producer.close();
      }
    }

    public static void main(String[] args) throws Exception {
      new Thread(new UserItemProducer(Parameter. KAFKA_TOPICS)).start();
    }
}
```

（3）OnlineRecommender.scala：作为 Kafka 的 Consumer，将 Topic 中的数据提取出来，通过本地两个线程进行数据的读取，间隔时间 2 秒一次；假设已经读取出某用户和其点击视频数据，然后可以结合离线部分我们得到的视频-视频相似度表，将和此用户点击视频相似的视频数据取出，推荐给此用户。

```scala
package com.sendto.recom

import com.alibaba.fastjson.JSON
import kafka.serializer.StringDecoder
import org.apache.spark.SparkConf
import util.{Parameter, RedisTool}
import org.apache.spark.streaming._
import org.apache.spark.streaming.kafka._
```

```
object OnlineRecommender {
  def main(args: Array[String]) {
//要读取的 BrockServer 和 Topic
    val Array(brokers, topics) = Array(Parameter.KAFKA_PATH, Parameter.
    KAFKA_TOPICS)

    //使用本地的两个线程、每隔 2 秒进行一次 Kafka 数据的消费
    val sparkConfig = new SparkConf().setMaster("local[2]").setAppName
    ("OnlineRecommender")
    val ssc = new StreamingContext(sparkConfig, Seconds(2))

    //从 Kafka 中读取数据
    val topicsSet = topics.split(",").toSet
    val kafkaParams = Map[String, String](
      "metadata.broker.list" -> brokers,
      "auto.offset.reset" -> "smallest")
    val messages = KafkaUtils.createDirectStream[String, String,
    StringDecoder, StringDecoder](
      ssc, kafkaParams, topicsSet)
     //将读取数据的第二部分,即用户点击视频行为取出
    messages.map(_._2).map{ event =>
      JSON.parseObject(event, classOf[UserItemEvent])
    }.mapPartitions { iter =>
      val jedis = RedisTool.getJedis
      iter.map { event =>
        println("UserItemEvent" + event)
    //将用户点击的视频 ID 取出,将它作为 Key 从离线时存储在 Redis 的视频-视频信息表中
    查出与其相似度高的视频,推荐给用户
        val userId = event.asInstanceOf[UserItemEvent].getUserId
        val itemId = event.asInstanceOf[UserItemEvent].getItemId
        val key = "II:" + itemId
        val value = jedis.get(key)
        jedis.set("RUI:" + userId, value)
        println("Recommend to user:" + userId + ", items:" + value)
      }
    }.print()
    //开始计算过程
    ssc.start()
    ssc.awaitTermination()
  }
}
```

18.6 小　　结

本章通过对已知的用户视频观看信息,构建了"视频-视频相似度"和"用户-视频相似度"模型表,并利用协同过滤推荐算法实现了用户的视频推荐,推荐更贴合消费者喜好的视频,这对视频平台提高播放量有重要作用。

第 19 章 电信离网用户挽留实战

在各大通信运营商持续增大网络建设规模的背景下，我国通信市场呈快速扩大趋势，而互联网和 ICT（信息和通信技术）行业的发展使移动通信网络的主要任务不断发生改变，用户需求的重要性逐渐增大，通信产业的竞争不断加剧。

某移动通信运营商针对近期用户流失严重的问题，希望通过数据挖掘建立流失用户预警名单，并将名单交由市场管理部，据此推出如研发新产品、用户关怀和提升服务质量等一系列针对性营销策略来维系和挽留客户。

本章中我们将采用数据挖掘的方法获取上述流失用户的预警名单。其中，数据挖掘标准流程包括商业理解（business understanding）、数据理解（data understanding）、数据准备（data preparation）、建模（modeling）、评估（evaluation）和部署（deployment）共 6 步，具体说明如下。

- 商业理解：进行业务理解，即用户的业务种类和消费行为，以及企业现有的商业服务和商业目标等。
- 数据理解：从商业目标的角度看数据，理解数据的意义，并根据不同的业务数据，选择不同的工具和方法进行数据采集。
- 数据准备：对企业的客户数据进行数据采集，并根据商业策略进行数据的整理、清洗和转换等，将原始数据转换为最终建模用的数据。
- 建模：根据业务特点和机器学习模型的优缺点，选择最佳模型进行建模，并在学习模型的过程中不断调整参数。
- 评估：对于我们拟合得到的模型，根据真实值对它的预测结果进行评估，并通过混淆矩阵和 ROC 曲线反映预测情况。
- 部署：进行项目部署，得到评分清单后推出客户维系方案，并在各渠道执行，最后将客户挽留结果返回给销售管理人员，继续数据分析和策略调整这一整个循环过程。

19.1 商业理解

对于商业建模，我们首先要明确企业的商业环境、商业目标和商业策略，然后确定挖掘目标，并将商业定义和目标转化成数据挖掘中的定义和目标，最后根据实际情况采集相关数据，并进行数据挖掘。

因为我们的目标是针对电信离网用户，所以根据电信行业的特征，我们的数据需求来源包括账单（每月生成）、详单（每次业务生成）、终端库（移动设备）、MR（手机信号）、客服（是否有投诉）和套餐等。数据需求的详细内容如图 19.1 所示。

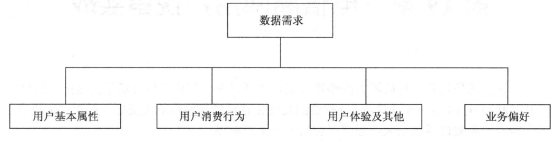

图 19.1　数据需求分类图

- 用户基本属性：包括年龄、是否是 VIP、是否属于集团网、付费方式、合约到期时间等。
- 用户消费行为：包括当月用户消费额、通话时长（长话、市话、漫游）、数据流量、短信条数、流量/语音/短信费用、增值费用。
- 用户体验及其他：包括用户感知评分、终端价位、电话集中度（网内拨打次数占比）、圈子（网内联系人比例）。
- 业务偏好：包括时间偏好、业务种类偏好、居住地等。

为了将此商业目标和数据挖掘的定义、计划和目标相结合进行数据挖掘，我们需要对用户的流失规则进行定义。比如，关于数据挖掘的定义，t 月正常，$t+1$ 月用户状态为注销，则视为主动流失用户；t 月正常消费且 $t+1$ 月零消费的用户可视为预流失用户；t 月出账且 $t+1$ 月不出账也可视为预流失用户。关于数据挖掘的结果，在模型中定义 $Y=1$ 则为流失用户，$Y=0$ 则为未流失用户，通过构建模型最终可以筛选出高概率离网客户。

19.2　数 据 理 解

在数据理解阶段，我们要从商业目标的角度看数据，进行数据的收集和数据的可用性评估。数据理解主要包括三部分：收集数据、了解数据和保证数据质量。

19.2.1　收集数据

收集数据时我们要考虑到用户流失收益，即用户之所以会流失，会有如卡费、话费、客户社交关系网可能会丢失等成本，但是用户的总体流失收益是大于流失成本的，如图 19.2 所示。

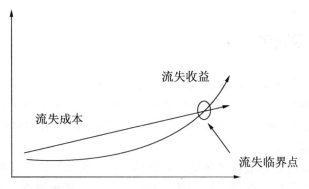

图 19.2 用户流失收益-成本图

举例来说，如果用户的套餐中不包含漫游优惠，但是该用户的漫游资费占总经费 80% 以上，而竞争公司的套餐中漫游费用低，即用户如办理竞争公司的套餐会有优惠力度大、费用低、流量赠送多等好处，那么此用户在成本一定的情况下离网收益会上升；又比如对于集团员工用户，因为电信集团对内部员工的优惠活动多，集团同事多为本集团用户，因此关系网复杂，离网的成本可能较高，直到用户辞职。

因此，对于离网用户的挖掘，我们需要找到影响用户离网成本和离网收益的特征变量，根据综合分析得到的离网成本和离网收益，考虑用户离网的可能性。

19.2.2 了解数据

首先要了解数据字段代表的意义，如在网时长较长，则可以推测此用户的关系网络复杂，离网成本高，离网可能性低；又比如用户的每月通话时长在增加，那么其离网可能性较低。收集数据的字段及说明如表 19.1 所示。

表 19.1 用户数据字段说明

字段名称	标 签	字段名称	标 签
subcriberID	用户编号	posTrend	用户通话是否有上升态势（1=是）
churn	是否流失（1=流失）	negTrend	用户通话是否有下降态势（1=是）
gender	性别（1=男）	nrProm	营销次数
age	年龄	prom	最近一个月是否被营销（1=是）
edu_class	教育等级：0=小学及以下，1=中学，2=本科，3=研究生	curPlan	统计开始时套餐类型（1=200分钟，2=300分钟，3=350分钟，4=500分钟
incomeCode	用户居住区域平均收入代码	avgPlan	统计期内使用时间最长的套餐
duration	在网时长（月）	planChange	是否更换过套餐（1=是）
feton	是否飞信用户（1=开通）	posPlanChange	统计期内是否提高套餐（1=是）
peakMinAv	最高单月通话时长	negPlanChange	统计期内是否降低套餐（1=是）
peakMinDiff	统计期内结束月份与开始月份通话时长增长量	call_10086	是否拨打过客服电话（1=是）

19.2.3 保证数据质量

在获取数据的过程中要保证数据的质量,这样后续的数据处理和数据建模才有意义。因此获取数据时应确保获得数据与字段的说明,以便进行重要变量的筛选,并确认字符编码,保证数据正常显示;确认数据的字段完整和格式正确,在读取数据时使用对应的方法读入,防止噪声数据产生。

采集数据完成后需要对数据进行处理和转换,将原始数据处理成最终建模所需要的数据。数据准备主要分为数据整理、数据清洗和数据转换,详细准备过程如下。

19.3 数 据 整 理

数据整理主要分为数据整合和数据过滤两步,详细介绍如下。

19.3.1 数据整合

数据整合是把不同数据源的数据进行汇总,形成可用于数据分析的表,主要包括横向连接(合并、追加)和转置的方式,具体整理方式如下。

- 合并、追加:通常是根据记录的 ID 向表中添加其他表中的字段和记录,进行表与表的数据合并,如图 19.3 所示。

ID	Age	Sex
1	20	1
2	35	0

ID	Call_Freq	Fee
1	245	143
2	78	56

ID	Age	Sex	Call_Freq	Fee
1	20	1	245	143
2	35	0	78	56

图 19.3 数据横向连接图

- 转置:对于每条 ID 数据,即每个用户,通常只记录一条观测值,因此需要把用户和对应多个产品的矩阵转置每个用户只一条观测值的矩阵。这里是将长表转化为宽表,将作业型表转化为分析型表,属于表内的数据合并。转置方式如图 19.4 所示。

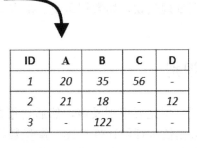

图 19.4　数据转置图

19.3.2　数据过滤

数据过滤是从业务、策略方面进行数据的筛选，排除实际生活中不正常的数据和字段，根据业务特点设置数据类型。主要包括记录选择、字段选择和设置字段类型等方法，具体处理如下。

- 记录选择：对宽表的行数据进行处理，筛选不正常消费用户，只分析正常的样本行记录，筛选结果如下。
- 基础筛选：去除上网卡用户、数据卡用户、非出账用户、企业用户。
- 套餐筛选：去除 M2M、无线座机、无线公话、客服专用、员工套餐等特殊用户。
- 非常规消费行为用户：去除语音（>5000 分钟）、流量（>15GB）、短信超常消费（>2000 条）的用户。
- 字段选择：对宽表的列数据进行处理，删除不需要的字段和冗余字段，筛选结果如下。
- 排除对分析无贡献、明显不需要的字段：营销网格、客户编码、表内记录编号、中继设备号、姓名等。
- 排除冗余字段：需要排除一个或几个相关性太高的列变量，包括通话开始/结束时间（原表已包含通话时长字段）、IMEI（原表包含 TAC）等。
- 设置字段类型：根据业务需求设置字段的类型，通常自动识别类型，也可人工调整。字段类型分类如下：
- 字段的值类型可分为字符串、整数型、浮点型和日期时间型。
- 字段的变量类型可分为离散、连续、有序和标识型。

19.4　数据清洗

实际应用中采集的数据通常不能保证精确性，存在噪声数据。产生噪声数据的原因有很多，如统计、调研类数据在录入信息时缺失或错误输入；账单信息等数据计算错误、设备故障等，电信级通信设备出错概率很小，但采集设备可能出现故障；还存在因为对数据理解不够导致数据处理失误，如数据重复、错误截断数据等问题，甚至还有一些数据分析工具导致的错误。

19.4.1 噪声识别

针对噪声数据，首先要识别噪声，然后再根据数据特点进行处理，识别方法如下：

（1）首先对数据值进行统计性描述，衡量数据的集中趋势、离中趋势、分布测度，那些与集中数据相距较远，即离中水平较大的样本有可能为离群值，其中，衡量集中趋势指标如下。

- 连续变量：均值、中位数、方差、标准差、峰度、偏度；
- 离散变量：众数、频次、频率等。

（2）根据图形直观、快速地对数据进行初步分析。

首先画出样本关于通话时长的直方图，查看是否有离群值，如图 19.5 所示。

由直方图 19.5 可知，图中右侧有一部分数据距样本集中水平较远，且数量很少，因此猜测为离群值。

然后通过样本盒须图分析 VIP 用户和通话时长之间的关系，如图 19.6 所示。

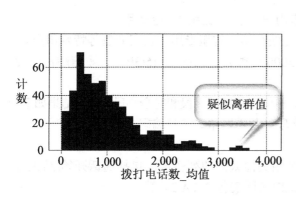

图 19.5　样本直方图

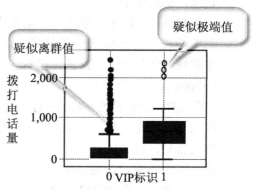

图 19.6　样本盒须图

由盒须图 19.6 可知，分类为 0 的样本中有一部分变量值在内限之外，外限之内，因此猜测为离群值；分类为 1 的样本中有一部分变量值在外限之外，离中水平大，猜测为极端值。

分析通信量和用户状态之间关系时用条形图展示，如图 19.7 所示。

由条形图 19.7 可知，右边部分的数据离中水平高，数量又极少，因此猜测为噪声数据。

分析样本在总体中所占比例时，我们利用饼图进行分析，如图 19.8 所示。

由饼图 19.8 可知，1 类和 2 类占比重较大且较均匀，而第 3 类数据数量极少，疑似错误值。

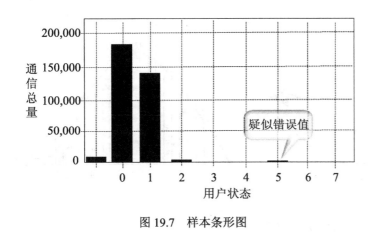

图 19.7 样本条形图

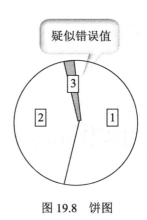

图 19.8 饼图

19.4.2 离群值和极端值的定义

对于图形中反映出的疑似离群值和极端值，我们要根据规则、定义来确定，离群值和极端值设置标准如下：

1．极端值

对于样本整体 5 倍标准差之外的数据，我们可以看作极端值，但极端值有时意味着特殊值，例如"特殊用户的超大额消费"，此时应重新理解数据。

2．离群值

- 利用平均值法设置标准，如平均值±3×标准差之外的数据看作离群值。
- 四分位数法设置标准，如四分位距 IQR=Q3-Q1，将 Q1-1.5×IQR～Q3+1.5×IQR 范围内的数据看作离群值，其中 Q1 是第 1 四分位数，Q3 是第 3 四分位数。

19.4.3 离群值处理方法

对于某些明显错误值，有修正和删除两种方法，修正即补充正确的数据，但因为我们处于流程后方的开发、建模步骤，因此通常选择删除错误值。而对于离群值，因为它是成批的数据，我们可以进行数据的转换来减小离群值的影响，处理离群值方法如下：

（1）视为空值。

（2）变量转换：当样本数据右偏时，可以对自变量数据取对数，使其服从正态分布，如图 19.9 所示。

（3）盖帽法：对于超过 Q1-1.5×IQR～Q3+1.5×IQR 范围的值（超出内限），我们取边缘值 Q1-1.5×IQR 和 Q3+1.5×IQR 的值替代超出内限的样本值，如 19.10 图所示。

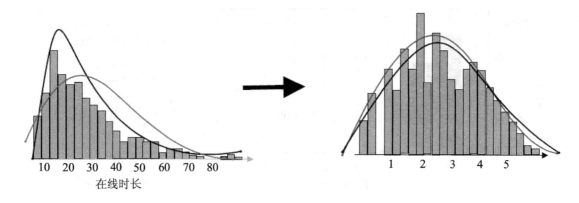

图 19.9　变量转换图

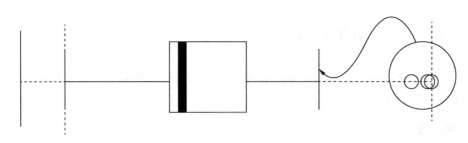

图 19.10　盖帽法转换图

（4）考虑分段建模或变量离散化：

分段建模是将重要变量进行分段，创建多个模型，使模型预测效果更好，如图 19.11 所示。

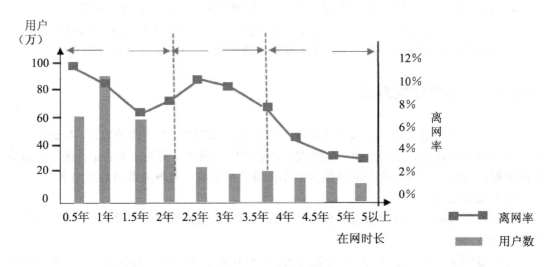

图 19.11　分段建模图

变量离散化是将连续变量 X 分为不同组，将连续变量转化为等级变量进行建模，如图 19.12 所示。

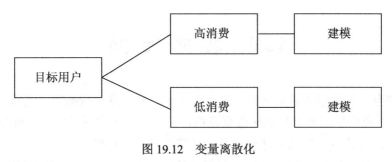

图 19.12　变量离散化

19.4.4　数据空值处理示例

前面介绍了数据整理和数据清洗的方法，下面介绍一个数据处理的例子。假设我们已经将收集的数据制成表，要对表中数据的空值进行处理，主要数据如图 19.13 所示。

ID	gender	年龄	教育程度	所在区域	峰值月时长	营销次数
1	0	22	3	郊区	1232	2
2	1	18	3	市区	522	2
3		45	3	市区	845	1
4	1		2		1321	1
5	0	15	2	市区	611	0
6	1	22	0		967	1
7	1	25	1	郊区	662	0
8				市区	10710	
9	0	27		市区	996	1
10	0	30	3	郊区	422	2
11	0	18	3		776	

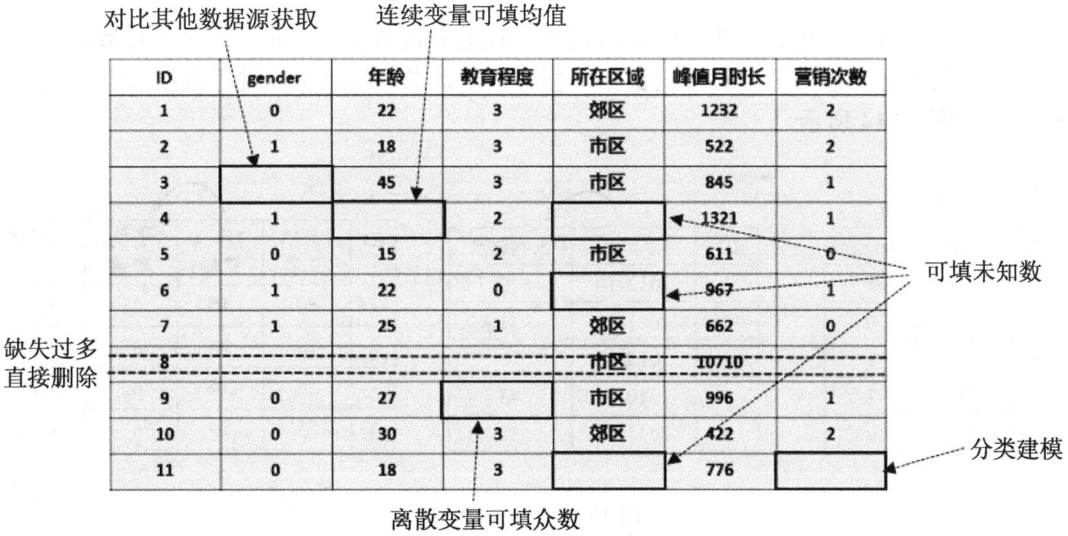

图 19.13　用户数据图

对于图 19.13 中的数据，第 8 行的样本的观测值缺失数目大于本行所有观测值的 50%，因此就删除此样本；对于列来说，如“所在区域”等离散变量的字段可利用填第三个分类（未知类）的方式补充数据，“教育程度”字段可填众数；像“年龄”这样的连续变量，可取均值或中位数；关于“营销次数”字段的数据补充，可将此“营销次数”缺失

变量看作 y，其余特征变量为 x，求其他特征变量作为 x 和 y 的关系来预测 y 从而得到缺失值。

19.5 数据转换

我们知道有一些模型如逻辑回归、决策树等不需要标准化数据即可建模，但是如神经网络和 KNN 等模型却需要对标准化数据建模，而在实际应用过程中，大部分的数据都是右偏数据，除此之外多数模型无法对字符型和连续型变量建模。

因此为了方便计算机处理、满足模型假设条件、降低模型复杂度和增强模型稳定性，我们在前面的错误值、缺失值等数据处理之后，进行一系列数据转换，常用方法包括衍生新变量（右偏数据取对数）、压缩分类水平数、数据离散化、变换哑变量、标准化和数据压缩等，具体方法如下。

19.5.1 变量转换

字符型变量无法进行建模，因此可转换为数值形式的变量，如存在时间点数据，转换为时长更具有分析意义。当对多列数据不便分析时，求均值水平、进行数据整合可以便于建模，如图 19.14 所示。

ID	性别	gender
1	男	1
2	女	0
3	男	1
4	男	1
5	女	0

ID	入网时间	入网时长
1	2012-10	20
2	2011-09	31
3	2012-12	18
4	2010-09	43
5	2011-08	32

ID	1月消费额	2月消费额	平均消费	增长率
1	30	46	38	53%
2	78	96	87	23%
3	28	108	62	243%
4	55	79	67	44%

图 19.14 变量转换图

19.5.2 压缩分类水平数

压缩分类水平可以降低模型复杂度，增强模型稳定性。我们根据专家的建议或公司实现目标来减少分类，降低了建模难度，且便于后期对结果的解释，具体操作如图 19.15 所示。

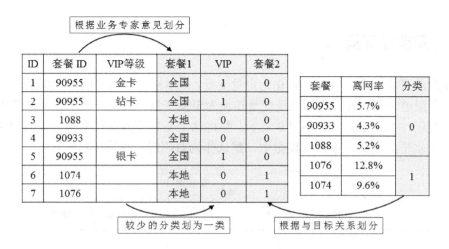

图 19.15 压缩分类水平图

19.5.3 连续数据离散化

离散化能降低数据的复杂度，支持许多无法处理数值型属性的分类算法，提高分类器稳定性和准确度，有助于解读最终结果。离散化分为等宽离散、等分离散和人工离散。

- 等宽离散：$Width=(max-min)/n$，保证样本分布不变，如果是右偏数据那么离散后仍为右偏的，如图 19.16 所示。

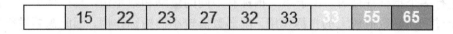

图 19.16 等宽离散图

- 等分离散：$Width=$样本数$/n$，分类后每类的样本数相同，如图 19.17 所示。

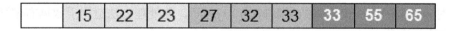

图 19.17 等分离散图

- 人工离散：手工分箱，根据业务需求进行分割得到不同样本，更易解释，如图 19.18 所示。

图 19.18 人工离散图

19.5.4 变换哑变量

哑变量的变换是将定性变量转化为定量变量,即本来是某用户的办理渠道表示为如营业厅、社会渠道和掌厅等,将此渠道变量转换为定量变量,某用户属于哪个办理渠道则此渠道用 1 表示,不属于的渠道都用 0 表示,并将这个变量作为虚拟变量进行回归建模,其中最后一个水平的哑变量不放入模型中,默认作为对照组,如图 19.19 所示。

ID	办理渠道	营业厅	10086	社会渠道	掌厅
1	营业厅	1	0	0	0
2	10086	0	1	0	0
3	社会渠道	0	0	1	0
4	掌厅	0	0	0	1

图 19.19 哑变量变换图

19.5.5 数据标准化

在进行多个指标评估时,由于不同指标有不同的量纲和数量级,直接使用原始数据进行分析会过于凸显数值较大的指标作用,削弱数值较小的指标作用,因此需要进行数据标准化,这对回归模型很重要,一般由软件自动执行,标准化方法如下:

(1)极值标准化

$$v' = \frac{v - \text{Min}}{\text{Max} - \text{Min}}$$

通过对原数据的线性变换将结果限制在[0,1]之间,此处理方法适用于数据右偏严重时。

(2)中心标准化

$$z' = \frac{v - \text{平均值}}{\text{标准差}}$$

利用均值和标准差进行数据标准化,得到的数据都服从正态分布,此处理方法适用于最大值和最小值未知或有超出取值范围的离群数据时。

19.5.6 数据压缩

我们在进行多个指标评估建模时,要保证 X 和 Y 相关性高、自变量 X 之间相关性低,最后得出的模型才能更准确、更有意义。相关性评估方法如下。
- X 和 Y 相关性评估：对于分类模型用卡方检验、方差分析方法；对于预测模型用 pearson 相关系数、spearman 相关系数进行分析。
- 自变量 X 之间相关性评估：利用主成分分析和因子分析方法进行评估。

19.6 建　　模

前面几节中我们已经熟悉了业务知识，并对收集的数据进行了整理、清洗和转换，接下来要进行数据建模。在建模之前我们先考虑选择什么样的模型更符合数据挖掘的目标。针对用户离网情况，一定有是否离网的结果，即一定有 Y，因此我们选择创建预测模型。

预测模型有分类模型和回归模型两种，分类模型又包括评分卡和分类器两种模型。其中，评分卡模型是指由业务人员根据情况主观定义一个规则作为分类依据，如 t 月出账而 $t+1$ 月不出账时定义为流失；分类器是以客观事实作为分类依据，如无人驾驶汽车识别信号灯时，得出客观的信号灯颜色情况，是客观、精确的情况。

本项目中由于用户的感受和行为的不确定性，我们无法精确判定用户是否离网，因此应选用评分卡模型，即我们根据用户特征和行为分析对用户的流失规则进行定义。如 t 月正常，$t+1$ 月用户状态为注销，则确定为主动流失；t 月正常消费且 $t+1$ 月 0 消费的用户可视为预流失用户；t 月出账且 $t+1$ 月不出账也可视为预流失用户；而在模型中我们定义 Y=1 则为流失用户，Y=0 则为未流失用户，据此通过构建模型最终可以筛选出高概率离网客户。

19.6.1 决策树算法概述

20 世纪 70 年代后期至 80 年代初期，Quinlan 发明了 ID3 算法，后来在 ID3 的基础上改进得到 C4.5 算法，后来又发布了 C5.0 算法。其中，ID3 是根据熵增益（Information Gain）作为决策树分割最优属性的评判指标，C4.5 和 C5.0 是根据熵增益率（Gain Ratio）求最优分割属性。

CART 算法在 1981 年被提出，它是根据最小基尼系数（Gini Index）求最优分割属性。

分类模型都是样本可分的，决策树正是基于属性分割后结果的纯净度原则，以树型结构组织的规则集合。决策树易于构建和可视化，能简约表示、易于执行，是一个有效的数据挖掘技术。

19.6.2 决策树的训练步骤

决策树的训练步骤如下。
（1）将所有训练集样本放入根节点。

（2）找到样本分割的最优属性，使分割后的正例/负例样本更集中，分割后此字段不再参与样本的分割。

（3）每个节点都按照最优分割属性进行分割，直到无法分割为止（用于分割的属性为 0 或节点中全部是正例/负例样本），最末端节点为叶节点。

（4）最终生成一棵决策树后将此模型应用到测试集中进行预测，并根据预测结果对模型进行评分。

19.6.3 训练决策树

在本例中我们需要创建客户流失决策树，根据用户的特征和行为判断用户是否流失。生成客户流失决策树包括获得训练集、训练决策树、产生预测集和获得预测结果，详细过程如下：

（1）获得训练集。

从历史数据中产生训练集，观察 Y 的情况，分析 X 的情况，如表 19.2 所示。

表 19.2 客户通信数据表

客户ID	曾经逾期	往来时长	通话趋势	无通话月	外网增多？	流失标志
张三	有	3个月	下降40%	3	无	1
李四	无	6个月	上升30%	1	无	0
...	...	...	...	...	...	...

（2）训练决策树。

对 X 和 Y 的数据进行建模，得到决策树，如图 19.20 所示。

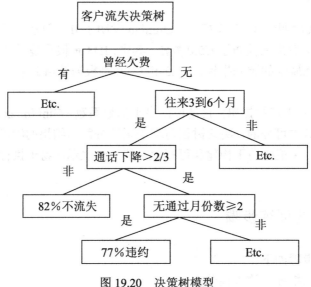

图 19.20 决策树模型

（3）产生预测集并得到预测结果。

根据现在的特征 X 利用决策树预测结果。

19.6.4　C4.5算法

C4.5 算法是根据熵增益率最大原则得到决策树模型，S 为样本集合，A 为样本的某一属性，$V(A)$ 为 S 中属性 A 的值域，Sv 是样本集中关于属性 A 值是 v 的样本集合。

样本分类熵：$Entropy(S) = \sum_{i=1}^{c} -p_i \log_2 p_i$

属性 A 的条件熵：$SplitInformation(S,A) = -\sum_{i=1}^{c} \frac{|S_i|}{|S|} \log_2 \frac{|S_i|}{|S|}$

属性 A 的信息增益：$Gain(S,A) = Entropy(S) - \sum_{v \in V(A)} \frac{|S_i|}{|S|} Entropy(S_v)$

属性 A 的信息增益率：$GainRatio(S,A) = \dfrac{Gain(S,A)}{splitInformation(S,A)}$

我们要根据属性的信息增益率最大原则进行决策树的分割。

- 分类变量分枝：对于分类变量可以有多分枝，当输入 4 个属性的数据时，依次遍历所有的组合形式，找到熵增益率最大的组合方式进行分枝，如图 19.21 所示。

```
A1 v.s. A2 v.s. A3 v.s. A4
A1、A2 v.s. A3 v.s. A4
A1、A2、A3 v.s. A4
A1、A3 v.s. A2、A4
……
```

图 19.21　分枝属性组合图

- 连续变量分枝：对于连续变量或等级变量只能出二分枝，先等宽分割为 50 组，然后依次取阀值分割为两组，以熵增益率最大的分割方式作为决策树分割节点。

1．连续变量的决策树分割实例

假设某样本集合共有 X_1 和 X_2 两个变量且都是连续变量，我们在决策树的节点分割时要先比较两个变量的优先级，即通过比较两个变量的分割值的最大熵增益率，选取有最大熵增益率的变量作为分割属性进行分叉。

对图 19.22 中的样本进行分割,先是各自变量内部做分割,选取熵增益率最大的点作为分割点,然后将不同变量分割点的熵增益率进行比较,熵增益率最大的变量作为首先进行属性分割的变量。

变量 X_1 中最大熵增益率的点为分割点 W_{10},增益率为 0.35,分割情况如图 19.22 所示。

变量 X_2 最大熵增益率的点为分割点 W_{20},增益率为 0.14,如图 19.23 所示。

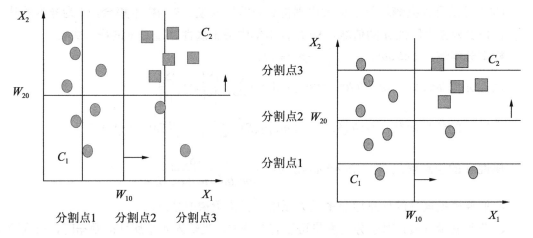

图 19.22　变量 X_1 最大熵增益率分割图　　图 19.23　变量 X_2 最大熵增益率分割图

比较两个属性各自的分割点熵增益率,X_1 中的分割点 W_{10} 熵增益率最大,因此首先取 W_{10} 的属性值进行节点的分割,即先用 W_{10} 分割;再根据剩余属性 X_2 的熵增益率最大值继续进行分类,即根据 X_2 的分割点 W_{20} 进行分割,如图 19.24 所示。

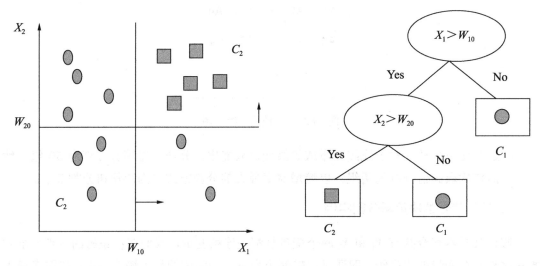

图 19.24　决策树分割图

2．分类变量的决策树分割实例

假设我们已经有一部分用户的样本信息数据，根据这些样本创建决策树模型，样本数据如表 19.3 所示。

表 19.3　用户样本信息表

记录ID	年　　龄	收　　入	性　　别	飞信用户	是否离网（Y）
1	青少年	高	男	否	1
2	青少年	高	男	是	1
3	中年	高	男	否	0
4	老年	中	男	否	0
5	老年	低	女	否	0
6	老年	低	女	是	1
7	中年	低	女	是	0
8	青少年	中	男	否	1
9	青少年	低	女	否	0
10	老年	中	女	否	0
11	青少年	中	女	是	0
12	中年	中	男	是	0
13	中年	高	女	否	0
14	老年	中	男	是	1

我们先计算样本关于年龄分类的熵增益率：

样本分类熵：$Entropy(S) = -\frac{5}{14}\log_2\frac{5}{14} - \frac{9}{14}\log_2\frac{9}{14} = 0.94$

在年龄属性值下的分类熵：

$$Entropy(青少年) = -\frac{2}{5}\log_2\frac{2}{5} - \frac{3}{5}\log_2\frac{3}{5} = 0.97$$

$$Entropy(中年) = 0$$

$$Entropy(老年) = -\frac{2}{5}\log_2\frac{2}{5} - \frac{3}{5}\log_2\frac{3}{5} = 0.97$$

进行年龄分类的条件熵：

$$SplitInformation = -\frac{5}{14}\log_2\frac{5}{14} - \frac{4}{14}\log_2\frac{4}{14} - \frac{5}{14}\log_2\frac{5}{14} = 1.577$$

关于年龄分类后的目标值信息增益：

$$Gain(S, AGE) = Entropy(S) - \frac{5}{14}Entropy(青少年) - \frac{4}{14}ntropy(中年) - \frac{5}{14}Entropy(老年) = 0.246$$

关于年龄分类后的熵增益率：

$$GainRatio = \frac{Gain(S, AGE)}{SplitInformation} = \frac{0.59}{1.577} = 0.156$$

以此类推，我们可以求出收入属性、性别属性和是否飞信用户属性分别对应的熵增益率。根据计算得知年龄属性的熵增益率最大，因此先根据年龄分类，然后从剩余属性集合中继续找出最优分割属性，其中，在中年样本中已进行完全分割，样本不纯度为 0，老年样本中进行飞信用户属性的分割熵增益率为 1，最大；在青少年样本中进行性别属性分割得到最大熵增益率 1。最终得到决策树，如图 19.25 所示。

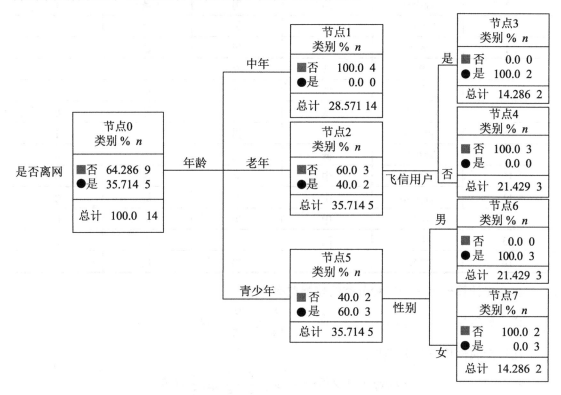

图 19.25 最终决策树模型

19.6.5 决策树剪枝

在决策树创建时，由于有很多噪声数据、离群数据和冗余信息，如果模型完美拟合训练集中的数据，我们会受到噪声数据的误导，即许多叶子节点反映的是训练数据中的特殊性、异常性。这种过分拟合的数据会影响模型的稳定性，所以我们通常使用统计度量剪掉最不可靠的分枝，剪枝后的树更小、更简单、更容易理解。剪枝分为预修剪和后剪枝，具体方法介绍如下：

1. 预修剪

预修剪的方法是控制决策树充分生长，通过事先指定一些超参数，如：
- 决策树最大深度，如果决策树的层数已经达到最大指定深度，则停止生长。
- 指定树中父节点和子节点的最少样本量或比例，对于父节点和分组后生成的子节点，如果节点的样本量低于最小样本量或比例，则不再分组；
- 树节点中输出变量的最小异质性减少量。如果分组产生的输出变量异质性变化量（分割后的样本特征差异程度）小于一个指定值，则不必进行分组。

2. 后剪枝

后修剪技术允许决策树充分生长，然后在此基础上根据一定的规则剪去决策树中那些不具有一般代表性的叶节点或子树，达到精简决策树的目的。后修剪技术是一个边修剪边检验的过程。在修剪过程中，应不断计算当前决策子树对测试样本集的预测精度或误差，并判断应继续修剪还是停止修剪。

CART 采用的后修剪技术称为最小代价复杂性修剪法（Minimal Cost Complexity Pruning，MCCP），是指随着决策树更加茂盛，模型精度不断增加，误判成本降低，但是决策树的复杂度增加，复杂成本增加，我们希望在误判成本和复杂度之间寻求一个平衡，得到较小的代价复杂度。

MCCP 有这样的基本考虑：首先，复杂的决策树虽然对训练样本有很好的预测精度，但在测试样本和未来新样本上不会有令人满意的预测效果；其次，理解和应用一棵复杂的决策树是一个复杂过程。因此，决策树修剪的目标是得到一棵"恰当"的树，它首先要具有一定的预测精度，同时决策树的复杂程度应是恰当的，如图 19.26 所示。

复杂度增加时，误差在减小，代价也在减小，但是复杂度过大导致模型泛化能力弱，因此在实际应用中我们只需要模型达到一定复杂度即可。关于模型的超参数（决策树层数、叶子节点数量），我们可以根据模型在测试集上的预测误差进行调整，如图 19.27 所示。

由图 19.27 可以看出，随着叶子数目增多，模型相对训练集误差降低、预测精度增大，相对于测试集的误差降低，但当叶子数目过多时，测试集精度反而下降，出现过拟合现象，因此我们要找到使测试集预测精度最高的叶子数目，并设置超参数创建决策树。

就机器学习算法来说，其泛化误差可以分解为两部分，即偏差（bias）和方差（variance）。偏差指的是算法的期望预测与真实预测之间的偏差程度，反应了模型本身的拟合能力；方差度量了同等大小的训练集的变动导致学习性能的变化，刻画了数据扰动所导致的影响。

当模型越复杂时，拟合的程度就越高，模型的训练偏差就越小，但此时如果换一组数据，可能模型的变化就会很大，即模型的方差很大，所以模型过于复杂的时候会导致过拟合，如图 19.28 所示。

当模型越简单时，即使我们再换一组数据，最后得出的学习器和之前的学习器的差别也不会太大，模型的方差很小，但因为模型简单，偏差会很大。

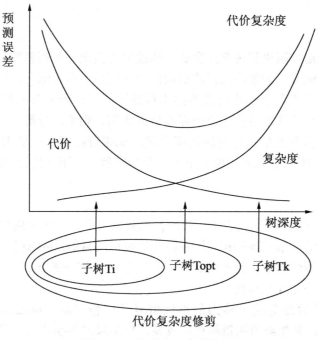

图 19.26 代价复杂度曲线图

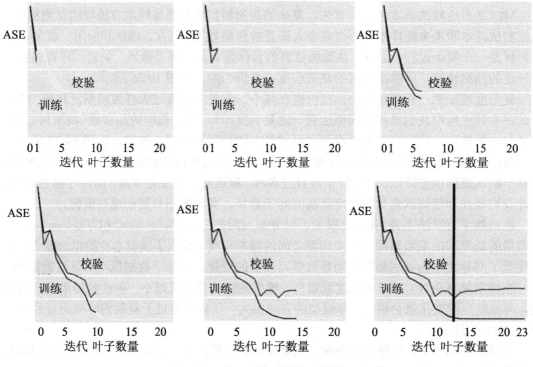

图 19.27 模型复杂度-误差图

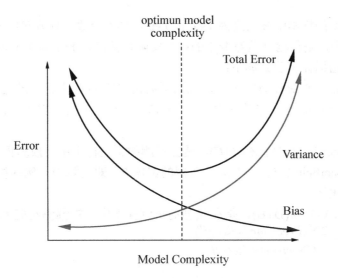

图 19.28　模型复杂度与偏差方差关系图

19.7　评　　估

对于分类器模型，有客观精确的结果，可以使用混淆矩阵进行评估；而对于评分卡模型，我们可以给定一个阈值，利用主观定义对样本进行分类后再计算混淆矩阵。

19.7.1　混淆矩阵

假设我们已经根据样本拟合出分类模型，我们通过评估模型的预测结果得到一个混淆矩阵，其中混淆矩阵如表 19.4 所示。

表 19.4　混淆矩阵计算表

混淆矩阵：每给定一个阈值，就可以做出一个混淆矩阵		打　分　值		合　　计
		反应（预测=1）	未反应（预测=0）	
真实结果	呈现信号（真实=1）	A（击中） True Positive	B（漏报） False Negative	A+B
	未呈现信号 （真实=0）	C（虚报） False Positive	D（正确否定） True Negative	C+D
合计		A+C	B+D	A+C+B+D

由表 19.4 混淆矩阵可知，它的行数据反映的是真实的目标值结果，例如第一行，即为

实际中分类为 Y=1 的样本中被预测为 Y=1 和 Y=0 的样本个数；列方向反映的是预测目标值结果，例如第一列，是预测分类为 Y=1 的样本实际目标值为 Y=1 和 Y=0 的个数。

混淆矩阵的指标计算公式如下：

- 正确率=（A+D)/(A+B+C+D)，是我们预测正例/负例样本正确的比例。
- 灵敏度（Sensitivity；覆盖率 Recall）=A/(A+B)，是指在实际的正例样本中预测正确的比例。
- 命中率(Precision/PV+)=A/(A+C)，是在预测的正例样本中，预测正确的样本的比例。
- 特异度(Specificity)=D/(C+D)，是负例的覆盖率，即实际为负例的样本中被预测到的样本所占比例。
- 负命中率(PV-)=D/(D+B)，是在预测的负例样本中，被预测正确的样本占比。
- $F - Score = \dfrac{2 \times (Precision \times Recall)}{(Precision + Recall)}$ 。

19.7.2 ROC 曲线

对于排序类（评分卡）模型，我们总是人为定义分类规则，如违约客户判定等。此时我们可以使用 ROC 曲线、K-S 曲线和 PR 曲线评估分类结果。

在使用 ROC 曲线对排序类模型进行评估时，我们人为给定一个阈值对排序类模型进行分类，得到确切的分类结果，并根据这个分类结果得到混淆矩阵，如图 19.29 所示。

Predicted Probability	True Class	TPR Sensitivity	Specificity	FPR 1-Specificity
0.90	1	0.09	3-1.00	0.00
0.80	1	0.18	3-1.00	0.00
0.70	0	0.18	0.89	0.11
0.60	1	0.27	0.89	0.11
0.55	1	0.36	0.89	0.11
0.54	1	0.45	0.89	0.11
0.53	1	0.55	0.89	0.11
0.52	0	0.55	0.78	0.22
0.51	1	0.64	0.78	0.22
0.51	1	0.73	0.78	0.22
0.40	1	0.82	0.78	0.22
0.39	0	0.82	0.67	0.33
0.38	1	0.91	0.67	0.33
0.37	0	0.91	0.56	0.44
0.36	0	0.91	0.44	0.56
0.35	0	0.91	0.33	0.67
0.34	1	1.00	0.33	0.67
0.33	0	1.00	0.22	0.78
0.30	0	1.00	0.11	0.89
0.10	0	1.00	0.00	1.00

图 19.29　样本分类图

根据上面这个分类结果得到混淆矩阵，如表 19.5 所示。

表 19.5 混淆矩阵

CM		真 实	
		0	1
预测	0	8	8
	1	1	3

遍历所有阈值，可以得到多个不同分类结果，进而得到多个混淆矩阵，再利用混淆矩阵计算得到召回率和特异度，如表 19.6 所示。

表 19.6 分类对应敏感度和特异度数据表

	阀 值	敏 感 度	特 异 度
1	0.96	0.003	0.998
2	0.91	0.022	0.997
3	0.89	0.038	0.996
4	0.83	0.082	0.989
5	0.75	0.173	0.959
6	0.72	0.227	0.948
7	0.69	0.280	0.929
8	0.61	0.438	0.846
9	0.60	0.467	0.834
10	0.52	0.670	0.721
11	0.48	0.730	0.666
12	0.41	0.829	0.517
13	0.36	0.880	0.412
14	0.35	0.882	0.399
15	0.30	0.908	0.323
16	0.21	0.953	0.200
17	0.17	0.967	0.152
19	0.11	0.983	0.092
19	0.05	0.991	0.048
20	0.00	0.998	0.005

根据计算得到的特异度和灵敏度画出 ROC 曲线，其中 ROC 曲线横轴为"1-特异度"，纵轴为灵敏度，如图 19.30 所示。

由图 19.30 可知，随着阈值降低，模型的灵敏度在升高，特异度在降低。

由图 19.30 我们知道 ROC 曲线是根据召回率和"1-特异度"作为坐标轴进行绘制的，其中横轴为 $FPR=1-FNR$（1-特异度），纵轴为 TPR（召回率），如图 19.31 所示。

以图 19.31 中画圈点为例，正例的正确命中概率（召回率）为 70%，反例的正确命中概率（1-特异度）为 1-20%=80%，虚线框住的面积中 AUC 越大，即曲线越靠近纵轴，模型越好；对角线代表随机猜测时的 ROC 曲线（此时模型效果最差）。

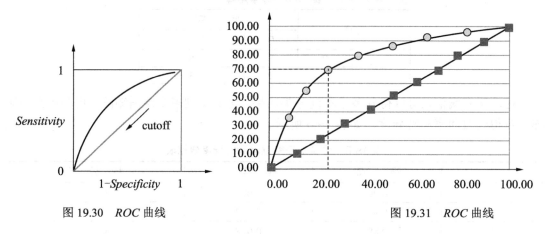

图 19.30　ROC 曲线　　　　　　　　　图 19.31　ROC 曲线

19.8　部　　署

部署过程主要为：建模人员将需求告诉数据处理及管理人员，数据处理人员对数据进行转换和加载，从数据仓库中返回数据集市给建模人员，建模人员对数据进行清洗、转换→建模→对模型进行评分→根据客户流失情况得到客户流失趋向评分清单→市场管理分析人员针对客户评分情况推出营销方案→通过客户关系管理系统在各个渠道执行→营销结果反馈给销售管理人员，继续进行数据的循环分析、营销策略调整。具体过程如图 19.32 所示。

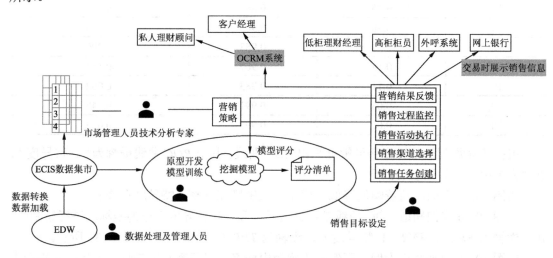

图 19.32　项目部署过程图

19.9 用户离网案例代码详解

根据数据挖掘标准流程，在进行数据采集后，还要进行数据准备、建模和评估等工作，这三部分工作我们通过代码来完成，具体步骤如下。

19.9.1 数据准备

首先读取客户信息的数据集，观察描述性变量 X 与 Y 是否相关，这里分别分析教育程度、通话最高峰值、在网时长和用户流失情况的相关性，并画出条形图和盒须图。

```
import pandas as pd
import numpy as np
import matplotlib.pyplot as plt
import seaborn as sns
churn = pd.read_csv('telecom_message.csv')    #读取已经整理好的数据
churn.head()

#分别分析教育程度、通话最高峰值、在网时长和用户流失情况的相关性分析
#多分类变量 edu_class 和二分类变量 churn 的相关关系用条形图表示
sns.barplot(x='edu_class', y='churn', data=churn)
#plt.show()

#连续变量通话最高峰值 peakMinDiff 和二分类变量 churn 相关关系用盒须图表示
sns.boxplot(x='churn', y='peakMinDiff', hue=None, data=churn)
#plt.show()

#在网时长 duration 也是连续变量，我们用盒须图寻找它和 churn 的相关关系
sns.boxplot(x='churn', y='duration', hue='edu_class', data=churn)
#plt.show()
```

得到结果如图 19.33 所示。

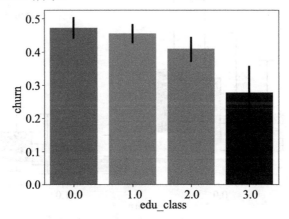

图 19.33 教育程度和用户离网情况条形图

由图 19.33 可知，当受教育情况变化时，用户离网概率不同，因此受教育程度和用户离网相关。

分析用户最高通话峰值和用户离网的相关情况，如图 19.34 所示。

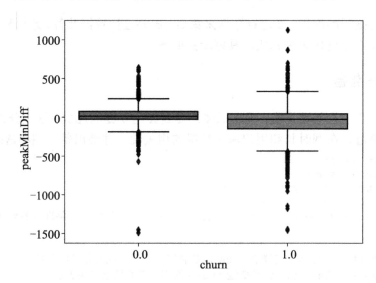

图 19.34　最高通话峰值和用户离网情况盒须图

由图 19.34 可知，对于离网用户（$y=1$）和非离网用户（$y=0$）的最高通话峰值差别不大，这个变量对离网与否影响不大，因此我们认为最高通话峰值和离网情况不相关。

分析关于用户的在线时长和用户离网的相关情况，其盒须图如图 19.35 所示。

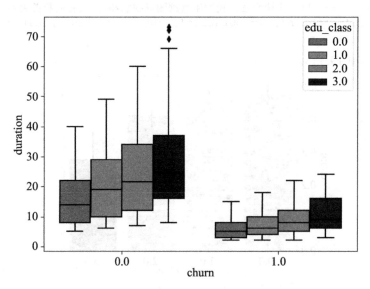

图 19.35　在线时长和用户离网情况盒须图

由图 19.35 可知，离网用户和非离网用户的在线时长差距较大，且受教育程度不同，时长也不同，由此可知在线时长和用户离网情况相关性较大。

19.9.2 相关性分析

前面粗略地分析了特征变量和目标值的相关性后，根据 apearman 相关系数矩阵选取相关系数大于 0.5 的变量进行查看，排除不相关字段，通过卡方检验，删除不相关字段和冗余字段。代码如下：

```
# ##筛选变量
#* 筛选变量时可以应用专业知识，选取与目标字段相关性较高的字段用于建模，也可通过分析现
有数据，用统计量辅助选择
#* 为了增强模型稳定性，自变量之间最好相互独立，可运用统计方法选择要排除的变量或进行变
量聚类

corrmatrix = churn.corr(method='spearman')   # spearman 相关系数矩阵，可选
pearson 相关系数，目前仅支持这两种,函数自动排除 category 类型
corrmatrix_new = corrmatrix[np.abs(corrmatrix) > 0.5]    # 选取相关系数绝对值
大于 0.5 的变量，仅为了方便查看
#   为了增强模型稳定，根据上述相关性矩阵，可排除'posTrend','planChange','nrProm',
'curPlan'几个变量
#print(corrmatrix_new)

# 连续型变量往往是模型不稳定的原因
# 如果所有的连续变量都分箱了,可以统一使用卡方检验进行变量重要性检验
#churn['duration_bins'] = pd.qcut(churn.duration, 5)
                                # 将 duration 字段切分为数量（大致）相等的 5 段
#churn['churn'].astype('int64').groupby(churn['duration_bins']).agg(['c
ount', 'mean'])

bins = [0, 4, 8, 12, 22, 73]     #根据需求将 duration 字段分为 5 段
churn['duration_bins'] = pd.cut(churn['duration'], bins, labels=False)
churn['churn'].astype('int64').groupby(churn['duration_bins']).agg
(['mean', 'count'])


# 根据卡方值选择与目标关联较大的分类变量，删除不相关字段和冗余字段
# 计算卡方值需要应用到 Sklearn 模块，但该模块当前版本不支持 Pandas 的 category 类型
变量，会出现警告信息，可忽略该警告或将变量转换为 int 类型
import sklearn.feature_selection as feature_selection
#将分类数据进行转换，并取一部分数据进行卡方检验，进行一系列变量的排除
churn['gender'] = churn['gender'].astype('int')
churn['edu_class'] = churn['edu_class'].astype('int')
churn['feton'] = churn['feton'].astype('int')
feature_selection.chi2(churn[['gender', 'edu_class', 'feton', 'prom',
```

```
                                   'posPlanChange', 'duration_bins', 'curPlan',
'call_10086']],
                      churn['churn'])         #选取部分字段进行卡方检验
# 根据结果显示,'prom'、'posPlanChange'、'curPlan'字段可以考虑排除
```

19.9.3　最终建模

把与目标值相关的、经过变换的变量数据抽取出出来作为测试集和训练集,进行建模。

```
# ## 建模
# * 根据数据分析结果选取建模所需字段,同时抽取一定数量的记录作为建模数据
# * 将建模数据划分为训练集和测试集
# * 选择模型进行建模

#  根据模型不同,对自变量类型的要求也不同,为了示范,本模型仅引入'AGE'这一个连续型变量
# model_data = churn[['subscriberID','churn','gender','edu_class',
'feton','duration_bins']]
model_data = churn[
    ['subscriberID', 'churn', 'gender', 'edu_class', 'feton', 'duration_
    bins', 'call_10086', 'AGE']]               #第二可选方案
model_data.head()

import sklearn.model_selection as cross_validation

target = model_data['churn']                    #选取目标变量
data = model_data.ix[:, 'gender':]              #选取自变量
#测试集和训练集划分
train_data, test_data, train_target, test_target = cross_validation.
train_test_split(data, target, test_size=0.4,
                                                           train_size=0.6,
random_state=12345)                             #划分训练集和测试集

# 选择决策树进行建模
import sklearn.tree as tree
#创建决策树,以信息熵作为分类属性标准,树的最大深度为8,最小为5
clf = tree.DecisionTreeClassifier(criterion='entropy', max_depth=8,
min_samples_split=5)   # 当前支持计算信息增益和GINI
clf.fit(train_data, train_target)               #使用训练数据建模

# 查看模型预测结果
train_est = clf.predict(train_data)                    #用模型预测训练集的结果
train_est_p = clf.predict_proba(train_data)[:, 1]   #用模型预测训练集的概率
test_est = clf.predict(test_data)                      #用模型预测测试集的结果
test_est_p = clf.predict_proba(test_data)[:, 1]     #用模型预测测试集的概率
pd.DataFrame({'test_target': test_target, 'test_est': test_est, 'test_
est_p': test_est_p}).T            # 查看测试集预测结果与真实结果对比
```

19.9.4 模型评估

建模后对模型进行评估:计算混淆矩阵和评估指标。

```
#模型评估
import sklearn.metrics as metrics

print(metrics.confusion_matrix(test_target, test_est, labels=[0, 1]))
                                                       #混淆矩阵
print(metrics.classification_report(test_target, test_est))
                                                       #计算评估指标
print(pd.DataFrame(list(zip(data.columns, clf.feature_importances_))))
                                                       #变量重要性指标

#查看预测值的分布情况
red, blue = sns.color_palette("Set1", 2)
sns.distplot(test_est_p[test_target == 1], kde=False, bins=15, color=red)
sns.distplot(test_est_p[test_target == 0], kde=False, bins=15, color=blue)
#plt.show()

fpr_test, tpr_test, th_test = metrics.roc_curve(test_target, test_est_p)
fpr_train, tpr_train, th_train = metrics.roc_curve(train_target, train_est_p)
plt.figure(figsize=[6, 6])
plt.plot(fpr_test, tpr_test, color=blue)
plt.plot(fpr_train, tpr_train, color=red)
plt.title('ROC curve')
#plt.show()
```

训练集和测试集的预测 ROC 曲线如图 19.36 所示。

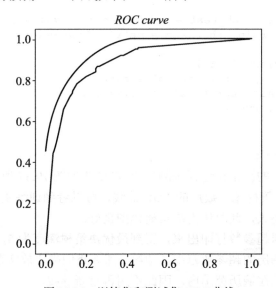

图 19.36 训练集和测试集 ROC 曲线

图 19.36 中下面的曲线为测试集，上面的曲线为训练集，我们看出测试集和训练集的 ROC 曲线有较大差距，且训练集的精度更高，因此我们猜测模型过拟合。利用 sklearn 的网格搜索功能，设置多个超参数进行遍历，从而找到最优的树的深度、最优分隔属性选择的指标和分割集数目，并利用最优参数重新创建决策树模型。

```python
# 这里表现出了过度拟合的情况
# 参数调优
# %%
from sklearn.model_selection import GridSearchCV
from sklearn import metrics

param_grid = {
    'criterion': ['entropy', 'gini'],
    'max_depth': [2, 3, 4, 5, 6, 7, 8],
    'min_samples_split': [4, 8, 12, 16, 20, 24, 28]
}
clf = tree.DecisionTreeClassifier()
clfcv = GridSearchCV(estimator=clf, param_grid=param_grid,
                    scoring='roc_auc', cv=4)
clfcv.fit(train_data, train_target)
# %%
# 查看模型预测结果
train_est = clfcv.predict(train_data)           #用模型预测训练集的结果
train_est_p = clfcv.predict_proba(train_data)[:, 1]
                                                #用模型预测训练集的概率
test_est = clfcv.predict(test_data)             #用模型预测测试集的结果
test_est_p = clfcv.predict_proba(test_data)[:, 1]
                                                #用模型预测测试集的概率
# %%
fpr_test, tpr_test, th_test = metrics.roc_curve(test_target, test_est_p)
fpr_train, tpr_train, th_train = metrics.roc_curve(train_target, train_est_p)
plt.figure(figsize=[6, 6])
plt.plot(fpr_test, tpr_test, color=blue)
plt.plot(fpr_train, tpr_train, color=red)
plt.title('ROC curve')
#plt.show()

clfcv.best_params_
```

利用最优参数创建的决策树进行预测的 ROC 曲线如图 19.37 所示。

图 19.37 为修改决策树超参数后的 ROC 曲线，可以看到训练集和测试集的 ROC 曲线基本一致，精度相差不大，此时认为决策树模型良好。

将改良后的决策树超参数打印出来，得到最优决策树深度为 5，样本分割数目为 24，这个参数如果在我们遍历的超参数范围的边缘，那么我们需要放宽范围，再进行最优参数选择，而此时最优参数在遍历范围内，因此不用再重新寻找。

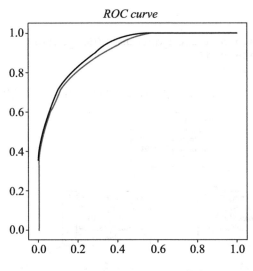

图 19.37 改进后的 ROC 曲线

求出最优参数后,利用这些参数创建决策树模型,并利用训练集训练此模型。代码如下:

```
clf = tree.DecisionTreeClassifier(criterion='entropy', max_depth=5, min_
samples_split=24)   # 当前支持计算信息增益和 GINI
clf.fit(train_data, train_target)   # 使用训练数据建模

# ### 可视化

# 使用 dot 文件进行决策树可视化需要安装一些工具:
#
# - 第一步是安装[graphviz](http://www.graphviz.org/)。Linux 可以用 apt-get
或者 YUM 的方法安装。如果是 Windows,就在官网下载 msi 文件安装。无论是 Linux 还是
Windows,装完后都要设置环境变量,将 graphviz 的 bin 目录加到 PATH,比如 Windows,
将 C:/Program Files (x86)/Graphviz2.38/bin/加入了 PATH
# - 第二步是安装 Python 插件 graphviz: pip install graphviz
# - 第三步是安装 Python 插件 pydotplus: pip install pydotplus

import pydotplus
from IPython.display import Image
import sklearn.tree as tree

# In[18]:


dot_data = tree.export_graphviz(
    clf,
    out_file=None,
    feature_names=train_data.columns,
    max_depth=3,
    class_names=['0', '1'],
    filled=True
```

```
)
graph = pydotplus.graph_from_dot_data(dot_data)
#Image(graph.create_png())
graph.write_pdf("churn.pdf")
```

最后,画出我们最终得到的决策树模型,这里只画出了决策树的三层,如图 19.38 所示。

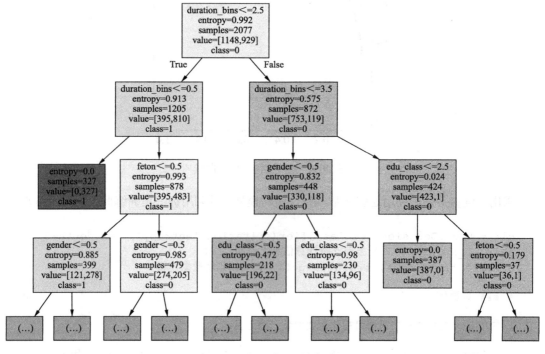

图 19.38　最优决策树模型

19.10　小　　结

本章主要介绍了数据挖掘标准流程中包括商业理解、数据理解、数据准备、建模、评估和部署共 6 个步骤在电信用户离网分析中的应用,并利用代码对处理后的数据实现了数据建模、评估和部署,最终得到用户离网预警清单,并据此清单提出营销策略,进行客户维系和挽留,起到了有效防止用户流失的重要作用。

推荐阅读

推荐阅读

Hadoop大数据挖掘从入门到进阶实战（视频教学版）

作者：邓杰　书号：978-7-111-60010-7　定价：99.00元

博客园资深博主、极客学院特邀讲师分享多年的Hadoop使用经验
全面涵盖Hadoop从基础部署到集群管理，再到底层设计等重点内容

本书采用"理论+实战"的编写形式，结合51个实例、10个综合案例及作者多年积累的一线开发经验，带领读者通过实际动手的方式提高编程水平。书中的所有实例和案例均来源于作者多年的工作经验积累和技术分享。本书提供近200分钟配套教学视频，手把手带领读者高效学习。

Spark Streaming实时流式大数据处理实战

作者：肖力涛　书号：978-7-111-62432-5　定价：69.00元

前腾讯优图实验室及WeTest研究员/现拼多多资深算法工程师力作
腾讯WeTest总监方亮与腾讯深海实验室创始人辛愿等5位大咖力荐
快速搭建Spark平台，从0到1动手实践Spark Streaming流式大数据处理

本书通过透彻的原理分析和充实的实例代码讲解，全面阐述了Spark Streaming流式处理平台的相关知识，能够让读者快速掌握如何搭建Spark平台，然后在此基础上学习流式处理框架，并动手实践进行Spark Streaming流式应用开发，包括与主流平台框架的对接应用及项目实战中的一些调优策略等。

Hive性能调优实战

作者：林志煌　书号：978-7-111-64432-3　定价：89.00元

作者曾经在国内互联网头部公司长期从事大数据项目的研发
百度无线搜索前负责人胡嵩、字节跳动算法团队技术总监丁锐等6位大咖力荐

本书旨在介绍如何进行Hive性能调优，以及调优时所涉及的工具。书中重点介绍了Hive性能调优所涉及的Hadoop组件和Hive工具。考虑到很多调优方法的着眼点有一定的相似性，这些调优方法可以适用于多个Hive版本，因此本书在介绍Hive的相关内容时会穿插Hive 1.x、Hive 2.x及Hive 3.x等多个版本的内容。